Advanced Combustion and Aerothermal Technologies

NATO Science for Peace and Security Series

This Series presents the results of scientific meetings supported under the NATO Programme: Science for Peace and Security (SPS).

The NATO SPS Programme supports meetings in the following Key Priority areas: (1) Defence Against Terrorism; (2) Countering other Threats to Security and (3) NATO, Partner and Mediterranean Dialogue Country Priorities. The types of meeting supported are generally "Advanced Study Institutes" and "Advanced Research Workshops". The NATO SPS Series collects together the results of these meetings. The meetings are co-organized by scientists from NATO countries and scientists from NATO's "Partner" or "Mediterranean Dialogue" countries. The observations and recommendations made at the meetings, as well as the contents of the volumes in the Series, reflect those of participants and contributors only; they should not necessarily be regarded as reflecting NATO views or policy.

Advanced Study Institutes (ASI) are high-level tutorial courses intended to convey the latest developments in a subject to an advanced-level audience

Advanced Research Workshops (ARW) are expert meetings where an intense but informal exchange of views at the frontiers of a subject aims at identifying directions for future action

Following a transformation of the programme in 2006 the Series has been re-named and re-organised. Recent volumes on topics not related to security, which result from meetings supported under the programme earlier, may be found in the NATO Science Series.

The Series is published by IOS Press, Amsterdam, and Springer, Dordrecht, in conjunction with the NATO Public Diplomacy Division.

Sub-Series

A.	Chemistry and Biology	Springer
B.	Physics and Biophysics	Springer
C.	Environmental Security	Springer
D.	Information and Communication Security	IOS Press
E.	Human and Societal Dynamics	IOS Press

http://www.nato.int/science
http://www.springer.com
http://www.iospress.nl

Series C: Environmental Security

Advanced Combustion and Aerothermal Technologies

Environmental Protection and Pollution Reductions

edited by

Nick Syred
Cardiff University, UK

and

Artem Khalatov
National Academy of Sciences,
Kiev, Ukraine

Published in cooperation with NATO Public Diplomacy Division

Proceedings of the NATO Advanced Research Workshop on Advanced
Combustion and Aerothermal Technologies: Environmental Protection and
Pollution Reductions
Kiev, Ukraine
15 –19 May 2006

A C.I.P. Catalogue record for this book is available from the Library of Congress.

ISBN 978-1-4020-6514-9 (PB)
ISBN 978-1-4020-6513-2 (HB)
ISBN 978-1-4020-6515-6 (e-book)

Published by Springer,
P.O. Box 17, 3300 AA Dordrecht, The Netherlands.

www.springer.com

Printed on acid-free paper

CONTENTS

PREFACE

The NATO Advanced Workshop "Advanced Combustion and Aerothermal Technologies: Environmental Protection and Pollution Reductions" was held in Kiev (Ukraine) from 15 to 19 May 2006 and was organized by the Institute of Engineering Thermophysics (Ukraine) and Cardiff University (UK). This Workshop based on the long-term collaboration between the Institute of Engineering Thermophysics and Cardiff University resulted in a first NATO Scientific Prize received by Professor N. Syred, UK, and Professor A. Khalatov in 2002, who served as Workshop codirectors.

The justification for this Workshop was based upon the perceived need for the bringing together of research in a number of combustion and aerothermal-related areas, so as to allow more rapid progress to be made. The primary Workshop objectives were to assess the existing knowledge on advanced combustion and aerothermal technologies providing reduced environmental impact, to identify directions for future research in the field, and to promote the close relationships and business contacts between scientists from the NATO and partner countries. This synergy in research and development is essential if advances in specific areas are to be widely utilized, whilst helping to cross-fertilize other areas and stimulate new developments. Of especial importance is the dissemination of concepts and ideas evolved in the aerospace industries into other related areas, whilst encouraging contacts, research exchanges, and inter-actions between engineers and scientists in the NATO and partner countries.

The Workshop program included keynote lectures, oral and roundtable discussions; the subjects addressed were as follows:

- A critical review of the use of swirling, cyclonic and vortex flow for stabilization of high-intensity flames and associated problems
- A critical review of the use of computational fluid dynamics (CFD) and advanced numerical modeling
- A review of the application of advanced diagnostics to flames and high-temperature systems
- A critical review of advances in gas turbine technologies and what directions future work needs to take
- Clean combustion of alternative fuels for industrial gas turbines
- Circulating fluidized beds for solid fuel combustion
- Minimization, utilization, and energy recovery from wastes
- Improvements to drying technologies through application of advanced combustion technologies
- Multifluid models of turbulent combustion
- Fuel cell power generation

These carefully chosen subjects were felt to cover the critical areas of combustion and aerothermal research and development where rapid progress is needed to improve performance in terms of efficiency, compactness, and emissions. This was a very timely and successful Workshop which brought together experts from the NATO and other countries to discuss increasing concerns about emissions.

The dramatic increasing price of fuels over the last 5 years means that every effort has to be made to spread best practice and cross-fertilize as many areas as possible for mutual benefit.

The editors take pleasure in acknowledging the cooperation and involvement of the NATO Science Committee, who have made this workshop possible. The outstanding work of the publishers, Springer, removed much of the onus for author and editor alike in the preparation of this book.

Finally, we express our profound gratitude to the authors of the individual papers, who have worked so competently and conscientiously to provide this extremely important collection.

CODIRECTORS

Professor	Syred N.	UK
Professor	Khalatov A.	Ukraine

INTERNATIONAL COMMITTEE

Professor	Dolinsky A.	Ukraine
Professor	Burdukov A.	Russia
Professor	Sudarev A.	Russia
Dr.	Spratt M.	UK
Dr.	Sy A. Ali	USA

LOCAL COMMITTEE

Dr.	Kuzmin A.
Dr.	Shevtsov S.
Mr.	Lisovsky A.
Mrs.	Shikhabutinova O.

POWER SYSTEMS ENGINEERING SCIENCE AND TECHNOLOGY

GENERATION AND ALLEVIATION OF COMBUSTION INSTABILITIES IN SWIRLING FLOW

N. SYRED[*]
*Cardiff School of Engineering, Cardiff University,
Queens Buildings, The Parade, Cardiff CF 24 3AA, Wales, UK,*

Abstract. Most gas turbine and other combustion systems use swirl as a method of flame stabilization. Swirling flows can in themselves generate several different types of flow and flame instability, as well as facilitating coupling with natural acoustic modes of a system. This paper reviews recent work in the area and describes how fluctuation in the structure of the re-circulation zone and tangential coherent structures (precessing vortex core [PVC]) can generate structures in other planes, capable of giving periodic heat release and excitation of combustion instability via the Rayleigh criterion. This is related to combustion-driven oscillations in gas turbine combustors. It is shown how damping of these tangential coherent structures can occur via certain modes of combustion whilst appropriate acoustic coupling can reexcite these phenomena. Methodologies for alleviation of these effects are described.

Keywords: combustion instabilities, swirling flow, precessing vortex cores, recirculation zones

1. Introduction

Lean premixed (LP) gas turbine combustion is the current approach employed to reduce pollutant emissions, particularly NO_x emissions. Some of the main advantages of LP operation are increased combustion intensity, shorter flame

---

[*]To whom correspondence should be addressed. Professor Nick Syred. Cardiff School of Engineering, Cardiff University, Queens Buildings, The Parade, Cardiff CF24 3AA, Wales, UK; e-mail: syredn@cardiff.ac.uk

N. Syred and A. Khalatov (eds.), Advanced Combustion and Aerothermal Technologies, 3–20.
© 2007 *Springer.*

lengths and better fuel burnout. However, the gains made by LP operating conditions can be accompanied by stability problems. Premixed flames are naturally more susceptible to static and dynamic instability due to a lack of inherent damping mechanisms. The absence of diffusive mixing increases the sensitivity of flames to acoustic excitation. If conditions favor instability, periodic fluctuations in the heat release will match the natural resonant frequency of one or more of the acoustic modes of the system geometry, self-exciting a thermoacoustic instability. The occurrence of thermoacoustic instability over a wide range of operating conditions in LP combustion continues to hinder the development of modern gas turbine combustors in both propulsion and power generation units.

A number of studies have been carried out to identify the various processes governing combustion oscillations. Keller et al. (1989, 1990) showed that a limit cycle can be divided into interrelated characteristic time lags representative of combustion chemistry, acoustic period, and flow dynamics, some typical figures are shown in Table 1. A series of parametric experiments in a gas turbine combustor by Richards and Janus (1997), Straub and Richards (1998) and Straub et al. (1998) have described the influence of various inlet parameters on reactant flow oscillations. An oscillating reactant flow can introduce variations to the reactant composition that is delivered to the flame front. The role of equivalence ratio fluctuations as a mechanism of combustion instability has also been treated by Lieuwen and Zinn (1998) and Mongia et al. (1998). The well-known Rayleigh criteria needs to be satisfied where the periodic heat release needs to be added as close as possible to the pressure peak of the oscillation for excitation to occur.

Recent studies by Broda et al. (1998), Venkataraman et al. (1999), Paschereit et al. (2000), Lee et al. (2000) and Giezendanner et al. (2003) have employed chemiluminescence and fluorescence techniques to study flame dynamics in swirl-stabilized flames. These studies have captured how the flame behaves over a limit cycle and have explored the mean periodic response of the progress rate variable and identified the location of the driving regions. Despite an absence of velocity information, most of these authors have referred to the importance of velocity oscillations around the flame zone as a key mechanism behind the observed flame response. There is an absence of detailed experiment and other data here.

Table 1 clearly shows the greatest similarity in timescales between convective processes and acoustic periods in the range of 100–500 Hz and indeed this can be easily extended to cover a much wider range of conditions. Convection effects can of course cover a wide range of different phenomena including convection of reactants in the inlets, convection of disturbances along a flame front or convection of vortical structures. Chemical kinetic effects can be seen to be important at higher frequencies. Although timing is important, it is not necessarily the only condition for excitation of oscillation. Although phasing of the unsteady heat release rate with the pressure is important, it must also be added at a rate that exceeds damping effects.

TABLE 1. Various processes in gas and liquid fueled combustors and associated characteristic times (Lieuwen 1999)

Process	Timescale (ms)
Acoustic disturbance of 100 Hz	10
Acoustic disturbance of 500 Hz	2
Chemical kinetic ignition delay, $\varphi = 0.7$	~1
Chemical kinetic ignition delay, $\varphi = 1$	~3
Convection of disturbance 100 mm at 10 m/s	10
Convection of disturbance 100 mm at 50 m/s	2
Evaporation of 10 μm hydrocarbon droplet	0.3
Evaporation of 50 μm hydrocarbon droplet	8
Growth rate of an acoustic disturbance 10 mm diameter liquid jet	125
Propagation of an acoustic disturbance 100 mm at 330 m/s (300 K)	0.3
Propagation of an acoustic disturbance 100 mm at 600 m/s (1,000 K)	0.17

Extensive investigations have been undertaken on the effect of various combustion parameters on the stability of combustion systems (Lieuwen and Zinn 1998; Lieuwen 1999) and it emerges that a very significant parameter is the fluctuation of equivalence ratio upon the reaction rate (Figure 1a) especially for ranges of equivalence ration <0.8 where LP combustors operate, this being derived from analysis of well stirred reactor simulations of the initial stages of gas turbine combustors.

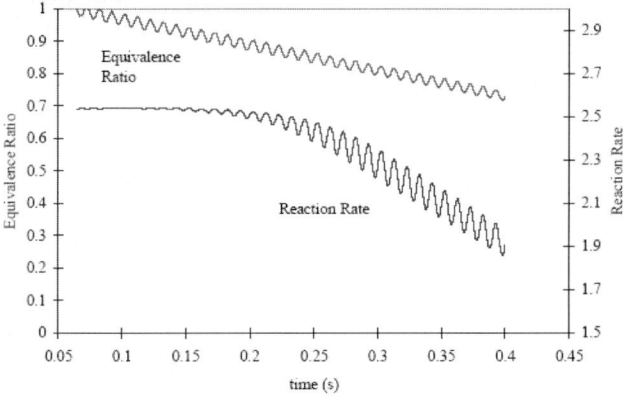

Figure 1a. Response of a well-stirred reactor to typical fluctuations in equivalence ratio

A typical gas turbine combustor flow is illustrated in Figure 1b. The formation of a central toroidal recirculation zone is evident, surrounded by an annular shear flow: this initial part of the flow can be easily perturbed as will be described later. Indeed this sensitivity to small perturbations allows coupling with acoustic oscillations without the presence of combustion. Syred (2006) reports the acoustic coupling of flows in large cyclone dust separators and long pipe runs at high pressure in process plants. Measure to reduce vibration and the acoustic coupling include the installation of carefully shaped center bodies in the exhaust of the cyclone dust separator and at the base of the long cone.

The most significant transient feature apparent within the computations is a precessing vortex core.

Normalised velocity

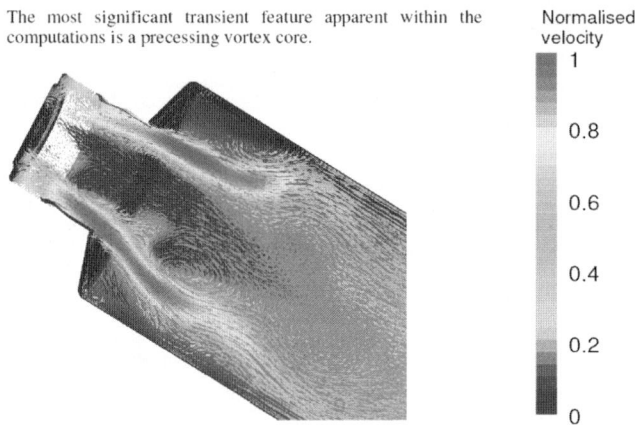

Figure 1b. Typical flow aerodynamics in a natural gas-fired swirl-stabilized gas turbine combustor. (From Turrell et al. 2004.)

The following discussion thus examines a series of studies which characterize oscillatory flow fields in swirl combustion system and seeks to analyze the underlying mechanisms of excitation. They comprise both experimental and predictive work and bring out the acoustic streaming or modification of the gross flow and flame characteristics which occur due to these oscillations. These results are then used to discuss the mechanisms of initial excitation followed by the features of the system which determine the stability behavior.

2. Swirling Flow and Combustion Instabilities

Swirl-stabilized combustion is widely used in gas turbines and many other systems. Stability considerations are crucial in high-intensity units. Figure 1a shows typical time-dependent flows generated by a Siemens swirl-stabilized gas turbine combustor as predicted by the CFX code (Turrell et al. 2004). The expected characteristics of an annular burning swirling flow leaving the swirler via an annular shear flow with a large central recirculation zone (CRZ) recirculating heat and active chemical species to the flame root is shown. Asymmetries in the CRZ are clearly shown with the formation of regular time-dependent axial radial eddies in the CRZ, whilst the presence of a precessing vortex core (PVC) is noted (Syred 1974, 2006). An external recirculation zone is also clearly shown (ORZ) as the swirling flow expands into the combustion chamber.

Swirl combustors and burners are usually characterized by the swirl number, S, defined as the ratio of axial flux of angular momentum to the axial flux of axial momentum, normalized by the exit radius of the system.

Although this parameter should be defined from experimental readings or measurements, it is commonly defined in terms of geometry (for further discussion refer to Beer and Chigier 1972; Syred and Beer 1974; Gupta et al. 1984).

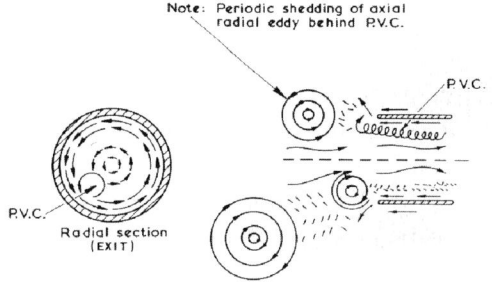

(a) Schematic of PVC (Syred 2006)

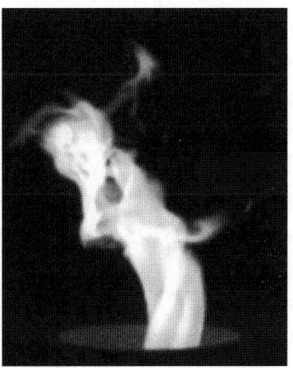

(c) Visualization of burning PVC

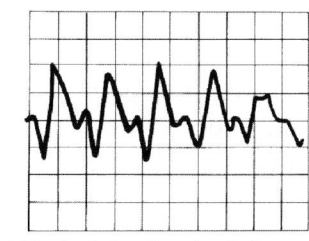

(b) Typical signal poducde by
the PVC via pressure transducer

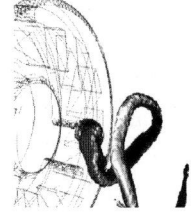

(d) LES visualization of PVC

Figure 2. (From Syred et al. 2006)

    The characteristics of swirl burners and combustors have been summarized
in Syred and Beer (1974) and Gupta et al. (1984). The occurrence of the PVC
and its associated instabilities have been summarized by Syred (2006). It was
shown that the formation of the CRZ is caused by the axial decay of radial
centrifugal pressure gradients, which in turn set up axial pressure gradients
sufficient to cause extensive central recirculation and CRZ formation. Often
associated with this is the formation of a PVCs, which occurs as the central
vortex core is perturbed by external disturbances i.e., hydrodynamic, acoustic.
The PVCs, appear to be linked to the CRZ, are wrapped around and distort it.
This is illustrated by Figure 2a, which shows a schematic of the PVC phenomena (Figure 2b), which shows the typical PVC signal from a pressure
transducer (Figure 2c), which presents visualizations of the PVC produced by
natural gas-burning fuel rich on its boundary and from LES predictions of
isothermal flow from a gas turbine swirler, presented by Figure 2d.

    In gas turbine combustors, the problem appears to arise from a threefold
interaction of combustion, system acoustics, and swirl flow aerodynamics. The
principal of acoustic streaming is well known in the areas of rockets and ram
jets, but its effect has not been considered to the same extent with swirl
combustion systems. Here, the acoustic forcing of the system interacts with the
flow and flame dynamics to actually maintain or increase the excitation and this
is illustrated by two examples; Figure 3 shows the first example (Rodriquez-
Martinez 2003, Syred 2006). Here, the swirl burner furnace system shown
(geometric swirl number 2.18) is set up in such a way that regular oscillations
of around 243 Hz occur, the oscillation being driven by a Helmholtz resonance
with the burner and neck forming the oscillator. Natural gas is used as fuel with
50% axial fuel injection and 50% premixed. A regular near sinusoidal pressure
oscillation is imposed on the system, shown by the pressure trace. Curve a
shows the cyclic variation of axial velocity over the averaged limit cycle, c, the
corresponding tangential velocity, and b and d, the corresponding directional
intermittency or percentage of time the flow is reversed. The oscillation has
significant effects on all the flow components. The axial velocity follows the
well-known pattern of a swirling shear flow entering the furnace and generating
a CRZ and an ORZ. Both the CRZ and ORZ vary considerably in size, extent,
and level of reverse flow velocity (and hence level of recirculation of hot active
products) over the oscillation cycle, and appear to be part of the mechanism
giving rise to excitation via the Rayleigh criterion. The boundary of the CRZ is
quite well marked by the yellow $\gamma = 50\%$ contour; it is clear that the visible
volume of the CRZ is varying by up to a factor of 5 over the phase angles
of 0–180°. Arising from this is the considerable variation of swirl number
throughout the oscillation cycle, clearly seen from Figure 3a and c where there

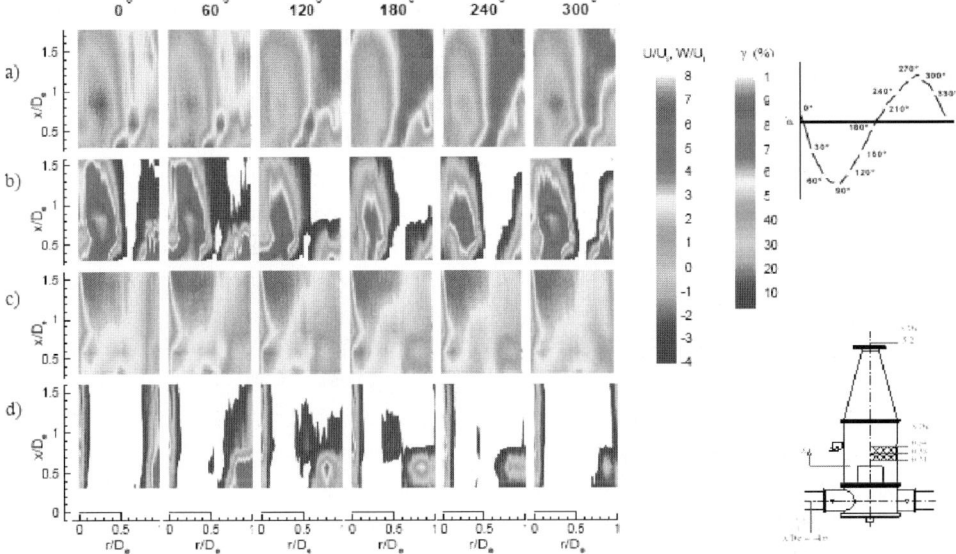

Figure 3. LDA phase-locked velocities in furnace of swirl burner/furnace system, excited at 250 HZ via Helmholtz resonance. (a) Axial velocity; (b) axial directional intermittency; (c) tangential velocity; (d) tangential directional intermittency. Mean inlet flow velocity $U_i = 4.3$ m/s: inlet flow rate air 0.468 m^3/s: $\varphi = 0.81$: natural gas injected 50% axially through base, rest premixed, no quarl

is substantially more variation in axial compared to tangential velocity for the various phase angles shown. Indeed, the CRZ is at its weakest between phase angles of 180° and 240° where axial velocity levels are highest, hence swirl numbers lowest.

The tangential directional intermittency levels are very high on the outside of the flow around the ORZ (Figure 3d), for much of the oscillation cycle indicating the presence of external helical flame structures on the edge of the flame. There is also some evidence of some PVC presence in the central region of flow. The axial direction intermittencies (Figure 3b), are also very high in the region of the ORZ and wall flow and again when considered with the tangential ones indicate the presence of helical or PVC type flow in this region. This is reinforced by reference to Figure 3e and f which replot the data. Figure 3e shows the phase-locked distributions of axial and tangential velocity in a tangential radial plane at $x/De = 0.58$ above the burner exhaust in the furnace. Here, it can be seen that the whole flow is rotating slightly off center or rotating eccentrically, and here the PVC effect in the center of the flow is small, but is amplified at the outside of the flow by the constraints applied by the furnace walls. Similarly, Figure 3f shows similar plots of the tangential directional intermittency at three different heights in the furnace. Very high levels of negative tangential directional intermittency are shown for a significant range of phase angles for $r > 0.8$ for $x/\text{De} = 0.31$ and 0.58, as well as in a small central region. The effect is reduced considerably by $x/\text{De} = 0.84$. Again, this reinforces the argument for helical structures on the outside of the flame.

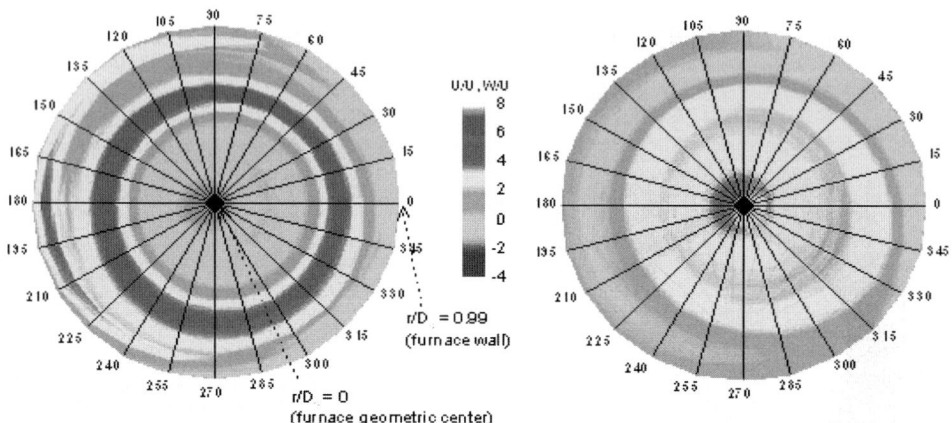

Figure 3e. Distribution of phase-averaged axial velocity (*left*) and tangential (*right*) in tangential radial plane for $x/\text{De} = 0.58$

The addition of an exhaust quarl, which nearly filled in the region where the ORZ forms, caused substantial reduction in the amplitude of the oscillation, primarily through the virtual elimination of the ORZ, illustrated by Figure 4. As can be seen the CRZ is much more regular in size (Figure 4a), whilst the levels of directional intermittency on the outside of the flow are much reduced (Figure 4b and d). However, there is still an underlying oscillation and this is returned to later. Figure 5a and b analyze the data from Figures 3 and 4 further by integrating the axial velocity profiles for each measured phase angle and measurement section in the furnace so as to give a volumetric flow rate at each section and phase angle. This can be used to give indications of the phase angle-related burning patterns in the system. Figure 5a clearly shows that most expansion of the flow occurs well downstream of the burner exit in the furnace for $x/\text{De} > 1.4$ and for phase angles between 150° and 270°, where axial velocity levels are highest in the furnace (average inlet flow 0.468 m³/s). Combustion appears to occur for all phase angles here. Closer to the burner exit for $x/\text{De} < 0.9$ volumetric flow rates are much less, hence the observed temperature levels, whilst just past the burner exit at $x/\text{De} = 0.31$ there is evidence of flow reversal back into the burner for phase angles between 330° and 90°, with very little combustion occurring. The flame then appears to reestablish itself close to the burner for phase angles between 120° and 300°, before lifting off again to a point further downstream. For the second configuration with the quarl there is much more uniformity of volumetric flow rate for all phase angles and measurement sections, whilst the flame is anchored much closer to the burner exhaust, as shown by comparing the volumetric flow rates at sections x/De = 0.58 and 0.84.

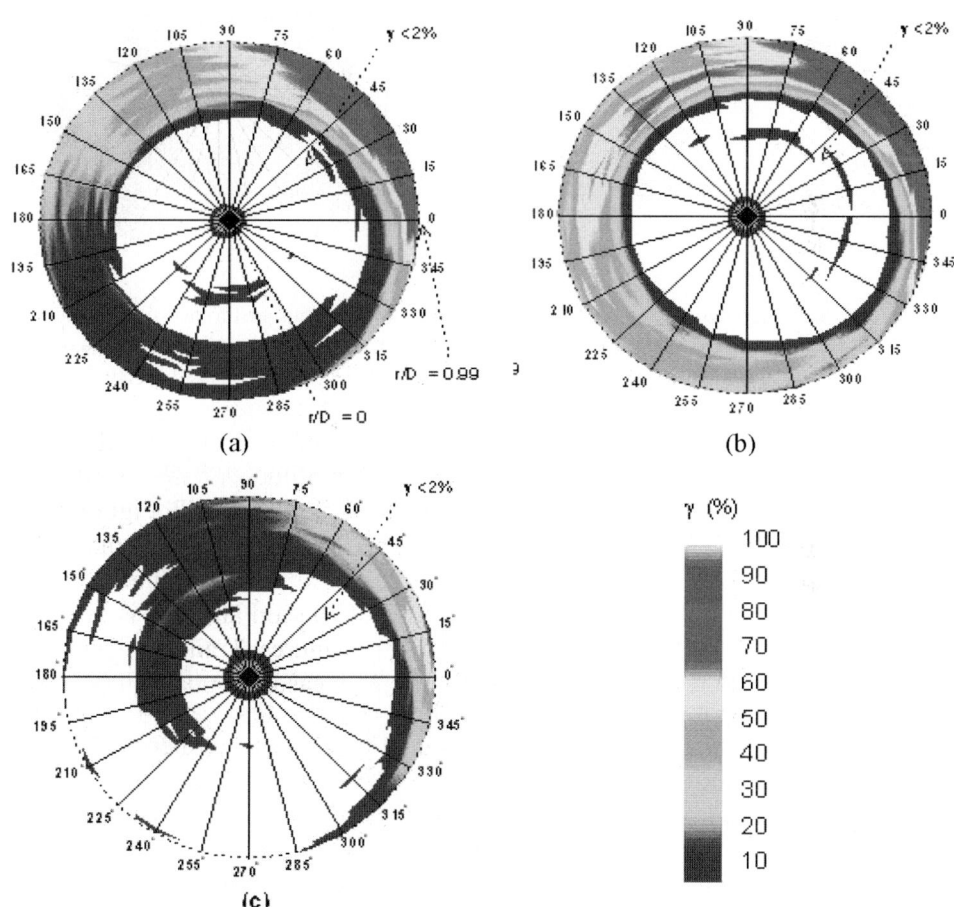

Figure 3f. Distribution of tangential intermittency at three heights in the furnace (a) x/De = 0.31; (b) x/De = 0.58; (c) x/De = 0.84

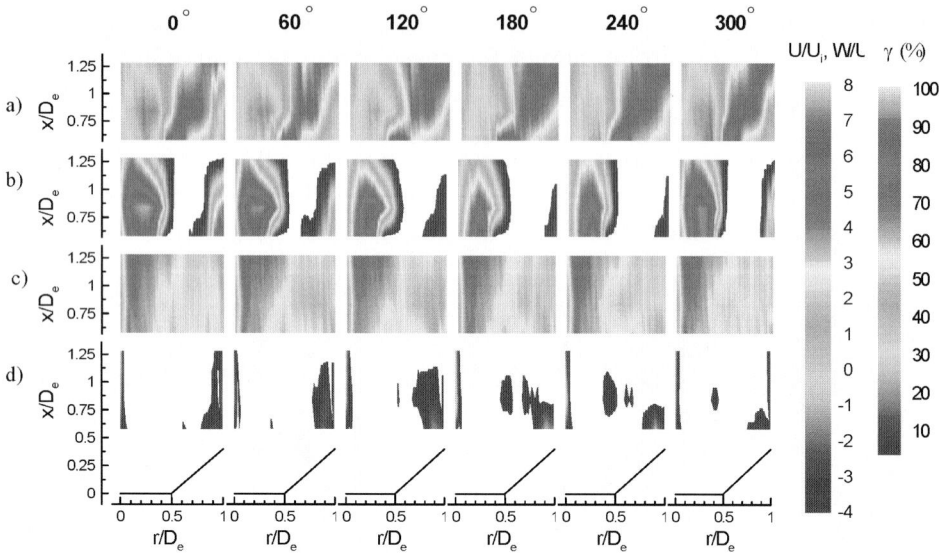

Figure 4. Phase-locked velocity measurements in furnace of swirl burner/furnace system, excited by 250 Hz Helmhlotz resonance. (a) Axial velocities; (b) axial directional intermittencies; (c) tangential velocities; (d) tangential directional Intermittencies. Cylindrical quarl inserted around burner exit. Other data as Figure 3

Schildmacher et al. (2002) have confirmed the above findings as to the variation in swirl number during a regular acoustically excited oscillation caused by a natural gas flame from a gas turbine swirler firing into a square enclosure. This is illustrated by Figure 6a and b. This system is similar to that used in Figure 3, although operating at higher intensity. In the reaction zone the swirl number varies between 0.1 and 0.9 for phase angles between 0° and 180°, which in many ways is similar to that indicated by Figure 3. Although the flow characterization is not so detailed as that shown in Figures 3 and 4, there are indications of considerable variation in axial flow, swirl number and hence CRZ size over the limit cycle of oscillation.

The second example of this phenomenon is also well illustrated by the work of Selle et al. (2006) who undertook a joint LES and Helmholtz analysis of rotating modes of oscillation in the natural gas flames produced by a Siemens gas turbine combustor firing into a square furnace. Although values of swirl number are not given, the device is stated to be of high "swirl" and thus comparable to other work discussed here. The 1,198 Hz helical structure

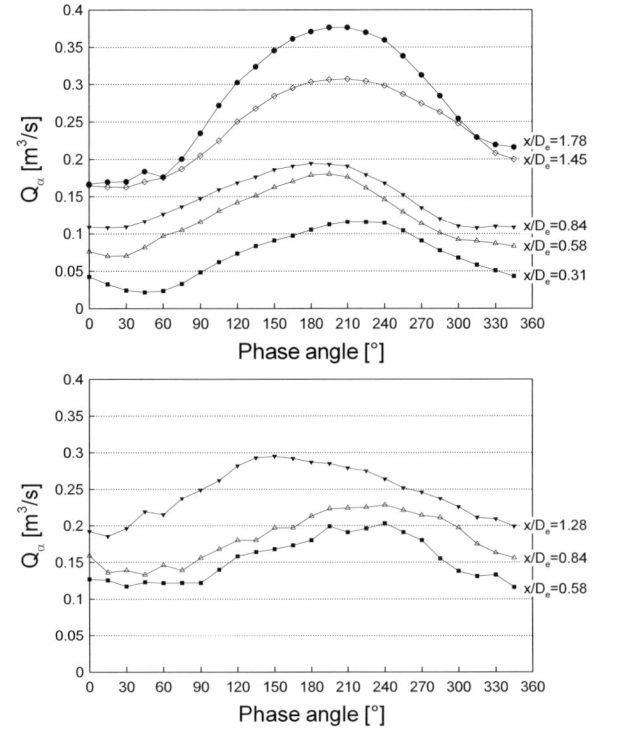

a) Distribution of Integrated Volume Flow Rate at each Phase Angle and Measurement Section, No Quarl, from Figure 3

b) Distribution of Integrated Volume Flow Rate at each Phase Angle and Measurement Section, with Quarl, as Figure 4

Figure 5. Distribution of volumetric flow rate at each measurement section and phase angle

observed in the flames was acoustically controlled and arose from the coupling of two transverse 1,192 Hz acoustic modes of the combustion chamber. This is shown in Figure 7 where the helical flame and flow shape is apparent. The work shows that in this combustion system the flame shape is strongly affected by acoustic velocity fluctuations at a frequency of 1,198 Hz. This mode is simply not forced by combustion. The acoustics also affect the flame itself leading to a self-sustained three-dimensional (3D) instability of helical shape. Critically, this occurs at the burner mouth, where the swirling flow is most sensitive to disturbances via coupling between the expanding swirling flow, associated centrifugal and axial pressure gradients, and formation of the CRZ and ORZ (Syred et al. 2006). The coherent structure is of spiral, tightly wrapped, PVC form and arises from the combination of two motions; the rotating mode and convection by the mean flow field. Selle et al. (2006) conclude that predictions of combustor flowfields need to take into account interaction of acoustics and turbulent flowfields. Other helical structures in similar gas turbine swirlers have also been reported (Schildmacher et al. 2002).

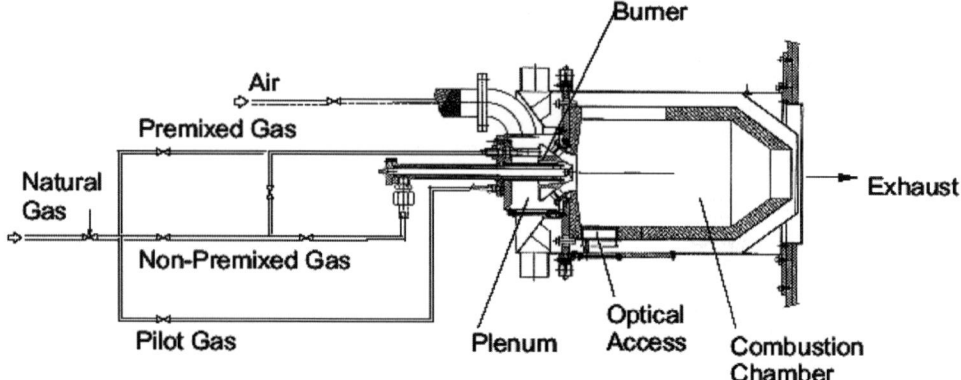

Figure 6a. Gas turbine swirler firing into combustion chamber with optical access

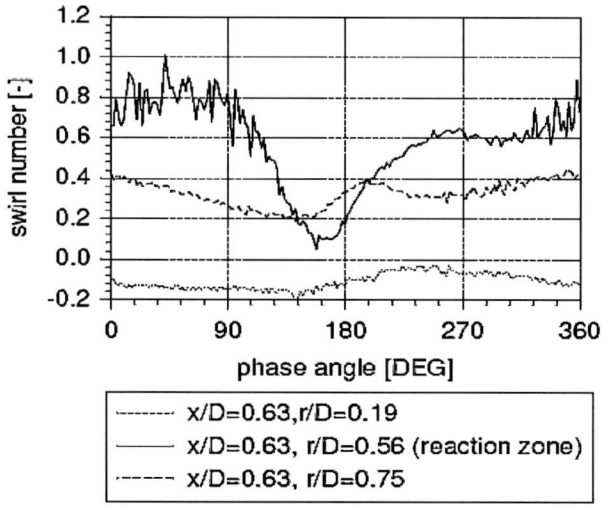

Figure 6b. Derived values of local swirl number as a function of phase angle over limit cycle oscillation

It is now pertinent to return to the work of Syred et al. (2002), Dawson et al. (2005) and Rodriquez- Martinez (2003) via reexamination of the data shown in Figure 3 and further data just above the furnace exit (Figure 8) from Syred et al. (2006) and Rodriquez-Martinez (2003). As has been discussed above, Figure 3 shows via the tangential directional intermittency measurements that there is tangential flow reversal on the outside of the flow near the furnace wall over part of the oscillation cycle, especially close to the region where the swirl burner fires into the furnace. When combined with high levels of axial directional intermittency in and around the same region, this again points to the

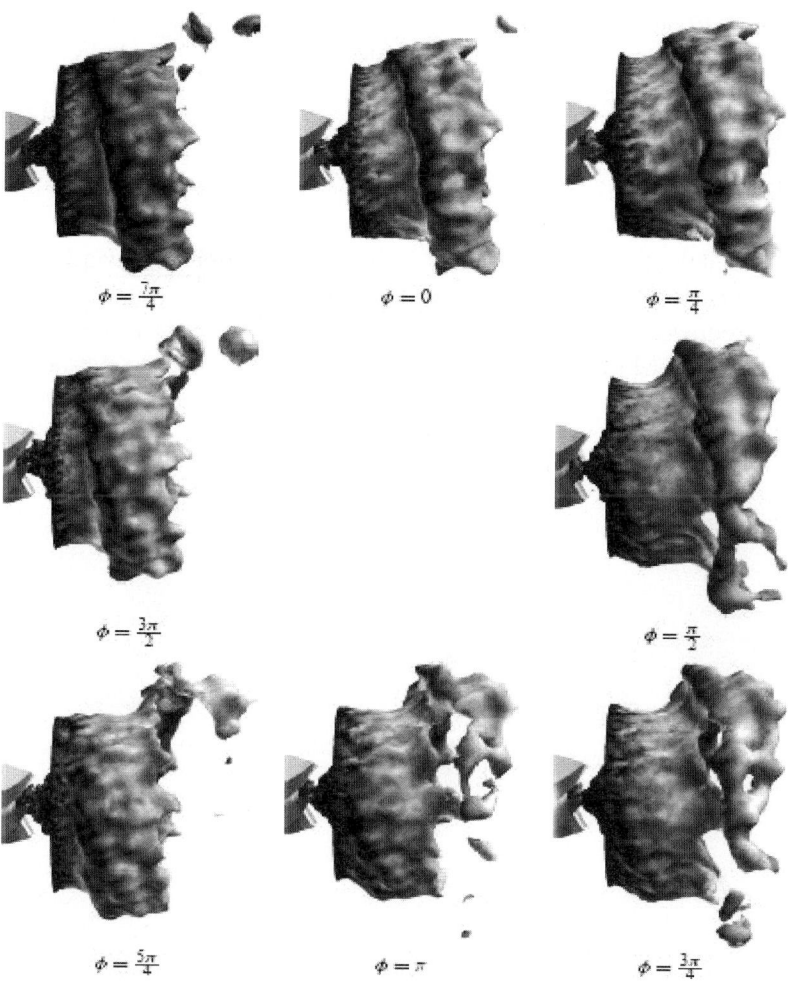

$\phi = \frac{7\pi}{4}$ $\phi = 0$ $\phi = \frac{\pi}{4}$

$\phi = \frac{3\pi}{2}$ $\phi = \frac{\pi}{2}$

$\phi = \frac{5\pi}{4}$ $\phi = \pi$ $\phi = \frac{3\pi}{4}$

Figure 7. Distortions produced in flame front by acoustic coupling of two transverse modes of the combustion chamber, resulting in a helical structure. The iso-surface is of the 1,000 K temperature

existence of spiral PVCs or helical structures close to the furnace wall. Again just above the furnace exit (Figure 8) the swirling flow is near symmetrical about the axis (Figure 8b), although the axial exhaust shows significant bias to one side (Figure 8a). However, the two sets of directional intermittency show different, but complimentary, pictures (Figure 8c and d). Tangential directional intermittency show a very extensive layer of values of $\gamma \sim 80$–90% close to the wall, for phase angles 90–240°, accompanied by an even thinner layer of axial directional intermittencies of $\gamma \sim 80$–90% for phase angles 105–240°. When combined with data from Figure 3f this can be interpreted as demonstrating the probable existence of tightly wrapped spiral PVCs or helical structures extending from the base of the furnace by the burner exit right the way to the furnace exhaust. Thus, the data indicates the existence of helical waves on the outer edge of these confined, combustion, swirling flows, excited by the interactions of the acoustics, combustion and swirling flow, the process being excited by wobble of the main vortex.

Unfortunately, the phase-locked technique used, although powerful, only locks onto a single frequency and thus information on other regular structures of different frequencies can be lost. Evidence from several sources (i.e., Syred 2006, Rodriquez-Martinez 2003) shows that there are often several harmonics

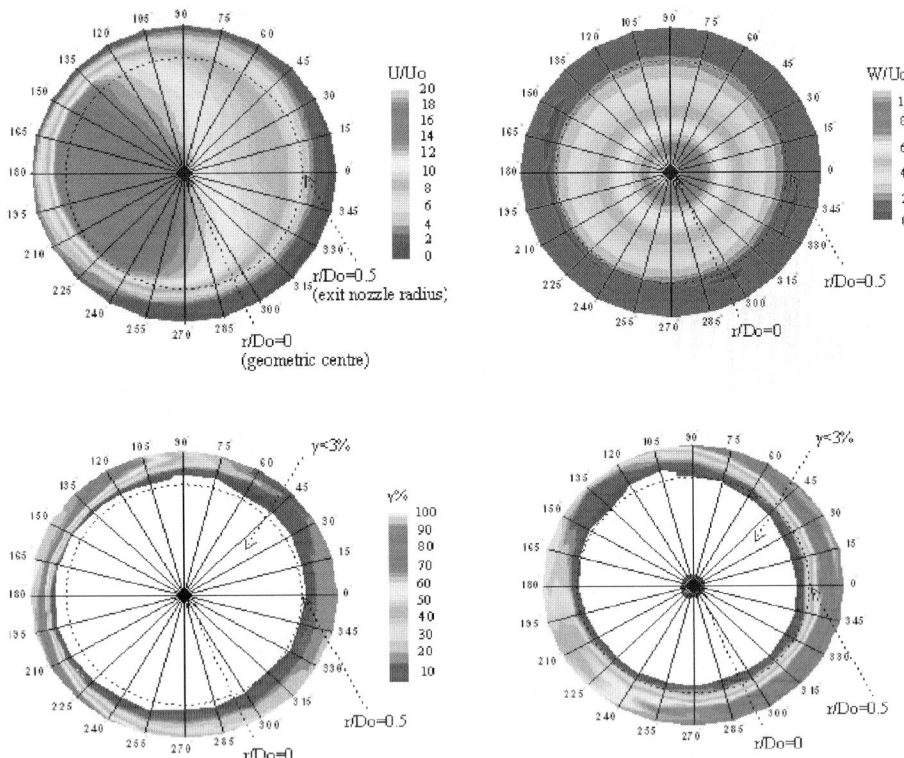

Figure 8. Phase-averaged (a) axial and (b) tangential velocities, directional intermittencies γ of the (c) axial and (d) tangential velocities just above the furnace nozzle exit ($x/De = 4.0$) for configuration 2a, as Figure 3a, sudden expansion swirl burner to furnace. Velocities normalized by the mean inlet velocity $U_i = 4.3$ m/s

of PVC or helical structures present and the flow actually jumps between states, i.e., single to double, etc.

It is now useful to look at the effect of configuration and equivalence ratio on oscillations in the swirl burner/furnace system shown in Figures 3 and 4. Figure 9a shows the effect of equivalence ratio upon pressure amplitude for the two configurations, 2a without the quarl, as shown in Figures 3 and 8 and 2b with the quarl (Figure 4). What is clear is that both configurations produce highest amplitudes for an equivalence ratio of about 0.65–0.7, with a rapid tail off in amplitude for $0.6 > \varphi > 0.9$. These limits correspond to frequency jumps from 250 Hz down to 50 Hz resonances, which are excited by other acoustic systems (travelling waves in the inlet pipes), Rodriquez-Martinez (2003). Thus, the ORZ clearly produces increased excitation of the Helmholtz 250 Hz oscillation over the equivalence ratio range $0.6 > \varphi > 0.9$, but does not remove the underlying cause of excitation.

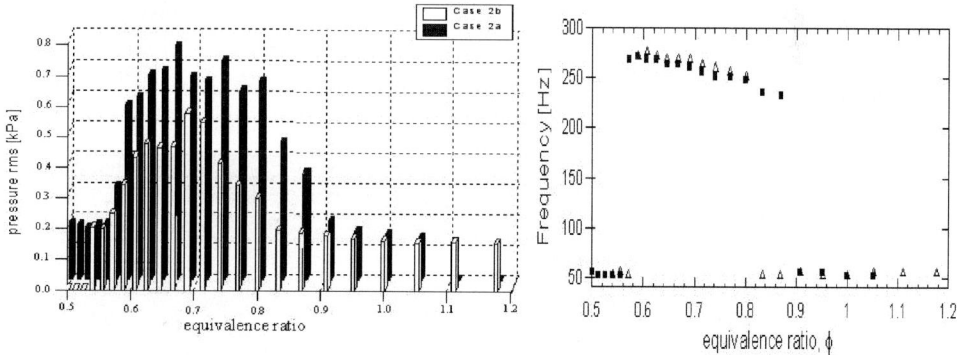

Figure 9. (a) Effect of configuration and equivalence ratio upon pressure amplitude; (b) effect of configuration and equivalence ratio upon oscillation frequency

Selle et al. (2006) fired their swirl combustor into a square furnace and undoubtedly there would have been regions where external recirculation zones would have formed, i.e., ORZs, possibly contributing to the excitation of combustion oscillations. Unfortunately, the work does not discuss the deformation of the CRZ over the oscillation cycle.

Rodriquez-Martinez (2003) and Froud et al. (1997) used the same swirl burner/furnace system as shown in Figure 3, but produced a wide range of stable oscillatory conditions by varying a wide range of parameters including inlet pipe length and configuration, furnace configuration, method of fuel entry, partial premixing, exhaust pipe length, and swirl number. System oscillations were typically in a frequency range of 40–60 Hz or more and corresponded to excitation either via Helmholtz resonance or via travelling acoustic waves in the inlet duct or exhaust pipe resonance. These low frequency oscillations could be of amplitudes higher than those produced by the 250 Hz oscillation (Figure 3), and very dependent on the system configuration. However, the common factor with the 250 Hz oscillation is the considerable variation in the size and shape of the CRZ over the limit cycle of oscillation, as well as the axial flow into the furnace and hence swirl number. Indeed reversed flow in the inlets was recorded by Rodriquez-Martinez (2003).

In all these studies of the swirl/burner furnace system of Figure 3 the common characteristic has thus been the variation in the size and shape of the CRZ and ORZ (where present), as well as the swirl number over the limit cycle of oscillation. Thus, this appears to be one of the major feedback mechanisms maintaining the amplitude of the limit cycle oscillation at high levels and is also in accord with typical convection and acoustic timescales.

However, what is still not clear is the initiation mechanisms for these oscillations and the role of the PVC or helical waves. One possible mechanism is the shedding of axial radial eddies by the PVC or by the helical structures on the edge of the flame as shown schematically in Figures 2a and 10. Figure 10

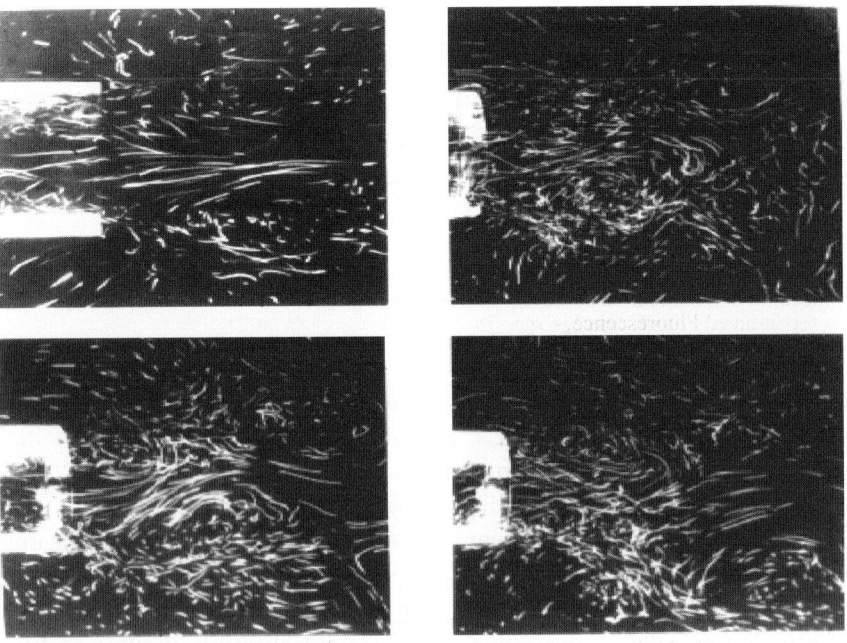

Figure 10. Four water model photographs of the exhaust of a swirl burner, $S = 1.8$, showing formation of axial radial eddies illumination via slit light in axial radial plane

shows four water model photographs of the flow leaving a model swirl burner, illuminated on its center line by slit light in the axial radial plane.

3. Conclusions

Thus, the work presented here appears to bring out two major findings:

Maintenance of the oscillations in swirl combustion systems appears to occur through amplification via variations in the size and shape of the CRZ and ORZ over the limit cycle of oscillation as this variation causes considerable variation in the flame holding capability of the system, allowing flame movement and on occasion near extinction to occur.

The initial excitation is more difficult to unravel, but does appear to be associated with the generation of axial radial eddies by the PVC, either internally in the CRZ or externally at the end of the shear layer.

Further detailed time-dependent experimental and predictive work is needed to resolve these important issues is needed.

References

Broda, J.C., Seo, S., Santoro, R.J., Shirhattikar, G., and Yang, V. (1998) An Experimental Study of Combustion Dynamics of a Premixed Swirl Injector, *Proc. Combust. Inst.*, **27**, 1849–1856.

Froude, D., Beale, A., O'Doherty, T., and Syred, N. (1997) Studies of Helmholtz Resonance in a Swirl Burner/Furnace System. Proceeding of the 26th International Symposium on Combustion, Pittsburgh, PA, The Combustion Institute, pp. 3355–3362.

Giezendanner, R., Keck, O., Weigand, P., Meier, W., Meier, U., Stricker, W., and Aigner, M. (2003) Periodic Combustion Instabilities in a Swirl Burner Studied by Phase-Locked Planar Laser-Induced Fluorescence, *Combust. Sci. and Tech.*, **175**, 721–741.

Keller, J.O., Bramlette, T.T., Dec, J.E., and Westbrook, C.K. (1989) Pulse Combustion: The Importance of Characteristic Times, *Combust. Flame*, **75**, 33–44.

Keller, J.O., Bramlette, T.T., Westbrook, C.K., and Dec, J.E. (1990) Pulse Combustion: The Quantification of Characteristic Times, *Combust. Flame*, **79**, 151–161.

Gupta, A.K., Lilley, D.G., and Syred, N. (1984) *Swirl Flows*, Abacus Press, Tunbridge Wells, England.

Lee, S.-Y., Seo, S., Broda, J.C., Pal, S., and Santoro, R.J. (2000) An Experimental Estimation of Mean Reaction Rate and Flame Structure During Combustion Instability in a Lean Premixed Gas Turbine Combustor, *Proc. Combust. Instit.*, **28**, 775–782.

Lieuwen, T. and Zinn, B.T. (1998) The Role of Equivalence Ratio Oscillations in Driving Combustion Instabilities in Low NO_x Gas Turbines, *Proc. Combust. Inst.*, **27**, 1809–1816.

Lieuwen, T. (1999) Ph.D. thesis, Georgia Institute of Technology, Atlanta, GA .

Mongia, R., Dibble, R., and Lovett, J. (1998) Measurement of Air-Fuel Ratio Fluctuations Caused by Combustor Driven Oscillations, *Proc. of the ASME/IGTI Turbo Expo*, ASME paper No. 98-GT-304.

Paschereit, C.O., Gutmark, E., and Weisenstein, W. (2000) Excitation of Thermoacoustic Instabilities by Interaction of Acoustics and Unstable Swirling Flow, *AIAA Journal*, **38**(6), 1025.

Richards, G.A. and Janus, M.C. (1997) Oscillations During Premix Gas Turbine Combustion, ASME Paper No. 97-GT-244.

Rodriquez-Martinez, V.M. (2003) Ph.D. thesis, Cardiff University, Cardiff, UK.

Roux, S, Lartigue, G., Poinsot, T, Meier, U., and Bérat, C. (2005) Studies of Mean and Unsteady Flow in a Swirled Combusator Using Experiments, Acoustic Analysis and Large Eddy Simulations, *Combust. Flame*, **141**, 40–54.

Schildmacher, K.-U., Koch, R., Krebs, W., Hoffman, S., and Wittig, S. (2002), Experimental Investigations of Unsteady Flow Phenomena in High Intense Combustion Systems, Proceedings of the Sixth European Conference on Industrial Furnaces and Boilers [ISBN 972-8034-05-9].

Selle, L., Benoit, L., Poinsot, T., Nicoud, F., and Krebs W. (2006) Joint use of Compressible Large-eddy Simulation and Helmholtz Solvers for the Analysis of Rotating Modes in an Industrial Swirled Buner, *Combust. Flame*, **145**(1–2), 194–205.

Straub, D.L. and Richards, G.A. (1998) Effect of Fuel Nozzle Configuration on Premix Combustion Dynamics, ASME Paper No. 98-GT-492.

Straub, D.L., Richards, G.A., Yip, M.J., Rodgers, W.A., and Robey, E.H. (1998) Importance of Axial Swirl Vane Location on Combustion Dynamics for Lean Premix Fuel Injectors, *AIAA Joint Propulsion Conference*, 13–15 July, Paper no. AIAA-98-3909.

Syred, N. and Beer, J.M. (1974) "Combustion in Swirling Flow: A Review". *Combust. Flame*, **23**, 143–201.

Syred, N. (2006) A Review of Oscillation Mechanisms and the role of the Precessing Vortex Core (PVC) in Swirl Combustion Systems, *Progress in Energy and Combustion Systems*, **32**(2), 93–161, ISSN 0360-1285.

Turrell, M.D., Stopford, P.J., Syed, K., and Buchanan, E. (2004) CFD Simulations of the Flow within and Downstream of High swirl Lean premixed Gas turbine Combustors, Proceedings ASME Turbo-Expo 2004, vol. 1, *Combustion and Fuels*, pp. 31–38, Education 2004.

Venkataraman, K.K., Preston, L.H., Simons, D.W., Lee, B.J., Lee, J.G., and Santavicca, D.A (1999) Mechansim of Combustion Instability in a Lean Premixed Dump Combustor, *J. Prop. Power*, **15**(6), 909–918.

MODERN TRENDS OF POWER ENGINEERING DEVELOPMENT

A. F. RIZHKOV*

V. E. SILIN

N. S. SHAREF

E. V. TOKAR
*Ural State Technical University-UPI, "Energy saving" Dept.,
Mira, 19, Yekaterinburg, 620002, Russia*

Abstract. This article considers world energy trends in use of solid fossil fuels, biomass, IGC cycle, and distributed generation (DG). Analysis of the usage of basic types of gasifiers is carried out and some characteristics of their maintenance are discussed. The dense bed gasifier allows the use of boulders of wet biomass residues providing terms for tar free wood gas. Its basic design and engineering are known, including from the field of blast furnaces, calcining kilns and other shaft units. These factors substantiate the selection of dense bed unit for use in DG. The main problems of biomass usage are underlined: nonordered size and high moisture fraction of the residue and tars in the producer gas. A new concept for obtaining tar free wood gasg is presented. Energetically (adiabatic firing temperature, effective output) and ecologically (CO, NO_x, PHC) characteristics of an IC engine supplied by wood gas are discussed. R&D carried out by USTU-UPI are presented, which was aimed at the creation of new gasifiers, pyrolisers and tar-free wood gas technologies.

Keywords: distributed generation, IGCC, fossil fuel, biomass, dense bed gasifier, tar-free wood gas, IC engine, adiabatic firing temperature

1. Global Trends in Power Engineering Development

According to the forecasts for global power engineering development, the rate of electric power demand will outstrip the rate of production. At the present stage, the factors that determine power engineering development are the

---

* To whom correspondence should be addressed. Aleksander F. Rizhkov, Ural State Technical University-UPI, "Energy saving" dept., Mira, 19, Yekaterinburg, 620002, Russia; e-mail: ensav@mail.ustu.ru

N. Syred and A. Khalatov (eds.), Advanced Combustion and Aerothermal Technologies, 21–30.
© 2007 *Springer.*

ratification of Kyoto Protocol by many countries and the appearance of new up-to-date power-generating equipment and technologies on the market. One can trace the following main tendencies in power production:

- Fuel–power balance structure has changed in favor of solid fuel usage, i.e., the preferred usage of coals, local power sources, biomass, and wastes.
- Power supply structure has changed and power markets are under liberalization process.
- New technologies of clean coal combustion with gas-generating equipment are being developed.

The tendencies are interrelated to a high degree. An example of such interrelation is the fact that the necessity to fire low-grade local coals and biomass stimulates the development of systems of local (on-site) power sources located near the fuel sources. Environmental safety and cost-efficiency requirements to such power sources necessitate the development of up-to-date technologies using solid fuel in the gas power cycle.

1.1 CHANGE OF FUEL AND ENERGY BALANCE

Coal is far inferior to natural gas and oil in its cost-efficiency and ecological impact. It is for this reason that the share of coal and other solid fuels in the general volume of power consumption was shortened drastically during the last 100 years. However, due to the decrease of oil and gas resources both the technological importance and usage of solid fossil fuels have expanded since the 1970s in other countries.

Solid fuels of adverse environmental impact such as sulfur coal and shale are used mainly in major power generation facilities where the usage of material-intensive and expensive technologies for emission control is expedient. For low-power distributed generation (DG) facilities a significant role will belong to renewable fuels of natural origin (biomass) as being the most preferable for environmental protection and providing simpler techniques of gas cleaning.

In some way or other, the problem of developing the power industry based on local fuel sources and biomass is being worked at in all countries of the world. Worldwide experience shows that woodworking produces over 50% of wood waste. More than 50% of this amount is used in power generation. Thus, in Denmark 25% of heat energy is produced by palletized wood fuel.

In industrial countries the transfer of regional power engineering to such fuels is aimed at providing the missing link in the power supply system, i.e., the implementation of distributed power generation systems based on local environment-friendly power sources in distribution networks. In countries with coal-based power engineering it will improve the efficiency and ecological safety of regional power systems.

In countries and territories with export-oriented economies (Iraq, Khanty-Mansi autonomous region in Russia) it will improve the standard of living and at the same time it will release high-quality fossil fuels for export in the amount of 1–2 t of conventional fuel/man-year. For countries and territories of import-oriented economies (Japan, northern Russian regions with season-dependent supplies) a wider use of local fuels and biomass will reduce power dependence and secure a better and safer power supply for the territory.

1.2 DISTRIBUTED POWER GENERATION AND LIBERALIZATION OF POWER MARKETS

Centralized production of recoverable power carriers, which is the basic system-forming principle, is the most cost-efficient procedure for a big Industrial region or a whole country, provided it is properly organized and up-to-date engineering decisions are made. The two energy production procedures, i.e., centralized generation at large-power stations (station-based power engineering) and DG at local installations (factory, settlement, cottage) will coexist in the world community in the near-term outlook.

Now, world power engineering faces the imminent problem of setting up a new power market. The pioneers in establishing the market relations in this sphere are Chili, Britain, and Norway. The processes of structural transformation in electric power engineering are taking place in Australia, Argentina, Brazil, Spain, Mexico, the USA, Sweden, Finland and other countries. Power Ministers of the European Community have adopted the directives on liberalization of relations in the European internal power market. The directives provide for changes in EC legislation and open a stepwise freeway for consumers to the electric power market.

The structure of world power markets is becoming orientated to the British liberalization model New European Trade Association (NETA), which creates a relatively flexible system of rates of charge and supply of electric and heat power to users. In Britain and in other countries that followed its lead, the government used to officially support the power market monopolies whereas now they have liberalized the markets to promote greater competition among suppliers.

While creating a competitive environment, Great Britain dealt with markets of two types based on the following systems: (1) the united system (the so-called British pool) (1990–2001); (2) the NETA (2001–present day). The united system allowed the creation of a free market with a centralized competitive price offer for producers with the contract price being preestimated and equal for everyone. NETA provided the opportunity to create a liberalized market based on private business relations between the consumer and the producer. The basic NETA principle is as follows: any persons who want to buy or sell electric power have the right to conclude any free contract agreements with each other.

NETA elaborators are sure that an immense share of electric power will be sold or bought through exchanges or a set of bilateral or multilateral contracts. The sellers of electric power include both the generating and marketing enterprises that produce and supply the energy to the final users and traders who are not natural persons. NETA provides a mechanism of close-to-real-time clearing and unbalance operations between contract and physical positions of persons who buy, sell, produce, and consume the electric energy.

An independent producer who can quickly change the ratio of heat and electric power supply to meet the ever changing consumers' demand on heat and electric energy has every opportunity to operate at a profit in new power market conditions and hence to be competitive. The producer who has such a capability is: (1) reliable (meets the forecasted demand) and (2) flexible (changes power generation according to the demand of local loads and the distribution network).

To achieve an adequate level of flexibility and reliability of operation, the independent producers with a combined heat/electric generation system should ensure a flexible ratio of heat and electric power supply at a general network level (together with steady producers) and execute continuous system control (metering and distribution). To achieve this goal, it is of primary importance to provide a flexible and reliable network which is the basis of a distributed power generation system.

Flexibility can be achieved by ensuring the grounds for quick connection of individual installations to the network or their disconnection. Reliability is based on the fact that the probability of simultaneous failure of several installations is low and that such installations operated as a group will satisfy variable consumers' demand within the entire network. Thus, a group of working installations is considered as a unified flexible and reliable generator of high capacity.

The experience of British reforms in power engineering has highlighted two procedures as a possible solution to the problem of flexibility and reliability: for small-capacity sources the installations with solid fuel elements (SOFC) are offered, for large-output sources the steam-gas plants –integrated cycle of coal gasification (IGCC) are preferable. The advantage of SOFC is high electric efficiency (at a level of 60%). SGP IGCC, with its dual-purpose produce, always operates at optimal mode with constant fuel consumption. Depending on electric energy demand, generator gas is supplied either to gas-fuel installations (ICE, GTU) or to liquid motor fuel producers.

The efficiency of power generation with various primary motors in large- and small-output power engineering is shown in Figure 1. While developing the fuel elements' processes and optimizing the design of small-output gas turbines, the prospective circuits of power sources of distributed power engineering can be arranged at these installations with a maximum electrical efficiency of up to 55–65%.

Figure 1 shows that small-power plants mainly use internal combustion engines and gas turbines. In the range of capacities equal to or below 30 MW the power sources with ICE are currently the least expensive, but according to British experts' estimates, the reliability of ICEs makes them less attractive for autonomous power sources working for a common network (for providing the required flexibility and high efficiency).

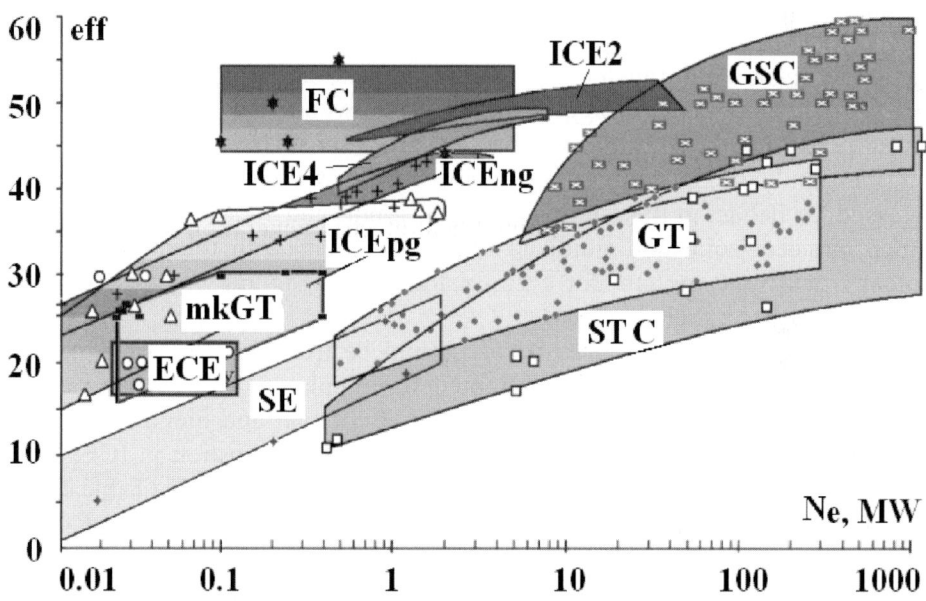

Figure 1. Electrical efficiency of various procedures: GSC – combined steam/gas turbine plant; GT – gas turbine plant; STC – steam turbine plant; ICEpg – internal combustion engine operating on producer gas; ICEng – internal combustion engine working on natural gas; FC – fuel cells; ICE2 – two-cycle diesel engines; ICE4 – four-cycle diesel engines; mkGT – microturbines; ECE – external combustion Stirling engines; SE – steam engines

1.3 DEVELOPMENT OF MODERN GAS-PRODUCING TECHNIQUES

In the USA, Europe, and Japan, large-scale international and national programs are being developed, aimed at the implementation of pollution-free solid fuel technologies for power generation in the gas-power cycle. In Russia, the plans for development of such technologies are included in the following materials on power strategy and program concepts, respectively; "Programs of renewal of major equipment at TPS, RAO 'UES of Russia' for the period until 2010" (technical retrofit of power units at condensing plants and TPS with different types of solid fuel gasification equipment, such as PFB) and "Power-efficient economy for 2007–2010 and prospects till 2015".

In distributed power generation a significant role belongs to local low-grade fossil fuels (peat, ligneous coals) and renewable fuels of natural origin (biomass). The main problems of gasification of such fuels are:

- CO emissions from gas-fuel installation (ICE)
- Caking

Due to pyrolysis with transfer through plastic state in the course of clinkering (for coals) at 400–500°C, or agglutination of fuel wood pieces as a result of tar, ballast increase in the shaft of solid-bed gas producer.

As a result of ash melting in the combustion zone (coals, peat), there is increased tar ballast in the range of 0.5–100 g/nm^3 and soot (to 5–10 g/nm^3) in known classical solid and fluidized bed installation (with tolerable pollutant content equal to 50 mg/nm^3 for a gas-piston ICE, 5 mg/nm^3 for a gas turbine, and 1 mg/nm^3 for a fuel element).

The problem of high CO emission should be treated as follows. In some countries, certain standards were adopted for regulating CO emissions from internal combustion engines working on generator fuel and biogas. The regulations were based on the experience of natural gas combustion in ICE. However, the composition of natural and artificial gases differs greatly (Table 1).

According to foreign studies a direct relation exists between the CO and CH_4 content in gas and the products of its combustion in an ICE, i.e., the higher the content in gas the higher it is in the combustion products. Hence, the actual CO emissions from a working ICE will always exceed the standards adopted in the EC. This means that the standards should be estimated with due account of actual fuel composition.

The problems of coking can be solved by maintaining the required lower temperature level in the combustion zone, giving priority to low-ash fuels with high-fusing ash (e.g., many-year cycle biomass) and by replacing the thermal carbon cracking and oxidation-free semiclinking processes with oxidation pyrolysis.

TABLE 1. Composition of fuel gases and products of their combustion in ICE

Component	Natural gas		Biogas		Generator gas	
	Content (%)	Content in combustion products (mg/m^3)	Content (%)	Content in combustion products (mg/m^3)	Content (%)	Content in combustion products (mg/m^3)
CH_4	90	1,000–1,500	50–65		1.0–1.5	250–500
CO		300		750–1,000	15–20	1,000–3,000
CO_2			35–40		8–10	250–650

To remove ballast components (e.g., tar and soot) from gas, the major power engineering firms and industries use well-designed secondary gas-conditioning procedures with proven workability and effectiveness. The standards for gas contaminants in generator gas for ICE and GTU are given in Tables 2 and 3.

TABLE 2. Requirements on gas quality for gas engines

Component	Max. permissible concentration (mg/m^3)	Experimental value (mg/m^3)
Particles	<50	<5
Tar	<100	<50

TABLE 3. Requirements on gas quality for gas turbines

Component	Concentration requirements
Particles	<1 ppm
Tars	5 mg/nm^3
HCl	<0.5 ppm
S (H_2S + SO_2, etc.)	1 ppm
Na	<1 ppm
K	<1 ppm
Other metals	<1 ppm

Generally, the block of power-generating plant with intracycle gasification includes at least three basic components: (1) gas generator or producer; (2) a system of gas conditioning to remove tars, usually "wet" procedures with solid particles removal from gas using fabric filters; and (3) gas-fuel power installation.

The efficiency of different installations for secondary gas cleaning is shown in Table 4. The principle of staged combustion executed at several installations minimizes flue gas losses by precise monitoring and control of the air flow factor at each stage. The gas conditioning system precedes the gas combustion stage and gas "load" on the latter is reduced. It can operate on a more concentrated gas flow and the general ecological efficiency of the station rises.

TABLE 4. Efficiency of various installations for secondary gas conditioning

Installation	Temperature (°C)	Efficiency (%)
Hose filter	200	Max. 25
Sand filter	10–20	60–95
Rotatory wet scrubber	50–60	10–25
Venturi scrubber		50–90
Electrostatic precipitator	40–50	99.3
Fluidized bed with catalyst	700	99.5–99.9

Therefore, in the past years, the so-called primary cleaning methods realized directly in gasification chamber, either by design changes or other measures, have been under development all over the world. A major advantage of such methods is that they do not use "wet" gas conditioning systems, which actually reduces the cost of 1 kW of installed capacity by a factor of 1.5–2 and improves environmental safety of the station. In this case the only thing needed is to remove ash particles from gas in a gas-cleaning system.

The gas pollutant content varies over a wide range, depending on the gas generator type (Table 5).

At present the following alternatives of primary gas-cleaning techniques are known:

- One-unit, for mini-TPS of electric output up to 500 kW, including

    - Direct process with steam-gas recirculation to high-temperature zone

- Downdraft process and its variations with two combustion zones (Russian and foreign two-zone type)

- Multiunits (decomposed in separate units), for mini-TPS of electric output over 500 kW

TABLE 5. Gas impurities at different gasifying installations

Gas generator	Max. temperature of pollutant formation (°C)	Pollutant components	Pollutant content (g/nm^3)
Direct technique	300–500	Primary tar > 80%	<150
		Secondary tar < 20%	
FB/CFB	700–800	Primary tar < 30%	10–30
		Secondary tar < 60%	
		Tertiary tar < 10%	
Downdraft low-temperature producer	1,200	Tertiary tar = 100%	Tars < 1–2
		Soot	Soot < 3
Flow-type, horizontal, slag-tap	>1,300	Soot	Soot < 3
Multizone	900	Tertiary tar = 100%	Tars < 0.05
		Soot	Soot < 0.5–2
Multistage	700	Tertiary tar = 100%	Tars < 0.005
		Soot	Soot < 0.4–0.5

The "Viking" gas generating installation at Denmark Technical University is an example of multiunits process.

Between the pyrolysis and gasification steps in this installation, the steam/gas products are partly oxidized by added air. This results in tar content reduction in volatiles by a factor of 100 and the developed heat energy is used for gasification. When partially oxidized products of pyrolysis pass through the coke residue layer, further tar content reduction occurs by a factor of approximately 100. The final tar content in the gas product is <15 mg/nm^3. The problem of soot formation in this installation is solved by secondary procedures.

The system has a well-developed regeneration system and tar content in generator gas is below 25 mg/nm^3. Tar content in cleaned gas (after particle removal) is about 5 mg/nm^3. In this two-stage gas generator the gas supply to the pyrolysis installation can be performed in two ways: by the recuperative principle through the wall of the worm feeder and by direct contact with superheated steam. In the latter case the superheated steam is directed to the coke residue layer for the purpose of steam gasification.

The developers of the technique estimate that the specific cost of power installation development, based on a two-stage gas generator, is equal to $1,100/kW of heat power. For two-stage installations with ICE, the specific cost of electric power is comparable to that of fuel elements and installations of nonfuel power generation ($3,000–$4,000/kW). Such installations become efficient above 2 MW in terms of the product-gas output (electric power of ICE-TPS over 600 kW).

2. Conclusions

Modern trends in power engineering development greatly depend on ever increasing imbalances, i.e., ecological imbalances in the global system "Nature-Civilization" and economical imbalances connected to nonuniform production and consumption of high-grade primary energy carriers. One of the most preferable solutions of eliminating these imbalances is to use local fuels in power generation, starting with widely distributed pollution-free renewable biomass.

In light of the above, comprehensive research and development studies of clean power gas production technologies for steam/gas and IGCC have been started abroad. In Russia, the introduction of up-to-date power supply technologies based on local fuels is hindered, to some extent, by the insufficient development of Russian equipment and the high price of foreign equipment.

The formation of complete and effective systems of distributed power generation in Russia can be a matter of some difficulty as it is not an easy task to ensure the reliable power supply in this country by abandoning the centralized development and execution of generation modes. The operating and dispatching system is also unavailable for microtrade which is required, for instance, by NETA ideology. In addition, the Russian simplified concept of bilateral contract does not comply with the above system.

INFLUENCE OF THE HEAT-AND-POWER PLANTS OF UKRAINE ON THE ENVIRONMENT AND PRIMARY WAYS OF POLLUTION REDUCTION

A. S. BABYAK
Ministry of Fuel & Energy of Ukraine

A. A. KHALATOV, S. V. SHEVTSOV
Institute of En gineering Thermophysics (IET), National Academy of Sciences, Ukraine

Abstract. Heat-and-power plants form the basis of this country's power engineering which in turn is the principal sector of the National Fuel & Energy Complex (FEC) and will do so for the foreseeable future. Apart from power, the heat and power plants (HPPs) also generate an unwanted by-product – toxic releases into the air basin of Ukraine in quantities beyond tolerable limits. In fact, the contribution of HPPs in this respect runs to 30% of the total emissions coming from stationary sources.

A threat to human health and environmental hazards are strongly amplified by the concentrated locations of HPPs in densely populated industrial and agricultural regions as well as by the presence in the emissions of such toxic compounds as sulfurous anhydride (70% of total emissions in Ukraine), nitrogen oxides, ash, and carcinogenic substances (benzopyrene, vanadium oxide, high-molecular compounds).

In an effort to reduce the adverse environmental impact the power engineering enterprises focus their activity on the retrofit of energy units, introduction of new technologies for solid fuel combustion, maintaining flue gas purifying facilities in proper working condition and other environmental protection measures.

Keywords: heat-and-power plants, environment protection, harmful releases, national energy program retrofit, advanced technologies

N. Syred and A. Khalatov (eds.), Advanced Combustion and Aerothermal Technologies, 31–45.
© 2007 *Springer.*

1. The Adverse Impact of Heat-and-Power Plants that Remain Within the Jurisdiction of Ukrainian Ministry of Fuel and Energy Upon the State of the Air Basin of Ukraine

Power engineering as the basic sector of the Fuel and Energy complex (FEC) meets this country's demands in electricity and partially in heat. The power engineering today and for the foreseeable future rests upon heat-and-power plants (HPPs) whose performance and development level are directly responsible for the volume of the polluting emissions into the air basin of Ukraine. The power engineering enterprises fall into the category of elevated-hazard objects and any emergency situation, breakdown or faulty performance may readily result in supernormative polluting emissions.

1.1 BRIEF CHARACTERISTIC OF THE MAIN POWER-GENERATING EQUIPMENT

The basis of this country's power engineering consists of 104 generating units with output capacity ranging from 150 to 800 MW. The units are installed at 14 HPPs and three central heat-and-power plants (CHPP). Arbitrary arrangement of the units in clusters of equal output capacity yields the following pattern:

- 8 units of 720–800 MW, total capacity 6.3 GW

- 42 units of 282–300 MW, total capacity 12 GW

- 5 units of 250 MW, total capacity 1.2 GW

- 43 units of 175–210 MW, total capacity 8.0 GW

- 6 units of 150 MW, total capacity 0.9 GW

The majority of HPPs were constructed in the 1970s and have already been in operation for 30 years or more. Relating capacity to construction date, the HPPs with units of 150 MW were commissioned between 1959 and 1964, units of 200 MW between 1960 and 1975, units of 300 MW between 1963 and 1988 and 800 MW units between 1967 and 1977.

Some data on power generating units of Ukrainian HPPs are given in Table 1.

The current HPPs are in a poor state and suffering from obsolescence. Thus, 46 units from the total number of 104 (34.8% of the entire generating capacity) have already exceeded the permissible level of service wear having seen 200,000 h in operation, 29 units (28.4%) have exceeded their limiting operational resource (170,000 h) and 17 units (27.4%) are at the end of their rated lifetime. Only 12 units (9.4%) still remain on the safe side.

Even the youngest of the HPPs, namely Tripolskaya, Uglyegorskaya, Zaporozhskaya, Krivorozhskaya and Ladyzhinskaya are 30 or more years old, while the "veterans" of power engineering such as Luganskaya, Zmyeyevskaya, Starobeshyevskaya and Slavyanskaya HPPs are well over 40 years old.

Presently, a dramatic drop in the use of generating capacities is being observed in this country: from total installed potential of 36 GW only 19 GW (53%) is practically used due to the wear of equipment and the lack of funds for purchasing fuel. The remaining nonused capacities are either in a state of

TABLE 1. Some data on power-generating units of Ukrainian HPPs. (From Енергетичні ресурси та потоки, Київ, 2003 (Energy resources and flows, Kyiv, 2003.)

Name of power plant	Number and rated capacity of units (MW)	Total number of units	Years of commissioning
Luganskaya	8 × 200	8	1961–1969
Starobyeshevskaya	10 × 200	10	1961–1967
Slavyanskaya	1 × 800		1967–1971
Uglyegorskay	4 × 300 + 3 × 800[a]	7	1972–1977
Kurahovskaya	1 × 200 + 6 × 210	7	1971–1975
Zuyevskaya	4 × 300	4	1981–1988
Pridnyeprovskaya	4 × 150 + 4 × 300	8	1958–1965
Krivorozhskaya	10 × 300	10	1965–1973
Zaporozhskaya	4 × 300 + 3 × 800[a]	7	1972–1979
Zmiyevskaya	6 × 200 + 4 × 300	10	1960–1969
Tripolskaya	4 × 300 + 2 × 300[a]	6	1969–1072
Ladyzhinskaya	6 × 300	6	1970–1971
Dobrotvorskaya	2 × 150	2	1963–1964
Burshtynskaya	12 × 200	12	1965–1969
Kharkovskaya CHPP	1 × 250[a]	1	1986
Kievskaya CHPP-5	2 × 250[a]	2	1974–1976
Kievskaya CHPP-6	2 × 250[a]	2	1982–1984
Total	29,810	104	

[a]Gas-and-mazut units, the rest are coal-dust-fired units

conservation or about to be removed from operation, since the cost of their maintenance degrades economic performance of the branch as a whole. The state of CHPPs is even worse. For example, the age of certain boilers at the CHPPs listed below cannot but impress: Lysychanskaya – over 55 years; Syevyerodonyetskaya, Nikolayevskaya, Simpheropolskaya – over 40 years; Dnyeprodzyerzhinskaya – over 70 years.

1.2 THE IMPACT OF POWER GENERATION ON THE STATE OF AIR BASIN

The share of harmful emissions produced by the Ukrainian power engineering enterprises that remain under the jurisdiction of the Ministry of Fuel & Energy (hereinafter Mintopenergo) runs to over 30% of total emissions from stationary sources in this country (Table 2).

TABLE 2. The HPPs share of emissions into the air basin of Ukraine

	Harmful emissions into atmosphere (1,000 t)			
	2000	2001	2002	2003
In all Ukraine[a]	3,959.4	4,054.8	4,075.0	4,087.8
HPPs[b]	1,265.876	1,381.707	988.827[c]	979.547[c]
Percentage	31.9	34.0	24.3[c]	23.06[c]

[a]From Довкілля України. Статистический сборник, Госкомстат, 2003 (Natural Environment of Ukraine. Statistical collection. Ukrainian State Committee for Statistics, 2003)
[b]Emissions of power engineering enterprises that remain under the jurisdiction of Mintopenergo (form of the statistical reporting 2ТП-air)
[c]Stock Company "Vostockenergo". No data since 2002

The existing distribution pattern of power-generating enterprises is characterized by highly concentrated areas in certain regions of this country. Thus, 11 high-capacity coal-fired HPPs from a total number of 14 are concentrated within the boundaries on Donyetsky coal basin. The degree of environmental pollution from the regions of Donyetsky, Lugansky, Dnyepropyetrovsky and Zaporozhsky by the enterprises of heat and power engineering which are not equipped with effective flue gas purification facilities is being further amplified by harmful releases from chemical and metallurgical plants. Considerable contribution to atmospheric contamination is also being made by the active coal-mining/processing enterprises through permanently smouldering mine waste heaps that have been accumulating for years.

In separate regions of Ukraine the atmospheric contamination caused by the Mintopenergo's HHPs is over 50% more intensive than that coming from stationary emitting sources taken together (Table 3).

TABLE 3. Toxic emissions into the atmosphere from HPPs in percentage of total emissions over the regions

Region (oblast')	Contribution of HPPs to the total quantity of harmful emissions into the atmosphere (%)			
	2000	2001	2002	2003
Lvovskaya	67.2%	64.1%	59.1%	55.4%
Stationary sources[a] (kT)	108.6	114.5	97.8	96.1
HPPs[b] (kT)	72.972	73.391	57.764	53.249
Ivano-Frankovskaya	84.3%	82.7%	84.95%	84.6%
Stationary sources[a] (kT)	141.0	143.8	149.0	181.3
HPPs[b] (kT)	118.860	118.938	126.581	153.452
Vinnitskaya	77.9%	76.8%	71.7%	75.7%
Stationary sources[a] (kT)	80.1	71.7	57.8	59.4
HPPs[b] (kT)	62.407	55.094	41.435	44.954
Kievskaya	79.8%	79.9%	83.7%	78.2%
Stationary sources[a] (kT)	80.8	87.4	93.0	75.3
HPPs[b] (kT)	64.496	69.811	77.854	58.899
Kharkovskaya	66.3%	66.3%	66.3%	66.9%
Stationary sources[a] (kT)	143.7	156.1	151.6	148.3
HPPs[b] (kT)	95.224	103.452	100.470	99.201
Donetskaya	27.4%	29.3%	17.1%	18.4%
Stationary sources[a] (kT)	1,590.0	1,588.7	1,580.7	1,576.8
HPPs[b] (kT)	436.171	464.898	269.533	290.231
Zaporozhskaya	34.9%	39.7%	40.5%	38.1%
Stationary sources[a] (kT)	231.2	233.3	233.5	235.8
HPPs[b] (kT)	80.676	92.534	94.483	89.765
Dnyepropetrovskaya	19.2%	20.6%	20.6%	18.7%
Stationary sources[a] (kT)	783.6	848.6	888.6	834.0
HPPs[b] (kT)	150.252	174.484	183.042	155.798

[a]From Довкілля України. Статистический сборник. Госкомстат, 2003 (Natural Environment of Ukraine. Statistical collection. Ukrainian State Committee for Statistics, 2003)
[b]Emissions from the enterprises of Mintopenergo's power engineering branch (form of statistical reporting 2TP-air)

An increased hazard to public health and the natural environment of Ukraine arising from HPP installations is due to their high concentrations in densely populated industrial and agricultural regions and their emissions of such toxic compounds as sulfurous anhydride (about 70% of total emissions in Ukraine), nitrogen oxides (about 50%), carbon dioxide, ash, and some carcinogenic substances (benzopyrene, vanadium oxide, high-molecular compounds)

The emissions from HHPs cause intensive contamination of atmospheric precipitation, soil, plants, and surface waters.

1.3 THE CAUSES OF NEGATIVE IMPACT ON THE AIR QUALITY

1.3.1 Equipment Aging and Poor Quality of Fuel

According to current regulations the thermal generation sector needs to be fast in its response to any variation in power demand, standard or nonstandard, including emergency cases.

The operation of coal/gas/mazut-fired generating units in variable regimes leads to a drop in efficiency and an increase in specific fuel consumption for the generation of electricity together with excessive fuel and power demand for the internal needs of the HPP as a whole. Furthermore, the following adverse processes are at work: intensive wear of steam boiler equipment and control facilities, thermal cycle variations, and thermal fatigue of metal parts and units. By way of illustration, it is worth pointing out that an increase of 17% in the specific consumption of equivalent fuel for the generation of electric power by HPPs since 1991 has been registered which is 372 g/Kw, while the corresponding world average against the figure is 315 g/kW. The same tendency holds true for other forms of operating costs.

As has already been mentioned, the main cause of intensive environmental pollution lies in the fact that HPPs are still in widespread operation despite having main equipment well past the middle of its rated service life span. Obsolete boiler units are not efficient, the combustion of low-grade raw coal produces ash that contains 20–25% of nonburned carbon which leads to low efficiency in using fuel resources and generates problems of ash stockpiling. The ash with high carbon content is not usable by the construction industry but is efficient as an environmental contaminator. For this reason the ash dumps of HPPs have outgrown safe dimensions and transformed themselves into a source of permanent worry and anxiety.

The aging of power-generating equipment of HPPs was also accelerated by the steady drop in the quality of coal: the ash content has now increased from 26% to 38%, while the calorific value has slipped from 5,511 to 4,805 kcal/kg (see Table 4).

TABLE 4. The characteristics of fuel

Year	Combustion heat (kcal/kg)	Ash content (%)	Moisture content (%)
1966	5,511	22.6–24.4	7.5
1970	5,245	24.6–26.8	8.3
1975	4,844	27.1–29.8	9.3
1976	4,838	27.2–30.0	9.3
1977	4,671	28.7–31.8	9.8
1978	4,504	30.3–33.7	10.2
1979	4,340	32.2–36.0	10.5
1980	4,261	32.8–36.8	10.7
1981	4,024	35.6–39.9	10.7
1990	4,416	31.6	9.9
1991	4,316	33.3	9.8
1992	4,155	34.2	9.8
1993	4,353	32.0	10.0
1994	4,250	33.3	9.9
1995	4,282	33.4	9.7
1996	4,139	34.2	10.0
1997	4,230	33.0	9.7
1998	4,266	33.0	9.5
1999	4,069	35.6	9.2
2000	4,069	36.2	9.1
2001	4,436	30.3	9.4
2002	4,997	26.2	9.5
2003	4,805	27.6	9.5

1.3.2 Ineffective Operation of Flue Gas Purification Equipment

Increased fuel-ash content has resulted in the overload of coal pulverizing systems, flue gas purification equipment (FGPE), accelerated erosion of heating surfaces, maneuverability loss, degradation of ecological performance and growing demand for ash-dump space. As a result, the majority of coal-dust-fired units suffered capacity loss and were relegated to a category of lower rating units.

The FGPE in today's heat-power engineering is comprised of electrostatic filter-type dedusters, wet scrubbers, and multicyclones. Installed at HPPs electrostatic filters have been of underrated capacity from the very beginning due to economy of space and funds. As a result a high cost is now being paid in terms of dramatic efficiency loss owing to small sizes, short residence time, high gas velocities, and various internal failures. Thus, with the initial dust content in flue gas at a level of 30–35 g/m^3 the concentration of solid particles after filter ranges from 800 to 6,000 g/m^3 against current norm of 100 mg/m^3 for power boilers.

The average efficiency of FGPE installed at HPPs ranges roughly from 86% to 98 % (Table 5).

TABLE 5. The average degree of cleaning flue gases from solid particles

Name of power plant	Average figure in years (%)			
	2000	2001	2002	2003
Starobyehevskaya	92.52	92.63	92.51	92.32
Slavyanskaya	98.58	97.52	97.46	97.25
Uglyegorskaya	98.36	98.36	98.27	98.33
Kurakhovskaya	96.71	92.05	Stock company	
Luganskaya	87.01	85.87	"Vostokenergo"	
Zuyevskaya	98.26	98.34	No data	
Prdnyeprovskaya	96.29	96.31	96.33	96.25
Krivorozhskaya	95.62	95.25	95.58	94.66
Zaporozhskaya	98.44	98.3	98.38	98.2
Zmiyevskaya	94.06	94.05	94.04	94.55
Tripolskaya	96.1	96.1	95.7	96.04
Ladyzhinskaya	97.75	97.63	97.72	98.13
Dobrotvorskaya	94.43	94.52	94.73	94.75
Burshtynskaya	94.47	94.42	94.26	94.64

It is beyond any reasonable doubt though, that the only open way toward the reduction of sulfur oxide emissions today lies in the use of clean coal with a sulfur content of less than 1%. Otherwise, advanced technologies for sulfur removal should be introduced, which are currently unavailable due to high cost.

As to the suppressing of nitrogen oxide emissions, the power-generating enterprises have resort only to technological measures which are of limited application.

2. Ways of Reducing the Air Pollution

2.1 RENOVATION OF POWER-GENERATING UNITS

By way of executing the President's Commission of 25.04.2001 and the corresponding Cabinet of Ministers Order of 06.05.2001 the Board of the Mintopenergo drew up and approved the Program for the modernization of Ukrainian HPPs No. 5 on 12.11.2002. The Program's strategy implies integrated modernization and development of heat power plants and is mainly oriented to maximize the use of coal coming from domestic deposits to minimize this country's dependence on imports, recruit modern technologies and equipment giving preference to those offering environmental friendly performance.

Under a deficit of working and investment capital it is easy on perfunctory consideration to fall into the trap of believing that the retrofit of HPPs is economically justifiable. However, the retrofit measures of extending power plant service life in the long run will only result in the accumulation of obsolete equipment, its progressive aging, performance degradation and elevated environmental hazards. Eventually, the expenditure required for this sort of patch-up activity will be comparable to that for a major modernization.

Rehabilitation of power generating units may be arbitrarily divided into three levels:

- First-level rehabilitation (low cost) is being practiced at Burshtynskaya, Krivorozhskaya, and other HPPs, and consists of extended overhaul and the modernization of individual pieces of equipment.
- Second-level rehabilitation (medium cost) gives satisfactory results close to European Standards in terms of technical characteristics, economic indicators, and environmental safety.
  Examples:
  > *Pridneprovskaya HPP.* Installation of steam turbine K-3106-23,5, increase of installed power from 275 to 310 MW, thus extending service life of power units to 20 years.
  > *Power unit of Luganskaya HPP.* Installation_of new-design furnace with gas-proof screen, modernization of steam turbine.
  > *Power unit of Zmiyevskaya HPP.* Introduction of a new technology to the Ukraine for the combustion of low-reaction anthracitic coal with high ash content.

- Third-level rehabilitation (high cost) implies total dismantling of the main and auxiliary equipment and its replacement on the basis of modern techno-logies meeting European requirements and standards.

Example:

> *Starobyeshevskaya HPP.* Implementation of a pilot project consisting of the installation of a 670/Th capacity boiler with circulating boiling bed at atmospheric pressure (CBB). The boiler will be fired by low-grade coal and the so far unused waste of coal, available in the quantity of 125 Mt. The introduction of new technology and equipment will make it possible to successfully handle the issue of hazardous waste disposal, obtain 130 kt of economical fuel and dramatically reduce harmful emissions, namely nitrogen oxides – five times due to reduced combustion temperature, sulfur oxides – 13 times owing to the binding action of limestone being added to the fuel, dust – 59 times (low content from small fractions in the fuel and improved performance of electrostatic filters). The concentration of hazardous components in flue gases entering the stack will not be above the following figures: nitrogen oxides – 200 mg/m^3, sulfur oxides – 200 mg/m^3, ash – 50 mg/m^3. The work started in 11,997 is being financed by the credit funds of EBRD (90%) and the earmarked fund of Stock Company "Donbasenergo" (10%).

The experience gained in the process of construction and exploitation of the 210 MW power unit with CBB will be transferred to the 200 and 300 MW power units as soon it is their turn to be modernized.

Measures planned for the immediate future:

- *Ladyzhynskaya HPP.* Adaptation to brown coal delivered by the DHC "Alexandriyaugol": modernization of fuel feed, processing facilities, and boiler equipment.
- *Dobrotvorskaya HPP.* Additional construction work at the generating set No 9 with desulfuring unit or the construction of new generating set No 10 following CBB technology as an alternative. The final decision will be taken following the detailed consideration of both versions. The investments volume may range between $100 and $170 million depending on the version adopted.
- *Mironovskaya HPP.* CBB-based modernization of the 115 MW power unit currently being completed. A 115 MW steam turbine has been delivered and is ready for installation.
- *Tripolskaya HPP, Zmiyevskaya HPP.* Pending modernization of 300 MW power units fired by run-of-mine anthracitic coal. Final modernized version

will be adopted following the completion of technological developments that are being carried out by the institutes of the National Academy of Ukraine and Kharkov Design Bureau "Energo".

2.2 REPAIRS OF FLUE GAS PURIFICATION EQUIPMENT

Growing concern about the state of the air compels power engineering enterprises to focus special attention on the maintenance of flue gas purification equipment.

To support the above statement described below are the relevant measures taken by several power plants in particular:

- *Zaporozhskaya HPP*. Medium repair of electrostatic filter of power unit No. 3 and running repairs at power unit No. 2 (2003–2004).
- *Burshtynskaya HPP*. Replacement of electrostatic filters of power unit No. 12 resulted in the reduction of ash emissions by a factor of 2.5. Presently, the ash content in flue gases leaving the stack is kept below 250 mg/m^3. A major overhaul of the electrostatic filters of power unit No. 8 caused ash emissions to reduce 3–4 times (absolute ash content less than 440 mg/m^3). Major overhaul and partial modernization of electrostatic filters of power unit No. 6 had the result of a twofold reduction in ash emissions (absolute ash content less than 670 mg/m^3). Nearing completion is a new electrostatic filter of Ukrainian make whose commissioning will bring about a 20-fold reduction in solid particles content in flue gases to the absolute expected level of 50 mg/m^3.

Similar activity is underway at Dobrotvorskaya, Ladyzhinskaya, Luganskaya, and other HPPs.

However, successful tackling of the environmental pollution issue does not solely reside in heavily relying on various kinds of purification equipment. A good solid line of attack on the problem may be drawn through the upgrading of fuel and reducing its specific consumption.

2.3 ENVIRONMENTAL PROTECTION MEASURES

On Mintopenergo's initiative the Cabinet of Ministers of Ukraine has issued a Decree No. 447 of 1.04.03. whereby up to 70% of funds from the State Fund for the Protection of Natural Environment is to be directed to the environmental protection measures at the HPPs of the Mintopenergo system. Guided by the Decree the Mintopenergo with the approval of Ministry of Nature drew up programs of environmental protection covering the time span of 2001–2005.

Environmental protection activity involved the following measures:

- *Tripolskaya HPP.* Modernization of electrostatic filters of power unit No. 1
- *Burshtynskaya HPP.* Construction of a flue gas purification installation at the 200 MW power unit (st. No. 11), refitting of a flue gas purification installation (st. No. 9)
- *Tripolskaya, Zmievskaya, Uglegorskaya HPPs.* Introduction of automatic emissions monitoring systems

It is pertinent to note that the implementation dynamics of joint environmental protection programs was worsened drastically following the adoption of State Budget 2006. This is because in the preceeding years the Ministry of Nature alone was in charge of the State Fund and the mechanism for funds allocation for environmental protection activity was well developed and refined. However, the State Budget 2006 has already identified seven fund allocators instead of the one previously and thus the power engineering enterprises have been left with practically no financial support in their effort to reduce the harmful impact on the environment.

2.4 PARTICIPATION IN THE EXECUTION OF WORK RELATED TO REDUCING THE EMISSIONS OF GREENHOUSE GASES

As far as Ukraine in general and the National Fuel and Energy Complex in particular are concerned, the key issues defined by the Kyoto Protocol are the following:

- Attraction of additional investments through joint introduction and sale of quotas
- Changeover of the branch to energy-promising and resource-saving technologies

From an economic standpoint the mechanisms of the Kyoto Protocol created for the first time the possibility of initiating normal market relations favoring the creation of high-liquidity ecological commodities whose efficient turnover will be profitable to the seller, the buyer, the economy and the natural environment.

In the Mintopenergo's backlog of investment projects related to the reduction reduction in the emissions of greenhouse gases there are, in particular, the following:

- *Burshtynskaya, Slavyanskaya, Krivorozhskaya, Luganskaya, Dobrotvorskaya, Pridnyeprovskaya, Zmiyevskaya HPPs* – Modernization of power units including the introduction of new technologies (coal combustion under pressure, coal gasification, etc.), construction of FGPE and others

- Introduction of cogeneration and turbo expander power units at the installations of power engineering and gas transport systems

3. Ecologization of Power Generation Branch

The perfection of economic incentives and other ways of salvaging our environment, as the ecological legislation becomes more stringent, is a necessary condition for steady econo mic and social development of the power industry of Ukraine. "The Power Strategy of Ukraine till 2030 and Long Term Outlook" starting from a draft version drawn up by the Institute of General Power Engineering of the National Academy of Sciences of Ukraine in 2003 and having survived a lengthy series of hearings in Parliament, Ministries, Departments, public discussions, and reinforcement by relevant proposals, emerged at last in its final version in the current year of 2006.

The Strategy's major goal implies alleviation of the injurious impact produced by the enterprises incorporated in the Fuel and Energy Complex on the environment on the one hand and the rational use of natural resources on the other.

To attain the goal the following priority lines of attack have been identified:

- Technical reequipment of the Production Complex by introducing the most up-to-date scientific achievements, energy and resource-saving technologies and environmentally safe technological processes
- Harnessing renewable energy resources
- Rendering harmless and turning to practical use all kinds of waste

The program of actions following these strategic lines must clearly specify provisions for counteracting the two types of environmentally harmful effects: those that may emerge under normal operating conditions due to the imperfection of equipment and technologies involved and the effects triggered by the violation of technical regulations and safety rules.

According to the long-term concept of extended development of solid fuel-based power engineering with the purpose of ecological improvement the following steps are imperative:

- Introduction of new technologies for low-grade coal combustion in boiler units with CBB and intracyclonic coal gasification, use of generator gas in steam and gas plants

- Application of modern high-efficiency FGPE on the operating power units and on those still in the process of construction
- Upgrading of solid fuel quality by reducing ash and sulfur content
- Development and introduction of FGPE thus bringing to a minimum the emissions of dust, sulfur, nitrogen compounds, etc.
- Implementation of solid waste utilization program meeting the demands of the construction industry
- Based on the principles of priority and economic feasibility the following steps of power engineering ecologization are identified:

  - First step (until 2010) – maximum attention to low and partially to medium cost measures
  - Second step (2010–2020) – implementation of primarily medium-cost and partially high-cost measures
  - Third step (2021–2030) – transition to high-efficiency measures based on advanced technologies

The implementation of key measures following the strategic lines of the ecologization planned to be completed by 2020 will result in considerable improvement of the state of the environment accompanied by a productivity increase within the FEC's branches and will thus create a favorable climate for Ukraine in meeting its commitments related to the protection of the natural environment.

References

1. Енергетична стратегія України на період до 2030 року та дальшу перспективу (Power strategy of the Ukraine for the period up 2030 and further perspective), Kiev, 2003, 364 pp. (in Ukrainian).
2. Енергетичні ресурси та потоки (Power resources and flows), under general reduction of Shidlovsky A.K., publishing house "Drednout" Ltd, Kiev, 2003, 468 pp. (in Ukrainian).
3. Довкілля України 2000 (Environment of the Ukraine 2000), statistical collection, Kiev, 2000, 285 pp. (in Ukrainian).
4. Довкілля України 2002 (Environment of the Ukraine 2002), statistical collection, Kiev, 2003. 309 pp. (in Ukrainian).
5. Довкілля України , (Environment of the Ukraine), statistical collection, Kiev, 2004, 263 pp. (in Ukrainian).
6. O. Liven. Экологические решения для предприятий ТЭК(Ecologic solutions for enterprises of FPC), journal "Power policy of the Ukraine", No. 12, 2004, pp. 72–73 (in Russian).

7. V. Sanzharskaya, V. Dubovik Научно-технический прогресс в электроэнергетике. (Scientific and technical progress in electric power engineering), "Power policy of the Ukraine", No. 9, 2005 pp. 76–81 (in Russian).
8. N. Borisov. Основные проблемы развития ТЭС в Украине и пути их решения на среднесрочную перспективу (The main problems of TEPS development in the Ukraine and ways of their solution for the middle-term perspective), Power engineering and electrification, No. 5, 2002, pp. 6–13 (in Russian).

TECHNIQUES TO LIMIT NO$_X$ EMISSIONS

L. SZECOWKA, M. POSKART
Czestochowa University of Technology, Faculty of Materials Processing Technology and Applied Physics, The Department of Industrial Furnaces and Environmental Protection, Al. Armii Krajowej 19, 42-200, Częstochowa, Poland, Tel.: (4834) 3250-723; e-mail: szecowka@mim.pcz.czest.pl

Abstract. Nitrogen oxides are one of most harmful components pollutants. It is possible to limit the negative influence of NO$_x$ concentration on the environment. This can be done by modifying the organization of the combustion process using so-called primary methods. Most published scientific reports reduce NO$_x$ by using only a single method. This work presents the results of nitric oxides reduction in the combustion process using simultaneous application of several methods. Most of the papers, devoted to this subject, includes both experimental and model research on the basis of comparative analysis. The tests were carried out in a quartz combustion chamber. Models of reduction in NO$_x$ emissions were formulated using the Chemkin program.

Keywords: combustion process, nitric oxides reduction, environment

1. Introduction

The progressive growth of industrialization has had an enormous impact on our environment. This is particularly true in the case of air quality. Nitric oxides are one of the main sources of atmospheric air pollution. They contribute to the formation of acid rain, smog and the greenhouse effect. It is, however, possible to limit the negative influence of toxic components on the environment. This can be done by modifying the organization of the combustion process using so-called "primary methods" [1–3].

The primary methods of NO$_x$ reduction are the methods which depend on the process temperature especially air staging and flue gas recirculation. The lowering of temperature significantly influences the decrease of NO$_x$ emissions.

N. Syred and A. Khalatov (eds.), Advanced Combustion and Aerothermal Technologies, 47–54.

Experimental research in the kinetics of the combustion process is difficult because of the complexity of the chemical reactions related with nitrogen. Numerical modeling helps engineers to optimize the operating conditions, reduce pollutant emission, investigate malfunctions in the equipment, evaluate different corrective measures and also improve the design of new boilers.

Submodels for simulating the in-furnace processes such as mixing, radiative heat transfer, and chemical kinetics have been under development. The development of the models depends on the availability of accurate and approximate experimental data for comparison. However, because of the expensive cost of measurements of the combustion and heat transfer characteristic and the limitation of a given geometry, time, and number of instruments and sills required, assessment of these models is still limited to laboratory scale [4, 5].

Many published works deal with both experimental research and modeling, particularly for comparative analysis. As a consequence of this, a series of models of nitrogen oxides emission reduction were created. These models involve primary methods of reduction such as fuel and air staging and combustion gas recirculation. Based on diffusion and heat exchange combined with chemistry of the process many of programs such as: FLUENT, KIVA, CHEMKIN, and others have been developed. In this paper results of complex research using several primary methods are presented [6–8].

This paper includes not only experimental research but additionally the numerical modeling of NO_x reduction for the single and simultaneous application of various primary methods.

2. Research

2.1 EXPERIMENTAL STAND

An experimental stand was built to determine the influence of the following "primary methods": air staging, reburning, and flue gas recirculation on NO_x emissions. The tests were carried out in a quartz combustion chamber with laboratory equipment to enable the measurement of all the thermal and chemical parameters of the process.

During the experiments, the distribution of the temperature and the NO_x concentration along combustion chamber and NO_x concentration at the end of chamber have been measured. The temperature profiles and NO_x concentrations with air staging, reburning and flue gas recirculation are thus described.

The experimental tests were carried out using the laboratory stand shown in Figure 1. The reaction chamber consists of three quartz segments each 1 m in length and 0.12 m in diameter. Individual segments are provided with measuring holes and supplying devices. The main burner is situated at the inlet into the chamber. The total length of the chamber together with the outlet is 3.8 m. Natural gas was burnt in the main burner. Natural gas was also used as the reburning fuel. The amount of feeding gas, air and recirculated flue gas was measured by rotameters. Pulsation disturbances, created by a mechanical pulsation generator, were introduced into the central segment. Pulsation disturbances frequency was determined via an electronic tachometer. Flue gas composition and temperature in the individual measuring points were determined with the use of a 360 Testo exhaust gas analyzer connected to the computer. Temperature was measured using separate thermocouples.

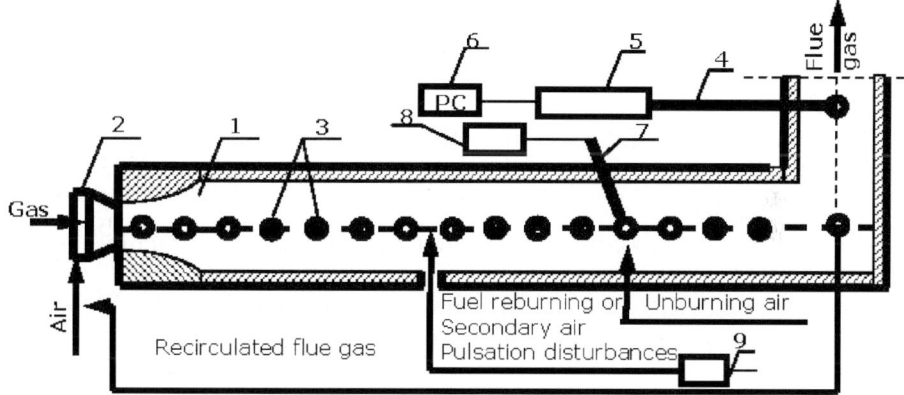

Figure 1. Laboratory for combustion tests: 1 – quartz chamber, 2 – burner, 3 – measurement points, 4 – probe, 5 – 360 Testo exhaust gas analyzer, 6 – PC computer, 7 – PtRh- Pt thermocouple, Pt thermocouple, 8 – measuring card, 9 – pulsation generator

All the researches were carried out at the same flow parameters. Preliminary measurements were made before each measurement series to check repeatability of results. This includes stabilization of air excess λ (1.07), and initial flue gas composition, mostly NO_x, at the 165 ppm level. Flue gas from the exhaust was recirculated and introduced into the combustion air of the main burner. Recirculation ratio r_c was 0.05–0.15, reburning fuel fraction r_b 0.05–0.17 based on the total amount of burning gas in the chamber. Secondary air fraction r_p is 0.11–0.26 based on the airflow into the combustion chamber. Pulsation disturbances (pd) were produced in the central segment of the chamber to intensify mixing where the secondary air or reburning fuel was introduced. Pulsation frequency was 30 Hz.

The study included single application of these methods and the following simultaneous measurement cycles: fuel staging with air staging, flue gas recirculation with air staging and flue gas recirculation with fuel staging, pulsation disturbances.

2.2 NUMERICAL STAND

The numerical modeling used 3.6.0 version of CHEMKIN. Scheme of the modeling chamber based on the laboratory stand is shown in Figure 2.

The model contained 126 reactions and 28 chemical compounds. The model of methane combustion was extended by NO_x formation reactions, reactions and the equilibrium constants have been taken from the Miller–Bowman model. All calculations were initially studied using two applications, Equil and Premix. The next step of the procedure was to prepare an input file containing the experiment conditions mass flux, pressure, temperatures profile and proper calculation grid.

The following quantities have been used:

- Mass flux 2.8–3.9 g/m^2s
- The temperatures profile taken from the initial experimental tests
- The initial equilibrium species concentrations using Equil to give the various fractions
- Estimated products of combustion calculated by Equil as a fraction rate

Numerical modeling for the reduction of NO_x emission by singly and simultaneous usage of the primary method was investigated as follows:

- Fue gas recirculation and reburning with pulsation disturbances
- Air staging and reburning with pulsation disturbances
- Air staging and flue gas recirculation with pulsation disturbance

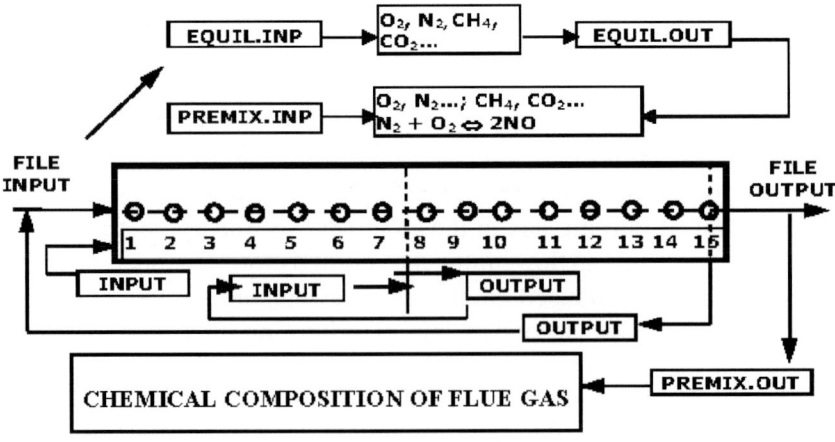

Figure 2. Schematic of the model

3. Results

Results of the experiments are presented in Figures 3–5.

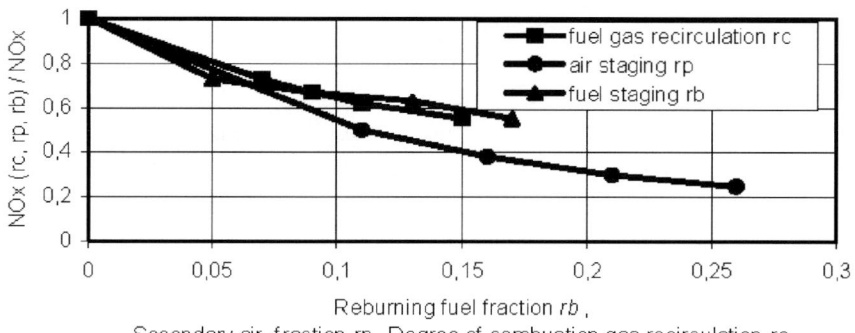

Figure 3. The impact of primary methods on NO_x concentration

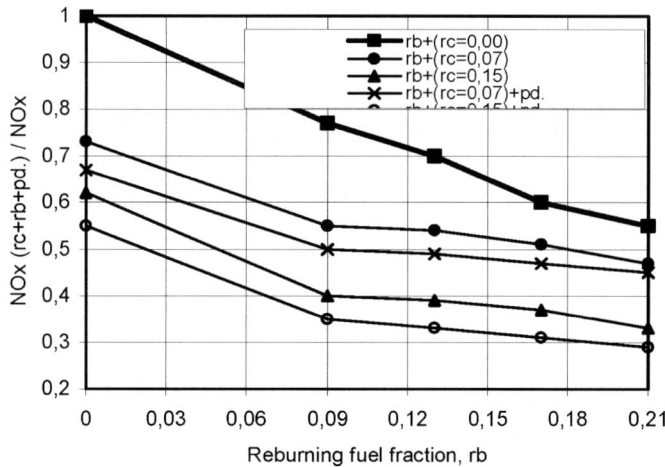

Figure 4. The impact of the flue gas recirculation, reburning and pulsation disturbances (pd) on reduction in NO_x concentration $\dfrac{NO_x\,(r_b + r_c + pd.)}{NO_x} = f(r_b)$

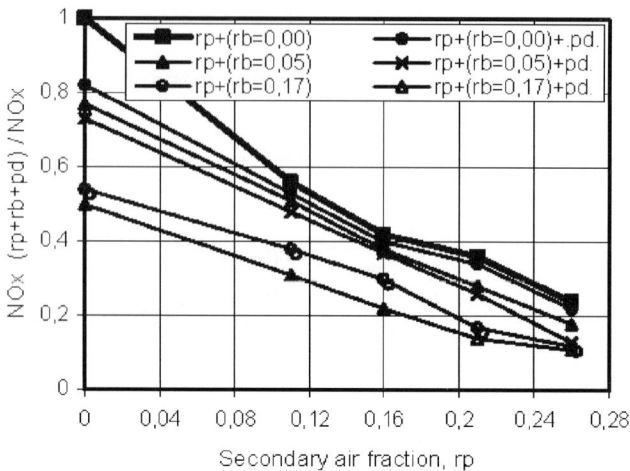

Figure 5. The impact of the air staging, reburning and pulsation disturbances (pd) on reduction in NO_x concentration $\dfrac{NO_x\,(r_p + r_b + pd.)}{NO_x} = f(r_p)$

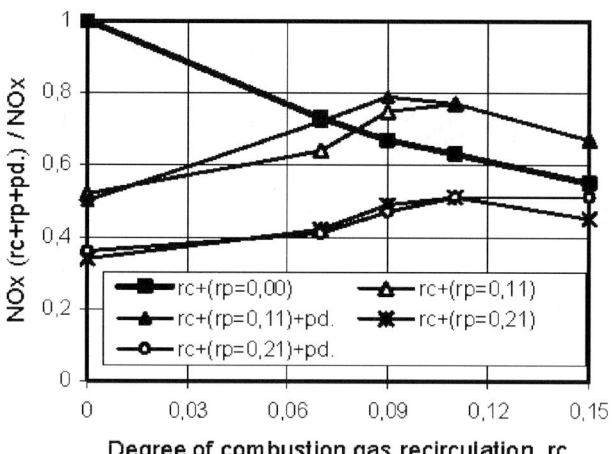

Figure 6. The impact of the air staging, flue gas recirculation and pulsation disturbances (pd) on reduction in NO_x concentration $\dfrac{NO_x\,(r_c + r_p + pd.)}{NO_x} = f(r_c)$

Results of the numerical modeling and experimental verification for singly and simultaneous application of primary methods are presented in Figures 7 and 8.

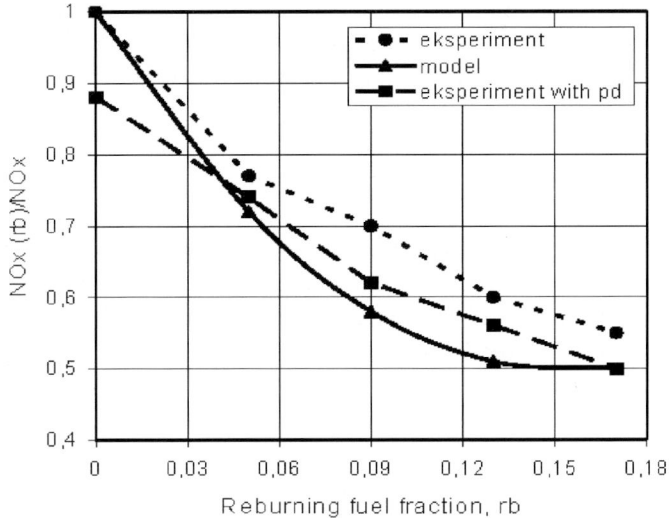

Figure 7. Comparison of the results from experiments and computer modeling during reburning with pulsation disturbances pd

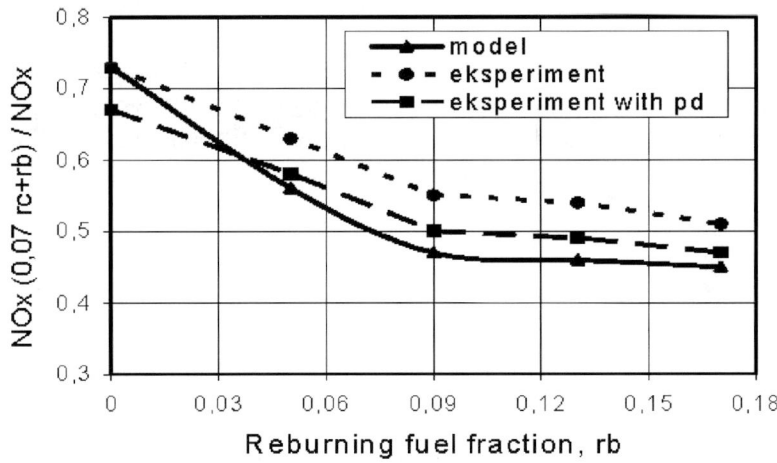

Figure 8. Comparison of the results from experiments and computer modeling during reburning and flue gas recirculation (for rc = 0.15)

4. Summary

Every primary methods has its own specific effectiveness. However, the highest efficiency was achieved by application of air staging (Figure 3). The final effect of NO_x reduction was up to 71%.

The simultaneous application of various primary methods causes additional increase in NO_x reduction in the certain systems. The highest efficiency was achieved by application of air staging and reburning (Figure 5) assisted by pulsation disturbances of the reactants. The final effect on NO_x reduction was 89%. An increased NO_x reduction by 14% in comparison to fuel staging and by 44% in comparison to reburning, was achieved. Simultaneous application of recirculation and reburning (Figure 4) was also very efficient, giving NO_x extra reduction of about 25% compared to individual application of each method.

The numerical calculations showed good agreement with the performed measurements for most of the analyzed single and simultaneous application of primary methods (Figures 7 and 8).

A good agreement between the numerical modeling and experimental results was obtained for the flue gas recirculation and also for the air and fuel staging. The pulsation disturbances additionally for reburning test intensified the process of NO_x reduction (Figure 7). It is essential to point out that the efficiency of the process resulting from the numerical calculation was always higher than the experiment because of assumptions involved in the ideal mixing of reacting substances. The good consistency of the results of the model used to the experimental tests suggests further work to extend the range of the experiments.

Best results overall were obtained for reburning and flue gas recirculation with pulsation disturbances. This showed higher level of effectiveness of NO_x reduction than the measurements obtained during the experimental research (Figure 8). This is due to the assumption of the ideal mixing of reactants in the model.

References

1. Jarosiński J.: Techniki czystego spalania", WNT, Wrocław, 1996.
2. Kordylewski W.: Niskoemisyjne techniki spalania w energetyce. Wrocław 2000.
3. Wünning J. A., Wünning J. G.: Flameless oxidation to reduce thermal NO-formation. Progress Energy Combustion Scientific, **23**, 1997, pp. 81–94.
4. Xu M., Azevado J. L. T., Carvalho M. G.: Modeling of the combustion process and NO_x emission in a utility boiler. Fuel **79**, 2000, pp. 1611–1619.
5. Romero C. E.: Reduced kinetic mechanism for NO_x formation in laminar premixed CH_4/air flames. Fuel, **77**, 7, 1998, 669–675.
6. Jones H. R. N., Leng L.: The influence of Fuel composition of CO, NO and NO_2 from a gas – fired pulsed combustor. Combustion and Flame, **104**, 1996, 419–430.
7. Li S. C., Williams F. A.: NO_x formation in two-stage methane-air flames. Combustion and Flame, **118**, 1999, 399–414.
8. Hill S. C., Smoot L. D.: Modeling of nitrogen oxides formation and destruction in combustion systems. Progress in Energy and Combustion Sciences, **26**, 2000, 417–458.

DEVELOPMENT OF ENERGY TECHNOLOGY FOR BURNING BROWN COAL WITH A SELECTION OF IRON LIQUID SLAG, TOGETHER WITH CARBON DIOXIDE UTILIZATION

M. J. NOSIROVICH
The Tashkent State Technical University, 700095, Tashkent, Universitetskaya str. 2, Republic of Uzbekistan

Abstract. The solution to the given scientific-technical problem is the creation of ecologically safe power plants (EcoPPs) with the removal of carbon dioxide emissions, which are dangerous for the Earth's climate and a sharp reduction in nitrogen oxides and sulfur oxides in flue gases. Of further benefit is the use of the by-products of the Angren brown coal (which contains 15% iron) carbon dioxide in the flue gases, strong hydrophobic silicates and ferroalloys in furnaces with liquid slag removal. As a solution to this problem we have investigated and developed systems and schemes employing gravitational separators of the liquid ash-slag wastes, coal-dust preparation and submission to a vertical prefurnace, submission of oxidant to the vertical prefurnace, drying of flue gases and derivation of liquid carbon dioxide. Investigations are ongoing.

Keywords: combustion, flue gases, brown coal, energy technology, gravitational separator, intensification, reduction of nitrogen and sulfur oxides, coal ash, hydrophobic silicates and ferroalloys

1. Introduction

In terms of fuels used for power engineering in Uzbekistan more than 80% is natural gas. Nowadays, a more balanced share of fossil fuels in this energy mix and an increase in the efficiency of their use is planned.

The increase of the share of coal in this power balance will lead to a rise of pollutant emissions into the environment, such as oxides of nitrogen and sulfur, carbon dioxide, ash, and others. It is caused by an absence of modern technology in the usage of fossil fuels, especially the most common energetic

N. Syred and A. Khalatov (eds.), Advanced Combustion and Aerothermal Technologies, 55–64.
© 2007 *Springer.*

fuel – low-calorie and high-ash brown coal. Following an increase in the volumes of coal combusted from 2.5 million tons in 2000 up to 12 million tons in 2010 the ecological damage for the region threatens to have a high cost.

Development of highly efficient and ecologically safe technologies in the usage of coal is a complex scientific-technical problem for the fuel and energy mix of the republic.

Characteristics of coal and a chemical composition of the ash and some results of the analysis are shown in Table 1.

TABLE 1. Coal characteristics and ash composition

Name	Characteristics	
	Design features	Actual middle value
Caloric value, Q (kcal/kg)	2,940	2,821
Structure of fuel on working mass (%)		
- Ash content, A	13.42	18.93–45
- Humidity, W	39.0	36.51
- Sulfur, S	1.34	1.85
- Nitrogen, N	0.38	0.2
- Carbon, C	36.42	33.56
- Hydrogen, H	1.82	1.69
- Oxygen, O	7.62	7.26

The ash has the following chemical composition: SiO_2 – 32.5%, Al_2O_3 – 21.5%, Fe_2O_3 – 15%, CaO – 24%, MgO – 3.5%, $(K_2O + Na_2O)$ – 3.5%. The high iron oxide content lead to surveys of ways of utilizing them.

One of main approaches to solving the given scientific-technical problem is the creation of EcoPPs with the full treatment of pollutant emissions – carbon dioxide (CO_2), sharp reduction of nitrogen oxides (NO_x) and sulfur oxides (SO_x) in the flue gases. It is thus essential to extract and utilize the useful by-products of hydrophobic silicates and Ferro-alloys from Angren brown coal, and the carbon dioxide from flue gases.

The aim of the investigation is to develop technologies for the effective combustion of coal with simultaneous extraction of the useful elements present in the slag composition. The most eligible technology for this purpose is that of operating the prefurnace with the highly effective cyclone-combustor for solid fuel [1, 4], coupled with application of an air or an air-oxygen blast and with liquid slag removal. Further developments include a gravitational separator for separation of an ash-slag melt to products and waste, fuel preparation and submission to the cyclone prefurnace, reception and passage of oxygen into the furnace and processing of flue gases with carbon dioxide capture. Some results from completed investigations are given below.

It is known that cyclone furnaces/combustors [3] use the aerodynamic advantages of a rotating airflow, which carries and supports the granulated fuel mixture. The fuel granules are blown into the unit at high speed, primarily to avoid deposition; larger particles stick in the liquid slag on the walls, gaining high residence time, whilst finer fuel granules burn intensively due to the aerodynamic characterdistics of the systesm. Thermal loading of the cyclone chamber reaches 3–6 MW/m^3. For comparison thermal loading of conventional pulverized coal Utility Boilers is 0.2–0.3 MW/m^3.

A model of a laboratory experimental cyclone prefurnace with a minor diameter of 600 mm has been built (which can be retrofitted to an existing boiler). For a better definition of the hydrodynamic and thermal processes occurring in the prefurnace, the gas field velocities of the cyclone chamber has been investigated. Distribution of tangential and radial velocity fields in the cyclone chamber with one- and two-sided admission of fuel granules in the cyclone chamber are shown in Figure 1.

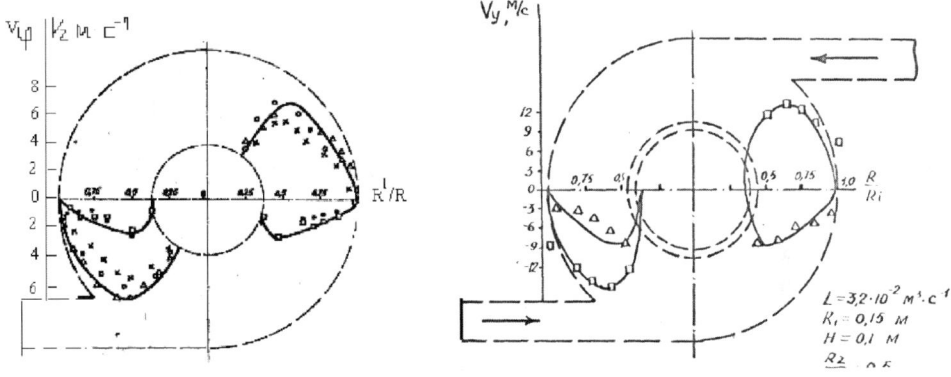

Figure 1. Radial and tangential speeds by cross section of the cyclone chamber. Δ – tangential speed, □ – radial velocity

Distribution of axial components of velocity in an axial radial plane of the unit at intervals from their base (top of diagram) is shown in a Figure 2. Curves characterize motion of gases in furnace chambers with two different designs of burners (here burners are used to fire vertically down into the cyclone chamber, well known in Russian work – editors comment). In the figure the velocities of the primary and secondary airflows at inlet are given (ω = axial velocity) For the two designs of inlet burners. Analysis of extensive velocity and other data demonstrates, that with change of the consumption of a gas–air mixture (i.e., heat input) and concentrations of the granulated coal the aerodynamic structure of a burning flame has virtually an even and constant nature. In the middle section of the furnace a developed, high velocity area of recirculation is observed. The main flows of the fuel and gas move towards the walls sides due to this recirculation zone. Their position in a furnace near an ionlet depends on the degree of swirl of the flows. For cylindrical nozzles and inlets the maximums velocities in the main flow are situated close to the axis of the burner and the flows start to diverge approximately 500 mm from it.

Cylindrical nozzles give less flame expansion than conical ones and thus the region where the flow reaches the walls is correspondingly more.

Length of the axial zone of recirculation at a ratio of speeds $\omega_2/\omega_1 \approx 1.6$ is $(1.5 \div 2)D_a$ for burners with conical entrances, and $(2 \div 2.5)D_a$ with cylindrical ones.

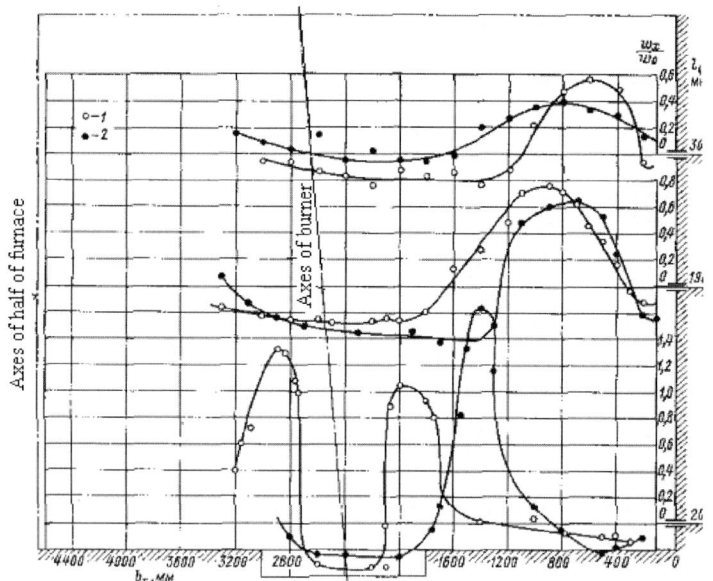

Figure 2. Fields of axial components of speed on a horizontal segment of a flame. 1 – utilization-blade burners with cylindrical entrances and nozzles ($\omega_2/\omega_1 = 1.5$; $\omega_2 = 30.4$ m/s; $\omega_0 = 28.5$ m/s); 2 – the same one with conical entrances and nozzles ($\omega_2/\omega_1 = 1.56$; $\omega_2 = 27.6$ m/s; $\omega_0 = 25.6$ m/s)

Results of calculations demonstrate that the design of the cyclone prefurnace gives better internal fly ash retention, not less than 80–90%. There is thus sharply reduced abrasive wear of heating surfaces of the downstream main boiler.

Good mixing enables combustion with fuel with small air-fuel ratios (α = 1.03 ÷ 1.1). There is a possibility of operation on dust from crushing operation or even on a crushed coal with sizes of granules of 1–5 mm, as the granules of fuel are retained by the liquid slag on the walls until it is burned completely.

Consider the mechanism for combustion of brown coal in a cyclone prefurnace [4].

During process of combustion of brown coal Fe, Si, and Mn pass into a liquid phase and in air the oxygen oxidizes iron

$$2Fe + O_2 = 2FeO + 54,4300 \text{ kJ} \tag{1}$$

Ferrous oxide is removed with the liquid slag, thus additives are used to promote the following reactions:

$$Si + 2FeO = Fe + SiO_2 + 370,000 \text{ kJ} \tag{2}$$

$$Mn + FeO = Fe + MnO + 125,650 \text{ kJ} \tag{3}$$

The oxides formed from the additives pass into the slag and due to the heat released by this oxidation reaction, the temperature of the prefurnace is increased by 300–400°C.

Hereinafter, there is an oxidation of carbon by following reaction:

$$FeO + C = Fe + CO_{gas} - 74,400 \text{ kJ} \tag{4}$$

As this process absorbs heat the, temperature in the cyclone prefurnace is a little reduced. After completion of process of burn-out of fuel, the slag which retains the iron oxide, is collected in the bottom of the prefurnace, where settling-out occurs prior to the subsequent separation of slag from Iron at a temperature of ~1,623 K. Metal sits on the bottom of a gravitational separator, and slag is collected at the top of the gravitational separator. After 300 s the slag is tapped off from the top of the molten metal, The metal then drains into an accumulator, where finally deoxidation of iron by ferrosilicon, ferromanganese and aluminium occurs via reactions:

$$2FeO + Si = SiO_2 + 2Fe \tag{5}$$

$$FeO + Mn = MnO + Fe \tag{6}$$

$$3FeO + 2Al = Al_2O_3 + 3Fe \tag{7}$$

There is a second way of deoxidizing iron in an accumulator, where it is necessary to support a temperature of ~1,800 K. For this purpose it is necessary to pass brown coal into the accumulator with a smaller quantity of oxygen giving coal gasification with the formation of hydrogen, which is then used to give the following reactions:

$$FeO + H_2 = Fe + H_2O \tag{8}$$

Thus, after chemical, thermal-technological and aerodynamic calculations of the process of combustion of brown coal with production of iron, the following scheme is preferred.

This consists of a vertical cyclone prefurnace with a gravitational separator and accumulator (Figure 3), and works as follows. From a fuel-preparation plant the coal dust passes to the cyclone prefurnace (3) by an air-dust pipe (1), where the oxidant (oxygen) (2) is added from an air-separation installation. In the cyclone prefurnace the process of combustion of brown coal whilst retaining Fe, Si, and Mn takes place. The process of combustion occurs at high temperatures owing to use of oxygen the and there is a fusion of the coal ash including iron, silicon etc. Owing to intensive swirl of the combustion gases the liquid slag is thrown to the periphery of the cyclone whereupon it flows own the walls to the bottom (6) and thence into a gravitational separator (7). Flue gases (5) have 80–85 % of the coal ash removed before passing into the main furnace of the boiler (14).

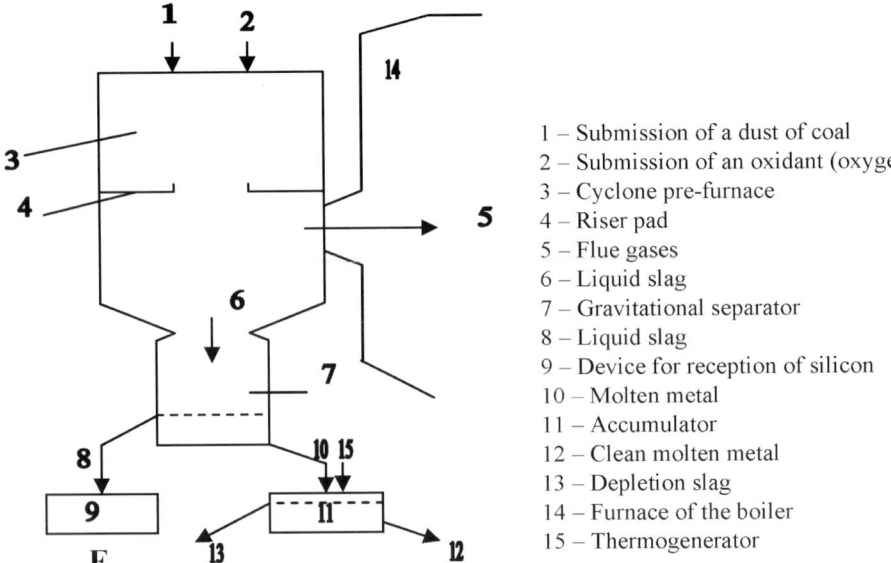

1 – Submission of a dust of coal
2 – Submission of an oxidant (oxygen)
3 – Cyclone pre-furnace
4 – Riser pad
5 – Flue gases
6 – Liquid slag
7 – Gravitational separator
8 – Liquid slag
9 – Device for reception of silicon
10 – Molten metal
11 – Accumulator
12 – Clean molten metal
13 – Depletion slag
14 – Furnace of the boiler
15 – Thermogenerator

Figure 3. Vertical cyclone pre-furnace with the accumulator

In the gravitational separator (7) settling-out of materials occurs and in a well-defined time the liquid slag (6), hydrophobic silicate (9) passes into the appropriate accumulator. The molten metal (10) passes into the accumulator (11), where a final deoxidation of iron takes place. Depleted slag (13) is let out through a notch, clean iron (12) is let out from bottom of the accumulator.

In the cyclone prefurnace the special devices are used, which results in strengthening of the swirl of the gas/fuel mixture, increase of convective diffusion of gas-air flows, and equalization of temperature of gases throughout the cyclone chamber.

The accumulator represents a vessel of cylindrical form manufactured from fire-clay or half-acidic brick. The molten slag continuously flows off through a riser pad of the cyclone prefurnace into the gravitational separator, thereafter into the accumulator. There is a notch for a descent of slag in mid part of the accumulator, and a notch for descent of metal in bottom of accumulator.

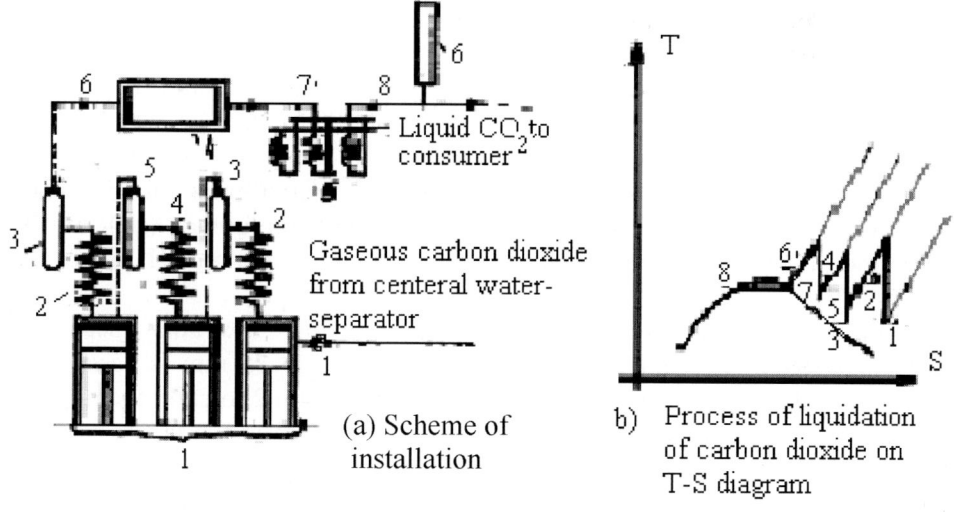

(a) Scheme of installation

b) Process of liquidation of carbon dioxide on T-S diagram

-.-   Carbonic gas at a I step of compression.
-..-   Carbonic gas at a II step of compression
-...- Carbonic gas at a III step of compression

Figure 4. Technological scheme of installation for production of liquid carbonic acid with the diagram of liquidation process

In Figure 4a scheme for the production of carbon dioxide is shown. Carbon dioxide is used as a raw for production of liquid carbonic acid. The carbon dioxide enters the suction inlet of a multistage compressor 1, where it is pressurized to a value which is determined by the temperature of the working medium of the associated refrigeration cycle. On the output leg of each stage of the compressor the water coolers 2 and separators of oil and fluid 3 are installed, which cools the gases and separates any microwater emulsion of fogs.

The compressed gas passes through dryer 4 and then goes to condenser 5. Carbon dioxide here, and the liquid carbonic acid through accumulators 6, then routed to a customer for storage of a liquid.

For operation of the scheme the calculations for each section of the installation have been done: heat exchangers, compressor, and condenser.

As a result, CO_2 emissions can be completely prevented. Pollutant emissions, such as the ash, the oxides of nitrogen and sulfur are reduced. It is possible to receive carbonic acid with purity of 95–100%. Strong hydrophobic silicates and ferro-alloys can be extracted from the ash.

The cost price of production of CO_2 does not exceed \$30/t, when the average global cost of industrial CO_2 is \$1,900/t.

On the basis of the developed technologies the common technological link of an ecological safe technology of combustion of Angren brown coal with an effective capture and useful use of the carbon dioxide and other by-products (Figure 5) is offered.

The given technology consists of five units intended for realization of separate technological processes, such as the unit of fuel preparation (I), unit of reception of oxygen (II), unit of the boiler with prefurnace (III), unit of reception of gaseous carbon dioxide (IV), and unit of reception of liquid carbon dioxide (V).

As a whole, the technology is operated as following – coal after a separation from metal additives and crushers with sizes of up to 25 mm is passed to a receiving bunker for crude coal 1, then into a gas stream 2 to enter the mill of a fan 3, where the flue gases consisting mainly of CO_2 from the last zone of the boiler. These flue gases are used for the drying of the wet crude coal. Granulated and dried coal dust passes via a coal handling duct into the cyclone prefurnace of the boiler 5, where the coal is fired. The oxidant (oxygen) enters separately. From the prefurnace the relatively clean flue gases (80–85% removal of ash) pass into the furnace of the boiler 11, and liquid slag collected in the gravitational separator 7. In the gravitational separator a settling-out of slag takes place and in set time the liquid slag passes out of the device for collection of hydrophobic silicate in accumulator 8. Molten metal enters into the accumulator 9, where a final deoxidation of iron with help of the high-temperature flue gases of a specific composition provided by thermogenerator 10. From the top of the accumulator the depleted slag passes down through a notch, and from bottom of the accumulator, the clean iron is collected.

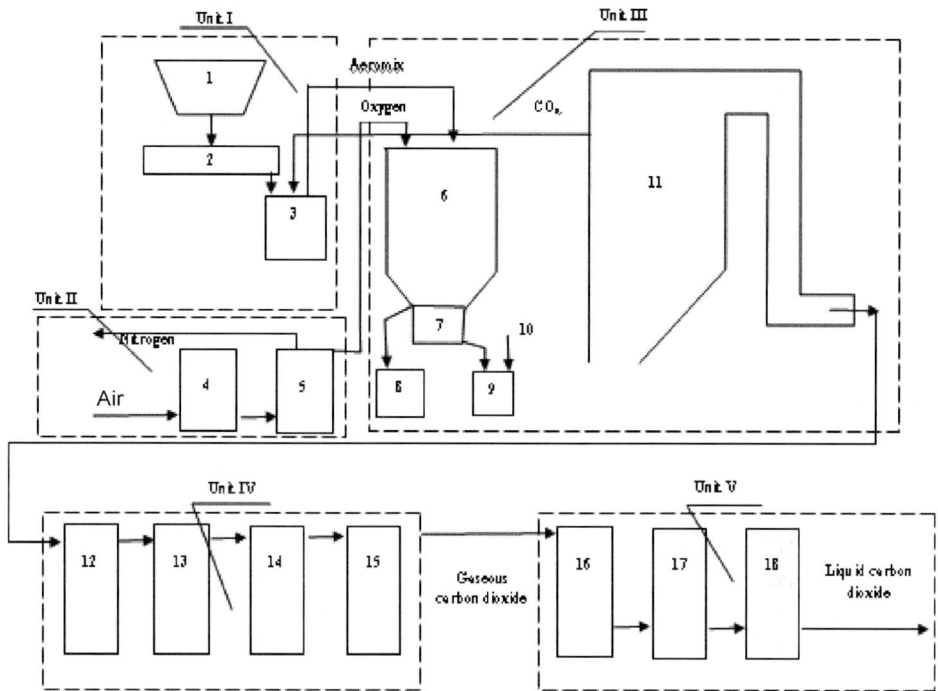

Figure 5. Technological scheme of ecological safe technology of burning of coal with effective selection and use of carbon dioxide and other by-products

Unit I   – Unit of fuel preparation

Unit II – Unit of reception of oxygen

Unit III – Unit of boiler with
           prefurnace

Unit IV – Unit of reception of the
           gaseous carbon dioxide

Unit V – Unit of reception of liquid
           carbon dioxide

1 – Bunker, 2 – feeder, 3 – fan mil, 4 – rectification colonna, 5 – heat exchanger, 6 – prefurnace, 7 – gravitational separator, 8 – device for receipt of silicon, 9 – accumulator, 10 – thermo-generator, 11 – boiler installation, 12 – scrubber, 13 – absorber, 14 – desorber, 15 – refrigerator, 16 – compressor, 17 – device for drying and cleaning of carbon dioxide, 18 – condenser for receipt of liquid carbon dioxide

The contaminated and high-temperature flue gases before entering into the absorption–desorption cycle, (for separation of CO_2) are routed to a scrubber 12, for cleaning from additives and cooling. They are then directed to absorber 13 for absorption of carbon dioxide, and then into the desorber 14 and finally into the refrigerator 15. Cleaned and cooled carbon dioxide is moved to the storage system for liquid carbon dioxide. This follows from the condenser, purification and drying plants. Received liquid carbon dioxide is routed to customers or to storage.

Other by-products of the given technology are the ferro-alloys and hydro-stable silicates. Taking into account the 15% iron in the ash structure Angren coal, on a new-Angren coal power plant it is possible to produce 450,000 t/year. Moreover, the application of this technology allows significant improvement in system ash collection and reduction in associated pollution, especially as there are many uses for slag wastes.

References

1. Akhmedov R. B. The patent of Russian Federation No. 2028542 on the invention Method of preparation of fuel to combustion and cleaning of combustion gases from pollutant emissions. Priority from 21.12.1990. Date of registration 9.02.1995.
2. Akhmedov R. B., Pojarov B. A., Grishutin K. S. Conception of use of fuel-energy resources on the basis of new technology of combustion of any kinds of organic fuels in a medium of a simulated oxidant. Papers of International conferences. Latvia. 13–17.05.1991.
3. Marshak Y. L. Furnace devices with vertical cyclone pre-furnaces Moscow: Energia, 1966.
4. Muhiddinov D. N., Babahodjaev R. P., Shaislamov A. Sh. Ecological perfect energy-technology of combustion of Angren brown coal. International Conference XX century in the History of Central Asia: Events. Tendencies. Lessons. Tashkent. 13 August 2004. pp. 26–27.

INTENSIFIED FLUIDIZED BED BURNING OF THE ANGREN
BROWN COAL CONTAINING AN INCREASED AMOUNT OF ASH

R. BABAHODZHAEV
Tashkent State Technical University, Power Engineering Faculty,
Thermal Power Engineering chair
Universitetskaya str. 2, Tashkent, Uzbekistan
Tel.: (99871) 3960340; e-mail: rachimjan@rambler.ru

Abstract. The intensified fluidized bed method is proposed to burn the crushed polydisperse low-grade brown coal. The principle involves the creation of a fluidized bed of coal through the use of intensive stirring and without the need for inert materials. The mode is jointly that of a fluidized bed and gushing layers with the movement of air at three levels. The primary air flow includes a proportion of recirculated flue gases to reduce the emissions of nitrogen oxides and prevent the melting of the coal ash particles. The addition of a suspension of hydrated lime or local quality clay to the layer binds the sulfur oxides, stimulating the granulation of the ashes and reducing the size of ablation. The tests were carried out in a quartz chamber and this topic continues to be the object of ongoing research.

Keywords: brown coal, reduces emissions, nitrogen oxides

1. Introduction

The largest coal deposit in Uzbekistan is that of Angrenskoye. The current known reserves of brown coal are about 2 billion tons, however, it is predicted that the total reserves are 5 billion tons [1]. Brown coal is burned through slot-hole burners on the coal-dust boilers TP-230-2 and TP-45 at the Angren Thermal Electric Power Plant. The flame is supported by a sulfurous boiler oil M-40. Sulfuric acid corrosion often causes emergency shutdowns, as well as increasing gaseous pollutant emissions. The low technical-economical characteristics have become the reference for the station, which has already fulfilled double its intended service life. To increase the overall performance and

N. Syred and A. Khalatov (eds.), Advanced Combustion and Aerothermal Technologies, 65–72.
© 2007 *Springer.*

decrease the emission of pollutants into the environment it is necessary to carry out a retrofit of the boiler. The retrofit is to be carried out for the minimum of costs through the use of domestic technology. The retrofit allows for a broad usage of coal in power engineering, an increase in profitability for the production of electric power, and the rational use of natural gas in the economics of Uzbekistan. The developed technology of burning the coal should be based on modern foreign technology. Such technologies are the VIR technology and fluidized-bed technology and are as follows.

VIR technology is based on a low-temperature burning, and is intended to burn coals of average caloric value in a pulverous condition [2]. The splitting of brown coal with a high initial ash content up into the required size results in the heightened wearing of mechanical and energetic equipment. A circulating fluidized bed is a large piece of equipment and requires the installation of an additional hot cyclone with return pipe lines, this complicates the design and increases the losses of heat in this part of the furnace [3].

Fluidized bed differs from the conventional methods of burning which have a high scale of isotherms within a layer, this is due to the stirring of the fragments of coal with an inert material. The residence time of granules in the combustion zone is regulated, and crushed coal of a broad class within the limits of 0.1–10 mm can be utilized. The granules' size depends upon the reactivity and ash content of the coal. However, the difficulty of maintaining the uniformity of the fluidized bed and the catching of the ash are problems with the fluidized bed method. Other problems include the heightened values from mechanical underfiring and the predilection of slagging of the ash. Some researchers and developers promote the use of a monodispersed structure of crushed coal with the presence of a layer of inert granules [4–6]. All these techniques are intended to broaden the application of the fluidized-bed technology in large power engineering.

An analysis of the performance of the technology for the burning of low-grade brown coals utilizing a fluidized bed is required. This paper demonstrates the development work carried out on an intensified fluidized bed system, and the investigations into its hydrodynamic parameters for the brown low-grade coal.

2. Research

The main properties of brown coal found in Uzbekistan are that it is characterized by a noncaking coke, a high outlet of volatiles (more than 40%), and a caloric value of wet ashless weight lower than 24 MJ/kg. The structure of fuel in relation to combustible and working weight by percentage is as follows: Carbon C^c = 76.5, hydrogen H^c = 3.8, nitrogen N^c = 0.4, oxygen O^c = 16.8, pyritic and organic sulfur $S^c_p + S^c_o$ = 2.5, volatiles V^c = 33.5, hygroscopic moisture W^{gi} = 11, working humidity W^w = 34.5, working maximum humidity W^w_{max} = 40, working ash content A^w = 14.4, maximum ash content A^d_{max} = 25, and a lowest caloric value Q^w_l = 3,210 kcal/kg. The apparent and bulk densities are ρ_{ap} = 1.28 t/m^3 and ρ_b = 0.8 t/m^3, and the values of the factors of abrasiveness and crushability are K_{abr} = 1.1 and K_{cr} = 2.1.

There are many different ash structures for brown coal, and it is these that determines the different melting temperatures. There are silicon dioxide SiO_2, aluminum oxide Al_2O_3, iron oxide Fe_2O_3, lime CaO, magnesia MgO, oxides of the alkali metals Na_2O_3 and K_2O, and also the sulfates and sulfides of sulfur present in the structure of ash of local brown coal. Coal contains between 2% and 20% clay, which in turn contained between 40% and 70% mineral substances. The temperature at the beginning of ash deformation for coal is 1,160°C. During the engine driven mining of the brown coal in Uzbekistan, it is common for the so-called third-order minerals to mix with the fuel, these artificially raise the initial general ash content of the brown coal up to 40%. The main impurities are earth silicon SiO_2, alum earth Al_2O_3, clay, the sulfides predominantly FeS_2, the carbonates and sulfates (i.e., $CaCO_3$, $MgCO_3$, $FeCO_3$, $CaSO_4$, $MgSO_4$), dioxides of metals, phosphates, chlorides, and the salts of alkali metals.

During the combustion of the coal, the ash content affects the scuffing of the burning granules with the fresh fuel. Accordingly, the reference speed of the fluidized bed will change. Therefore, the design of the cross-flow of the gas-distribution grating, and the hydraulic resistances of the layers must all be appropriate for the fuel before the implementation of this process. Also, it must be ensured that the pollutant emissions into the environment are within the required values. For the purpose of this investigation the low-grade brown coal was crushed up into a polydisperse condition, with granule sizes of 3–4 mm, as the mobility of small-sized granules between larger granules can cause the flow to be nonsteady.

For the combustion of the local low-grade brown coal, the intensified fluidized bed methodology was proposed. The fundamental design work of the test facility and its components was carried out by Profs. B. A. Baskakov and A. A. Belyaev (Russia), and is based on a classic spouting bed [4, 6, 7]. A schematic of the design is shown in Figure 1. The granulated polydisperse coal is loaded into the furnace (shown by 1), and is subject to intensive stirring by a mixture of air and flue gases. The intensive stirring of a layer allows the creation of a flow of coal. For this purpose a suitable geometrical shape for the appartus was created through the use of boundary walls II. To manage the flow within the fluidized bed, a special configuration of a gas-distribution grating I is used. The use of a number of different sections for the combustion of coal using the intensified fluidized bed method ensures that the required time for the complete burning of the fuel granules is maintained. The sectioning is implemented with a serial arrangement of furnaces, with the fuel medium moving in direction 2. Through the bottom of the conical vessels the fluidizing agent – mixture of air (3) and flue gases (4) enters, the arrangement allows the adjustment of the speed of the fluidizing agent for each vessel to take into account the dispersed structure of the granules. The upper, wider, section of the conical vessels has integrated protective walls (III), to prolong the thermal processes. The hot flue gases (6) exit through the top of the furnace vessels. The mass of ash and unburned granules pass from section to the next through pouring thresholds (IV), the heights of which are to be determined by calculation and experimentation. Ideally, only ash should be left to leave the last vessel section. The relatively heavy granules of the mineral additives fall through the gas-distribution grating of the first section (5).

The use of several different intensified fluidized bed vessels allows the adjustment of the heat loading of the boiler over a broad interval, and provides uniformity of the fluidized bed with more effective stirring. This results in a reduction in electrical energy requirement, at the expense of replacing a general draft-forced fan for a small adjustable unit. The gas-distribution grating is the shape of a truncated cone with a small slope of a vertical axis, similar to that of water-wheel blades. The grating generates a twist in the gas flow, providing uniformity and intensive stirring of the coal layer. The impurities, such as earth silicon, will in part act as inert materials in the intensified fluidized bed.

The constant addition of a limestone suspension or local clay into the coal layer effectively moistens the solid fragments, helping to maintain the necessary operation temperature of the layer between 850 and 950°C. It is known, that at such temperature the emission of nitrogen oxides is reduced, and also the emission of SO_x is suppressed through chemical changes. A further decrease in the emission of nitrogen oxides is achieved by supplying the required quantity of air into the furnace by three methods. Firstly, the primary air is supplied under the gas-distribution grating with between 6% and 22% recirculated flue gases. The exact quantity of recirculated gases depends upon the depth of the intensified fluidized bed, the sizes of the granules, and the technical-economic parameters of the boiler. The temperature of the primary air is established depending on the mode of the intensified fluidized bed. The second part of the air enters through the top of intensified fluidized bed. The third, relatively small proportion of air enters the convective section of the furnace to ensure the abundance of air in a ratio up to 1.07–1.11.

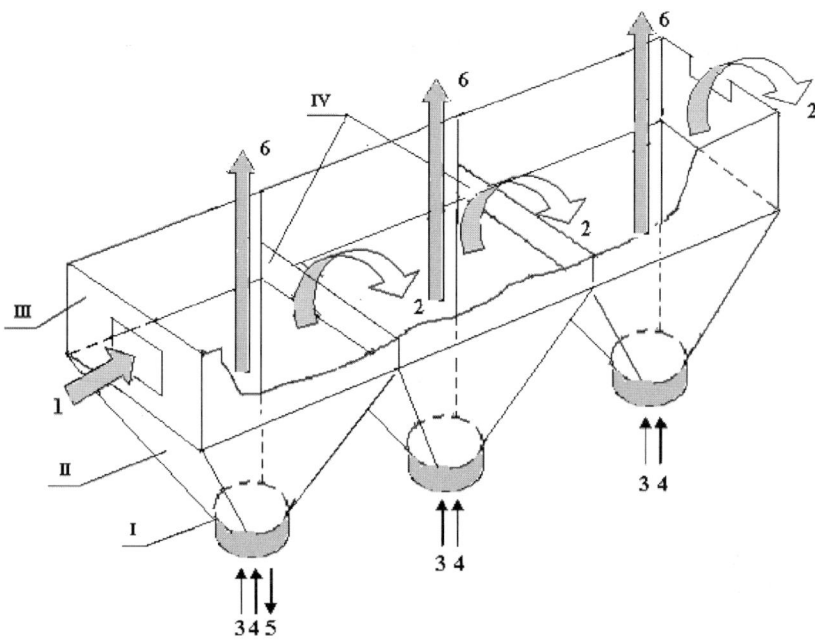

Figure 1. The schematic of the system with intensified fluidized bed, for the burning of low-grade brown coal: I – gas-distribution gratings, II and III – protective walls of the vehicle, IV – pouring thresholds, 1 – submission of coal, 2 – moving of firm medium, 3 and 4 – entrance of air and flue gases, 5 – outlet of heavy granules, 6 – moving of flue gases

The use of hydrated lime or a local clay can boost the granulation of ash, which improves the properties of the solid waste and allows them to be used as raw materials in construction, it also reduces wear. The local clay can be used instead of the limestone, as a cheaper alternative [8, 9]. According to some data, the possibility of using rather than disposing of the solid waste produced by thermal power plants (TPPs) results in a reduction in the price of the electric power by up to 15%, severely influencing the profitability of production.

Hydrodynamic investigations the intensified fluidized bed were conducted on a laboratory facility, under cold conditions. The experimental vessel incorporating the intensified fluidized bed was constructed from a transparent glass. The vessel represents a truncated cone with a lower diameter of 0.10 m, an upper equivalent diameter of 0.7 m, and a height of 1.0 m. The supply of the initial polydisperse coal was through the top of the vessel. The air enters from below, through the gas-distribution gratings with and without an induced twist. The solid granules are moved by gases in a rotary motion upwards as a result of the intensive stirring of the bed. At the expense of the twist in the flow of gases there is a considerable decrease detected in the thickness of a spouting bed.

Thus, the developed bed (located between a fluidized bed and aero-fountains in layers) is called as an intensified fluidized bed.

Some results from a series of experiments are given in Figures 2 and 3. The measurements were taken by pulse handsets, fixed to a micromanometer. The curves shown in Figure 2 were recorded at a bed depth of 350 mm in a quiet bed. From these results, it is possible to define the angle of twist of the gas-air flow with the least hydraulic resistance.

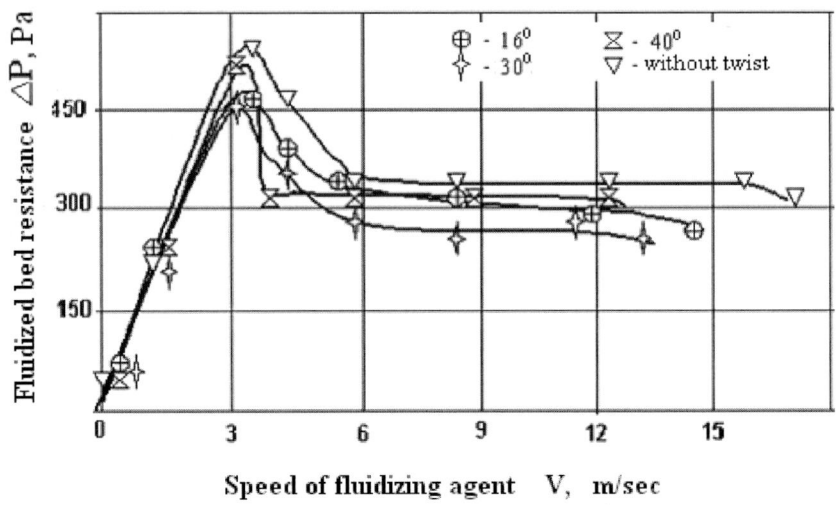

Figure 2. A bed resistance at miscellaneous angles of twist of gas-air flow

The bed resistance (without any twist of the gas-distribution flow) for several depths of a quiet bed in reletion to the speed of the fluidizing agent is given in Figure 3. It is visible from Figure 3 that an increase in the depth of a quiet bed results in an increase in the hydraulic resistance. However, an increase in the depth of a quiet bed results in an increase in the heat loading of the vessel. The optimum value of the depth of a quiet bed of incinerated coal will be determined on the basis of technical-economic calculations.

The stability of the bed depends on its depth, the physical characteristics of the solid and gas phases, and the geometrical parameters of the unit [10]. Because of the high initial humidity of coal (up to 40%), at the beginning of the process there are only a few, small gas channels and dead regions within the bed are detected. During the process of removing the free moisture, the adhesion–cohesive properties of the bed change, and it transforms into an intensively "boiling" condition. Changes in the design of the gas-distribution grating and the conical vessel shape were carried out to aid in the process removing the gas channels and dead regions that occur at the beginning of the process. Through these changes, an improvement in the initial bed performance was achieved, for only a small increase of hydraulic resistance of the gas-distribution grating.

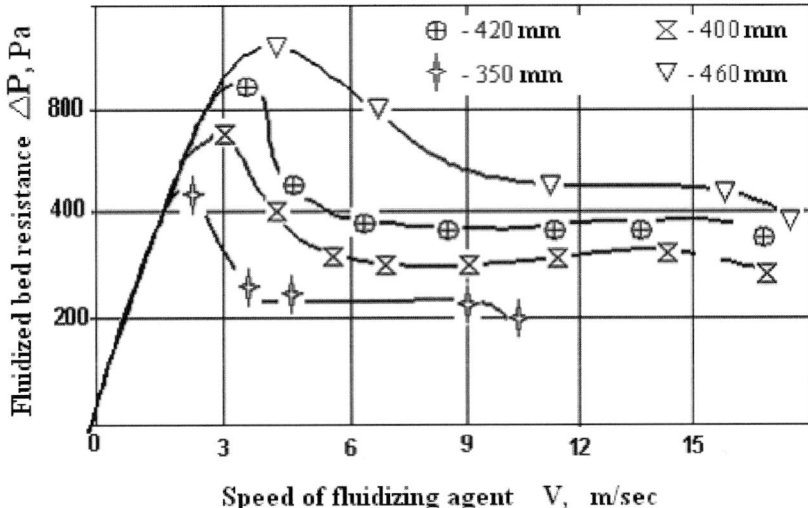

Figure 3. A bed resistance without twist of the gas-distribution flow at different altitudes of the bed

An investigation of the operation of the proposed design under laboratory conditions revealed the optimum settings and design characteristics. The depth of a quiet bed in the vessel should be between 0.35 and 0.45 m, the speed of the primary air that enters the vessel should be 8–10 m/s, the porosity of the bed must be between 0.65 and 0.85, the angle of the cone at of the bottom of the vessel should be 75–80° from the horizontal, the useful cross-flow of the distribution grating should be 12–15%, and the attitude the of upper equivalent diameter to the lower should be 9–11.

3. Summary

The intensified fluidized bed technique has been shown to effectively combust low-grade brown coal through the use of layer stirring, without the support of a boiler oil flame. An increase in depth of the quiet layer in an intensified fluidized bed results in an increase of the critical speed of the gas-air flow and the reduction in the range of working speeds (Figure 3).

Calculation results demonstrate that the heat loading of furnace applying the intensified fluidized bed technique can be higher than 10 MW/m^2.

On completion of the continuing research work, it will be possible to specify the design changes for the renovation of currently used boiler units. According to the market specialists, the payback period for carrying out such modifications would be between 2 and 4 years. The achievement of the desired technological advances it is anticipated would guarantee payback times of under 2.5 years.

References

1. Klimenko A. I. and others: Solution of problems of energy savings in a coal industry. Gornik, Bulletin, Uzbekistan. 2004. Volume 1, pp. 8–12.
2. Finker Z. B.: Experience of retrofit of boiler of the PC-10 CHP "Yavojno-2" (Poland) with transfer them to vortical combustion of stone coals. Thermal power engineering. 2000. No. 11, pp. 50–55.
3. Vikhrev Yu. B.: Development and experience for high-power boiler with a circulating fluidized bed. Power Engineering. 2004. No. 7, pp. 23–25.
4. Belyaev A. A.: Combustion of high-ash fuels in furnaces with a fluidized bed industrial boilers. The manual. Moscow, Publishing house Moscow Power Engineering Institute. 2004, 72 pp.
5. Munts B. A., Filippovskiy N. F., Baskakov A. P.: Furnace with a fluidized bed as object of regulation. Thermal power engineering. 1998, No. 6, pp. 15–19.
6. Baskakov A. P., Mantsev B. B., Raspopov I. B.: Boiler and furnaces with a fluidized bed. Moscow. Energoatomizdat, 1995.
7. Romankov P. G., Rashkovskaya N. B.: Drying in suspension. Leningrad, Publishing house Chemistry. 1979. 272 pp.
8. Babakhodjaev R. P.: Energy saving on Angren Thermal Electric Power plant (Uzbekistan) by replacement of a source of generator gas of a underground gasification on on-ground gas. Fifth International Association of Central Asian Studies Conference. 13–14 August 2004, Uzbekistan, Tashkent, pp. 23–25.
9. Babakhodjaev R. P.: Possibilities of application of a fluidized bed at combustion of brown coal on Angren Thermal Electric Power plant. International scientific – practical conference: INNOVATION-2004. Proceedings of the scientific articles. Tashkent. 2004, pp. 105–107.
10. Kubin M.: Combustion of firm fuel in a fluidized bed. Moscow. Energoatomizdat. 1991, 144 pp.

THERMODYNAMICS OF MODERN LOW-EMISSION (LOW-NO$_X$) PROCESSES IN THE COMBUSTION OF HYDROCARBON FUELS

B. SOROKA[*]
*Gas Institute, Nat. Ac. of Sciences, Degtyarivska 39,
Kiev 03113, Ukraine*

Abstract. This paper focuses on the results achieved through the addition of previously unaccounted for processes to existing calculation procedures relevant to combustion units as a whole and low-emission boiler furnaces in particular. It concerns the processes accompanying the use of NO$_x$ reduction facilities. Numerical simulations have been performed with respect to the chemical reactions of natural gas and/or methane–air mixtures in the recirculation of flue gases (combustion products). It is concluded that measures taken to reduce NO$_x$ emissions cause soot formation due to the fact that they involve the conversion of natural gas with steam, carbon dioxide or a combination of the two. The dependence of the thermodynamic equilibrium soot concentration on both the controlling temperature and concentration process conditions have been determined, based on the results of numerical simulations.

Keywords: flue gases recirculation, low-emission burner, natural gas conversion, nitrogen oxides reduction, numerical simulation, soot formation, thermodynamic equilibrium

1. Introduction

The basic principles of burner development associated with the reduction of pollutant effluent from fuel combustion, including those relevant to low NO$_x$ and ultra-low NO$_x$ burners, have been systematically considered by the author and various coworkers previously[1,2]. The establishment of the processes

---

[*] To whom correspondence should be addressed. Boris Soroka, Gas Institute, Nat. Ac. of Sciences, Degtyarivska 39, Kiev 03113, Ukraine; e-mail: soroka@kievweb.kiev.ua

N. Syred and A. Khalatov (eds.), Advanced Combustion and Aerothermal Technologies, 73–85.

involved in and the mathematical modeling of low-emission combustion, which up until now have been neglected by power engineers is considered below. Particular reference is made to soot formation in low-emission (low-NO_x) combustion through the various approaches to this process utilized in actual burner facilities.

The world-famous low-emission combustion FLOX system has several poorly understood specific features[3], including the conversion of natural gas through its reaction with the combustion products.

Defining the reactions between hydrocarbons and calculating the amount of prompt NO_x formation is highly significant because the most common combustion scheme (sequence of stages) utilized is the preliminary burning of rich mixtures of gas and oxidant, followed by secondary combustion of the rich after-mixture.

2. Evaluation of Soot Formation in Modern Low-Emission Gas Combustion Systems

2.1 MODERN APPROACHES TO AND THE BURNER SYSTEMS OF LOW-EMISSION COMBUSTION WITH RESPECT TO SOOT FORMATION

Any combustion process subject to a lack of oxidant within the primary combustion mixture is accompanied by higher soot formation the lower the primary excess air (the higher the ER – equivalence ratio).

The increase of soot formation caused by the recirculation of 300–400°C flue gases (combustion products) into the burning zone for the purpose of reducing NO_x formation is relatively well understood[4]. The process is accomplished by bleeding off the combustion products downstream of the principal smoke exhauster or from the boiler intermediate flue ducts. Recirculating gases are then directed into the furnace or into the combustion air flow by means of a special recirculation induced draught fan or are redirected to the suction section of the combustion air fan.

Sigal[5] asserts that the main techniques of NO_x reduction, namely those mentioned above, i.e., staged combustion and recirculation of flue gases, along with other examples, including water or steam injection, lead to a rise in soot formation. Furthermore, additional soot formation takes place in connection with the flame behavior, which determines the arrangement of secondary flows and local vortexes, causing an increase in the residence time within the zones of initial and incomplete combustion. The soot particles are detected as the illuminated ones in the photo of the flame organized by gas combustion with supplementary slot burner installed at the cold funnel-shaped section of the boiler of the PTVM-100 furnace at Darnitsa HPP-4 in Kyiv, Ukraine.

Precision distribution of gas and air flows within burners is of great importance from the standpoint of power efficiency support and NO_x effluent minimization, with particular reference to furnace facilities of low capacity[6]. This is demonstrated by induced smoking caused by short residence time in small furnaces and by soot formation caused by rich burning mixtures (sub-stoichiometric firing conditions).

Burner manufacturers have recognized the effectiveness of FGR in controlling NO_x and several new burner designs incorporate the benefits of FGR[5–9]. In the parallel-flow burner design (Figure 1), the primary and secondary airstream velocities are designed to aspirate hot furnace flue gas into the smaller annular space between the primary and secondary airstreams[6]. Depending on the specific burner design, approximately 20–30% of the fuel gas is recirculated in this manner. Initially, the aspirated flue gas acts as a dividing layer between the primary and secondary airstreams. Later, as the fuel and air mix, the flue gas serves to lower both the temperature and the O_2 concentration in the primary flame zone.

In the fuel-induced recirculation burner design, an "eduction" system actually draws the fuel gas, mixes it with the fuel gas at the nozzle, and propels it into the combustion chamber at the point of ignition. NO_x levels as low as 17 ppm has been reported in commercial boiler tests.

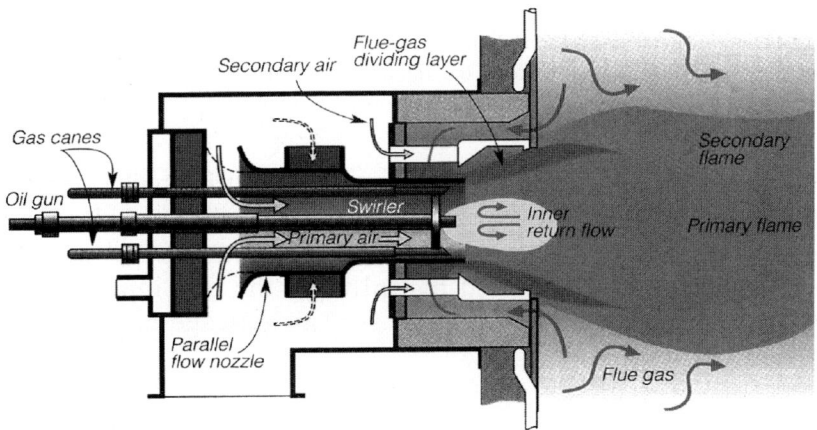

Figure 1. Double-fuel low-emission burner with combustion air twist and staging by fuel gas recirculation both into the burner itself and outside, within the zone along the flame axis [6]

Another approach to incorporating FGR into the burner design relies on rapid mixing rather than air staging (Figure 2)[6]. The burner operation is controlled by regulating air flow, flue gas, and fuel flow rates with exterior control valves. The design features no moving parts, eliminates the need for burner adjustments, and reportedly provides a stable and compact flame at all loads and avoids tube impingement.

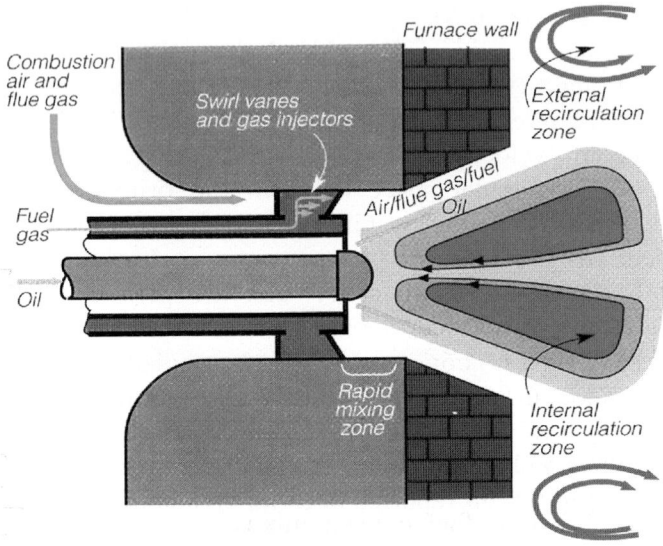

Figure 2. Double-fuel low-emission burner with numerous flows of recirculating combustion products (flue gases) within zone along the flame axis (internal), outside the flame (external) along with combustion products feed into the air flow[6]

A burner concept, developed for small industrial boiler applications, features both air and fuel-staged combustion. Jones[6] describes the burner as being arranged with both internal and external gas injection nozzles. Fuel staging is achieved by splitting the gas input between the internal gas nozzles located in the primary air passage and the outboard external gas nozzles[5].

2.2 THERMODYNAMIC ANALYSIS OF SOOT FORMATION

It is important to bear in mind the energy losses associated with the appearance of soot in flue gases (combustion products) as well as the complications introduced by dispersed particles afterburning over the active combustion zone.

In connection with above-mentioned case, the soot formation boundaries represent the area of practical importance with respect to low-emission burning. The most simplified approach to solving the problem involves calculating the state of thermodynamic equilibrium of the reacting systems.

According to preliminary calculations, soot formation occurs for any of the most important examples of low-emission fuel combustion: two-stage process, recirculation of flue gases (combustion products), water and water vapor injection into the combustion components, as well as by a combination of the listed approaches.

Among the examples of low-emission burning of interest here is the reaction of the fuel–oxidant mixture with the recirculating combustion products, including:

- Complete combustion products (external recirculation)[9,10]
- Intermediate species, where the depth of reaction performance is determined by the initial (or total) value of excess air factor α_Σ being fed through the burner (internal recirculation)
- Intermediate or final products determined by the first stage of combustion (e.g., – rich combustion products – internal recirculation)[7]

The results of analysis performed using the original computer code "FUEL" developed under the guidance of Soroka are presented below.

Three cases of fuel or fuel–oxidant mixture reacting with recirculating combustion products were considered:

1. Methane interaction with the complete combustion products CP^* (λ_Σ, T_{CP}) – reaction $CH_4 + \beta \cdot CP$ (λ_Σ, T_{CP}) → *Initial data for computation:*

- Preheated methane temperature T_{in} (CH_4) = 298, 600, 900 K
- Combustion products recirculation ratio β^{**} = 0.2, 0.5, 1.0, 3.0
- Combustion products temperature T_{CP} = 600, 900, 1,200, 1,600 K
- Excess air factor for recirculating gases λ_Σ = 1.05

2. Methane interaction with incomplete combustion products CP^{***} (λ_1, T_{CP}) – reaction $CH_4 + \beta \cdot CP$ (λ_1, T_{CP}) → *Initial data for computation:*

---

[*]Recirculating gases (agent).

[**]Under recirculation ratio β is assumed the value of ratio of volume or mole flow rates of CP to that of methane CH_4.

- Preheated methane temperature T_{in} (CH$_4$) = 298, 600, 900 K
- Combustion products recirculation ratio β = 0.2, 0.5, 1.0, 3.0
- Combustion products temperature T_{CP} = 900, 1,200, 1,600 K
- Excess air factor for recirculating gases λ_1 = 0.6

3. Methane–air mixture ($F + O_x$) interaction with the complete combustion products CP (λ_Σ, T_{CP}) – reaction CH$_4$ +2αO$_2$+7.52αN$_2$ + $\beta\cdot$CP (λ_Σ, T_{CP}) → *Initial data for computation:*

- Initial preheated methane–air mixture ($F + O_x$) temperature T_{in} (CH$_4$ + 2αO$_2$ + 7.52αN$_2$) = 298; 600; 900 K;
- Excess air factor of methane–air mixture ($F + O_x$) λ = 0.5
- Combustion products recirculation ratio β = 0.2, 0.5, 1.0, 3.0
- Combustion products temperature T_{CP} = 600, 900, 1,200, 1,600 K
- Excess air factor for recirculating combustion products λ_Σ = 1.05

The composition of recirculating gases under consideration meets fixed values λ_Σ and temperature T_{CP}. The process is considered to take place within the perfectly stirred reactor (PSR) under adiabatic conditions in accordance with each of the three cases mentioned above.

2.2.1 Case 1: reaction CH$_4$ + $\beta\cdot$CP (λ_Σ, T_{CP}) →

The computation has been carried out in two stages:

- Composition of recirculating gases in thermodynamic equilibrium were defined by varying T_{CP} and λ_Σ in terms of the limits mentioned above
- Reaction temperature T_{react} and composition of conversion products in thermodynamical equilibrium at T_{react} were defined for the reaction CH$_4$ + βCP(λ_Σ, T_{CP}) while varying the values of initial natural gas temperature T_{in}(CH$_4$) and of recirculation ratio β.

[***]Local composition, particularly reaction products after the first stage of combustion – for the case of two-stage or multistage combustion.

The results of computations are presented in Tables 1–3. Table 1 summarizes the composition of the recirculating gases. Tables 2 and 3 summarize the equilibrium reaction temperatures T_{react} and compositions of the reaction products at T_{react}, including the amount of soot formation. In each table content of gaseous components is given in terms of volumetric percentage and the content of soot particles is given in terms of mass concentration, i.e., kg of C per 1 kg of reacting mixture.

TABLE 1. Composition of recirculating combustion products CP, in equilibrium, with ,methane as fuel and air as oxidant. Excess air excess factor $\lambda_\Sigma = 1.05$

Component	T_{CP} (K)			
	600	900	1,200	1,600
O_2	0.988	0.9879	0.9859	0.9667
H_2	0	0	0	0.0017
OH	0	0	0.0002	0.0109
H_2O	18.011	18.011	18.0115	18.004
N_2	71.932	71.9325	71.93	71.909
NO	0	0.0002	0.0041	0.0397
CO	0	0	0	0.0018
CO_2	9.069	9.0684	9.0683	9.0662

The considered reactions are found to be endothermic across the whole range of parameters varied, i.e., in this case, the methane conversion reactions are characterized by the following;

$$CH_4 + H_2O \longrightarrow CO + 3H_2 - 209 \text{ kJ} \tag{1}$$

$$CH_4 + CO_2 \longrightarrow 2CO + 2H_2 - 252 \text{ kJ} \tag{2}$$

or in accordance with similar reactions, for example,

$$CH_4 + 2H_2O \longrightarrow CO_2 + 4H_2 - 165 \text{ kJ} \tag{3}$$

Under these conditions the reaction temperature $T_{react} < T_{CP}$ (Tables 2 and 3). Moreover, the clear correlation $T_{CH4} < T_{react}$ is changed to the opposite inequality: $T_{CH4} > T_{react}$ by initial methane jet preheating or by high local temperature of CH_4 at the entry to the reaction section.

TABLE 2. Equilibrium reaction temperature and combustion products composition, $\beta = 0.2$

Characteristic	T_{CP} (K)		
	900	1,200	1,600
T_{in} (CH$_4$) = 298 K			
T_{react} (K)	409.3	442.7	484.1
$H_2 \cdot 10^2$	22	51	129
H_2O	3.41	3.40	3.39
N_2	12.23	12.21	12.16
$NH_3 \cdot 10^2$	1.5	1.9	2.4
$CO \cdot 10^4$	0	0	0.2
$CO_2 \cdot 10^2$	0.6	0.8	1.1
CH_4	84.12	83.85	83.13
$C_2H_6 \cdot 10^4$	0.4	0.7	1.2
[C] (kg/kg)	0.016	0.017	0.020
T_{in} (CH$_4$) = 900 K			
T_{react} (K)	703.2	710.9	721.3
$H_2 \cdot 10^2$	2,652	2,831	3,079
H_2O	2.88	2.85	2.80
N_2	10.61	10.50	10.34
$NH_3 \cdot 10^2$	4.07	4.05	4.01
$CO \cdot 10^4$	292	347	436
$CO_2 \cdot 10^2$	2.91	2.98	3.09
CH_4	59.89	58.24	55.96
$C_2H_6 \cdot 10^4$	6.33	6.39	6.44
[C] (kg/kg)	0.118	0.126	0.137

TABLE 3. Equilibrium reaction temperature and combustion products composition, $\beta = 3.0$

Characteristic	T_{CP} (K)		
	900	1,200	1,600
T_{in} (CH$_4$) = 298 K			
T_{react} (K)	708.4	764.9	830.4
$H_2 \cdot 10^2$	1,526	2,276	3,131
H_2O	11.55	9.94	7.55
N_2	53.49	51.17	48.26
$NH_3 \cdot 10^2$	3.74	3.4	2.8
$CO \cdot 10^2$	24	76	22.4
$CO_2 \cdot 10^2$	159	181	186
CH_4	17.83	13.53	8.75
$C_2H_6 \cdot 10^4$	1.1	1	7
[C] (kg/kg)	0.064	0.079	0.093
T_{in} (CH$_4$) = 900 K			
T_{react} (K)	775.7	822.4	880.8
$H_2 \cdot 10^2$	2,423	3,033	3,673
H_2O	9.59	7.87	5.45
N_2	50.70	48.62	46.11
$NH_3 \cdot 10^2$	3.3	2.9	2.2
$CO \cdot 10^2$	92	199	424
$CO_2 \cdot 10^2$	184	188	161
CH_4	12.69	9.29	5.83
$C_2H_6 \cdot 10^4$	1	8	5
[C] (kg/kg)	0.086	0.091	0.096

For a given recirculation ratio β of the complete combustion products CP, the content of solid soot in the equilibrium reaction products increases proportionally with the equilibrium reaction temperature T_{react} irrespective of the cause of the rise in reaction temperature T_{react}: either by an increase of the initial CH$_4$ temperature or by a rise in the temperature of recirculating gases T_{CP}.

In turn, for reaction temperatures T_{react} below 700 K, soot concentration [C] decreases with increasing recirculation ratio. Hence, under conditions of low reaction temperatures T_{react} it is possible to increase soot growth [C] by increasing the value of β.

2.2.2 Case 2: reaction $CH_4 + \beta \cdot CP$ (λ_I, T_{CP}) $\rightarrow$

This case was investigated in the same sequence as the previous example, i.e., the composition of recirculating gases was initially computed (Table 4) followed by the concentration of conversion products, including the content of soot particles. Because $\lambda_I < \lambda_\Sigma$ (generally, if $\lambda_I < 1.0$) the content of reduction components (H_2, CO) was much higher than in the previous case. Moreover, the total content of reduction components ($H_2 + CO$) was higher than the oxidizing components ($H_2O + CO_2$) while the value of λ_I under consideration ($\lambda_I = 0.6$) was assumed for conditions of T_{CP} varying in the range: $T_{CP} \in \{900; 1,600 \text{ K}\}$.

TABLE 4. Composition recirculating combustion products CP in equilibrium, with methane as fuel and air as oxidant. Excess air factor $\lambda_1 = 0.6$

Component	T_{CP} (K)		
	900	1,200	1,600
H	0	0	0.002
H_2	15.47	13.3	11.674
H_2O	10.734	13.119	14.74
N_2	60.45	60.28	60.284
NH_3	0.006	0.001	0.000
CO	5.09	7.77	9.39
CO_2	8.11	5.53	3.91
CH_4	0.14	0	0

The results demonstrate that in case of CH_4 conversion with the intermediate reaction products of complete combustion or with the combustion products of a rich primary mixture ($\lambda_I = 0.6$ – see Tables 5 and 6) the conclusion outlined for the case of CH_4 reacting with the complete combustion products given above (Section 2.2.1) is conserved. However, for $\lambda_I = 0.6$, in the case of substantial recirculation ($\beta = 3.0$) soot formation decreases with increasing recirculating gas temperature (Table 6 – the case of $T_{react} = 898.5$ K).

TABLE 5. T_{react} (K) and relevant composition of combustion products (%, vol.) by reaction $CH_4 + \beta \cdot CP$ (λ_1, T_{CP}) under $\beta = 0.2$

Characteristic	T_{CP} (K)		
	900	1,200	1,600
T_{in} (CH$_4$) = 298 K			
T_{react} (K)	421.4	463.5	510.5
$H_2 \cdot 10^2$	30	83	215
H_2O	5.43	5.32	5.34
N_2	10.30	10.25	10.17
$NH_3 \cdot 10^2$	1.5	2	3
$CO \cdot 10^4$	0	0	0
$CO_2 \cdot 10^2$	1.6	2.3	3.2
CH_4	83.94	83.56	82.29
$C_2H_6 \cdot 10^4$	0. 45	0. 91	1.71
[C] (kg/kg)	0.024	0.025	0.030
T_{in} (CH$_4$) = 900 K			
T_{react} (K)	705.8	715.9	728.5
$H_2 \cdot 10^2$	2,710	2,949	3,254
H_2O	4.54	4.39	4.33
N_2	8.903	8.761	8.599
$NH_3 \cdot 10^2$	3.73	3.7	3.6
$CO \cdot 10^4$	490	600	800
$CO_2 \cdot 10^2$	7.3	7.3	7.8
CH_4	59.30	57.19	54.33
$C_2H_6 \cdot 10^4$	6.3	6.4	6.5
[C] (kg/kg)	0.127	0.137	0.152

TABLE 6. T_{react} (K) and relevant composition of combustion products (%, vol.) by reaction $CH_4 + \beta \cdot CP$ (λ_1, T_{CP}) under $\beta = 3.0$

Characteristic	T_{CP} (K)		
	900	1,200	1,600
T_{in} (CH$_4$) = 298 K			
T_{react} (K)	723.7	789.7	857.6
$H_2 \cdot 10^2$	1,753	2,690	3,567
H_2O	16.39	13.14	9.36
N_2	44.82	42.26	39.32
$NH_3 \cdot 10^2$	3.5	3.1	2.4
$CO \cdot 10^2$	48	165	455
$CO_2 \cdot 10^2$	345	367	348
CH_4	17.30	12.35	7.60
$C_2H_6 \cdot 10^4$	1.1	1	0.7
[C] (kg/kg)	0.073	0.087	0.092
T_{in} (CH$_4$) = 900 K			
T_{react} (K)	783.9	838.2	898.5
$H_2 \cdot 10^2$	2,599	3,343	3,986
H_2O	13.54	10.43	6.96
N_2	42.48	40.20	37.64
$NH_3 \cdot 10^2$	3.1	2.6	1.9
$CO \cdot 10^2$	152	349	719
$CO_2 \cdot 10^2$	374	357	291
CH_4	12.70	8.86	5.42
$C_2H_6 \cdot 10^4$	1	0. 8	0.5
[C] (kg/kg)	0.086	0.092	0.089

2.2.3 Case 3: reaction $CH_4 + aO_2 + 3.76aN_2 + \beta \cdot CP$ $(\lambda_\Sigma, T_{CP}) \rightarrow$

Preliminary calculations of the composition of CP (λ_Σ, T_{CP}) for $\lambda_\Sigma = 1.05$ (see Table 1) have been used for the next stage of calculations using the set of values for β and T_{CP} mentioned above. The combustible mixture to be converted is assumed to have the following local characteristics:

- Temperature T_{loc} $(F + O_x) = 298, 600, 900$ K, where $F \equiv CH_4$ (methane, natural gas); O_x – air $(O_2 + 3.76N_2)$

- $a = 2\lambda_{loc}$, where λ_{loc} is the local value of excess air factor at the point where the conversion process of the primary reacting mixture with the recirculating flow of complete combustion products takes place; $a = 0.2$, 0.5, 1.0, 2.1.

Tables 7 and 8 present the results of the interaction of primary combustible mixture with recirculating flow of the complete combustion products.

Unlike the previous case of gas conversion, in this case the reaction is found to be exothermic and is accompanied by temperature rise:

TABLE 7. T_{react} (K) and relevant composition of combustion products (%, vol.) by reaction $CH_4 + 2\lambda O_2 + 7.52\lambda N_2 + \beta \cdot CP(\lambda_\Sigma, T_{CP})$ under $\lambda = 0.5$ and $\beta = 0.2$

Characteristic	T_{CP} (K)		
	900	1,200	1,600
T_{in} ($CH_4 + 2\lambda O_2 + 7.52\lambda N_2$) = 298 K			
T_{react} (K)	1,461	1,500	1,555
$H \cdot 10^4$	4	7	15
H_2	16.16	16.04	17.95
H_2O	11.20	11.32	103.6
N_2	58.77	58.77	66.38
$NH_3 \cdot 10^4$	3.5	3.1	3
CO	10.81	10.93	12.50
CO_2	3.06	2.95	3.17
CH_4	0	0	0
T_{in} ($CH_4 + 2\lambda O_2 + 7.52\lambda N_2$) = 900 K			
T_{react} (K)	1,859	1,894	1,947
$H \cdot 10^4$	220	300	440
H_2	15.33	15.28	15.20
H_2O	12.02	12.09	12.14
N_2	58.75	58.75	58.74
$NH_3 \cdot 10^4$	1.2	1.1	1
CO	11.64	11.69	11.75
CO_2	2.24	2.19	2.13
CH_4	0	0	0

TABLE 8. T_{react} (K) and relevant composition of combustion products (%, vol.) by reaction $CH_4 + 2\lambda O_2 + 7.52\lambda N_2 + \beta \cdot CP$ (λ_Σ, T_{CP}) under $\lambda = 0.5$ and $\beta = 3.0$

Characteristic	T_{CP} (K)		
	900	1,200	1,600
T_{in} ($CH_4 + 2\lambda O_2 + 7.52\lambda N_2$) = 298 K			
T_{react} (K)	1,077	1,258	1,508
$H \cdot 10^4$	0	0	7
H_2	12.94	12.07	11.33
H_2O	7.94	8.81	9.55
N_2	68.53	68.53	68.52
$NH_3 \cdot 10^4$	14	5	2
CO	6.42	7.29	8.03
CO_2	4.17	3.30	2.56
CH_4	0.001	0	0
T_{in} ($CH_4 + 2\lambda O_2 + 7.52\lambda N_2$) = 900 K			
T_{react} (K)	1,201	1,381	1,630
$H \cdot 10^4$	0	1.3	025
H_2	12.31	11.66	11.09
H_2O	8.58	9.23	9.80
N_2	68.53	68.53	68.53
$NH_3 \cdot 10^4$	7	3	1
CO	7.04	7.70	8.27
CO_2	3.54	2.89	2.32
CH_4	0	0	0

$$\begin{cases} T_{react} > T_{CP}; \\ T_{react} > T_{in} \end{cases} \tag{4}$$

There is one exception to this outcome, where $T_{react} < T_{CP}$. This occurs at a comparatively low initial temperature of the gas–air mixture, i.e., 298 K, and a high T_{CP}, i.e., 1,600 K. However, in all cases for constant values of T_{in} and T_{CP}, the higher the recirculation ratio β, the lower the value of T_{react}.

Soot formation was not determined for any of the initial temperatures T_{in} or recirculation ratios β considered.

3. Conclusion

Combustion technology is a conservative area of engineering research and development. Nevertheless advanced approaches have been affected in low-emission fuel burning, based on the following methods: two-staged combustion process, recirculation of combustion products and flue gases, water or vapor injection into the burning components (fuel, oxidant), as well as a combination of the listed methods.

The task of this paper has been to evaluate how these new combustion procedures intended to reduce NO_x formation, influence the main characteristics of the combustion process, i.e., temperature, combustion products composition, including hydrocarbon and soot pollutants fraction.

Based on the results of both experimental and theoretical-computative studies and analysis of current data in the literature, the specifics of soot formation are determined by two basic schemes of gaseous hydrocarbons combustion; full preliminary mixing of the initial components (combustible + oxidant) or if no preliminary premixing occurs, simultaneous mixing and burning (diffusion combustion)[12].

The contribution of thermodynamics and chemical kinetics to the formation of soot particles have been compared for various schemes of hydrocarbon fuels combustion, Thermodynamic analysis has been found to accurately represent the combustion process by determining the influence of the composition of the initial mixture and the process parameters on both the reaction products yield and the soot formation.

The correlation between the techniques for reducing NO_x emissions from fossil fuel combustion, i.e., staged burning, recirculation of the complete and noncomplete combustion products, water or water vapor injection into fuel – with the formation of soot and other carbon-containing substances has been considered within the framework of this paper. It has been demonstrated by means of calculations of thermodynamical equilibrium within corresponding systems that the recirculation of the combustion products with the initial reactants (fuel) induces the endothermic reactions of carbon dioxide, water vapor and mixed conversion, which accompany soot formation.

References

1. Б.С. Сорока, Промышленные печи в проблеме снижения выбросов оксидов азота, *Известия ВУЗов «Энергетика»,* No. 4 (июль-август), с.51–66 (2004).
2. Б.С. Сорока, К.Е. Пьяных, В.А. Згурский, А.П. Апальков. Комбинирование способов снижения образования оксидов азота при горении – основное направление обеспечения экологических нормативов, *Экотехнологии и ресурсосбережение,* No. 5, с.60–69 (2000).
3. J.G. Wunning, "FLOX^{R}" – Flameless Combustion, *Thermoprozess Simposium* (2003), VDMA, 19 pp.
4. А.И. Щелоков, В.С. Щеглов, Обзор методов сжигания выбросов оксидов азота при сжигании газообразного топлива за счет рециркуляции дымовых газов, В кн.: *Энергосбережение в городском хозяйстве, энергетики, промышленности, Материалы Четвертой Российской научно – технической конференции,* Ульяновск (24–25 апреля 2003 г.), с. 311–313.
5. И.Я. Сигал, Влияние методов снижения NO$_x$ на концентрацию оксидов азота в продуктах сгорания и КПД котла (отчет Т15). Проект УНТЦ №2248, Киев (in Russian, unpublished), с. 3 (2004).
6. C. Jones. NOx emissions control: Small boilers poses great challenges, *Power,* 12. 34–41 (1994)
7. W. Leuckel. Combustion Fundamentals and Concepts of Advanced Burner Technology: Keynote Lecture, *6th European Conference on Industrial Furnaces and Boilers*: Estoril – Lisbon (02–05 April 2002), INFUB, 36 pp.
8. B. Soroka, K. Pyanykh, V. Zgursky, M. Khinkis, H. Abbasi, J. Rabovitser, Mathematical modeling of low-emission combustion processes basing upon Monte-Carlo procedures, *Preprints of 5th European conference on industrial furnaces and boilers* (11–14 April 2000), Espinho-Porto (Portugal), Vol. II, 12 pp.
9. M. Flamme, Neuester Stand der NO$_x$-Minderung am Beispiel von Industriebrenners, *GWF Gas Erdgas* 134(1993), Nr. 10, p. 533–541.
10. M. Flamme, H. Kremmer, NO$_x$ Emission and Reduction Potential from Industrial Gas Burners with Air Preheat Temperatures of Up To 1,000°C, *Natural Gas Policies and Technologies European Conference.* Part II: Infrastructures and Technologies Conference Proceedings, (14–16 October 1992), Athens, Greece, pp. 136–144.
11. *Общая химическая технология*/Под ред. проф. И.П. Мухленова, Ч.II, Высшая школа, Москва, с. 287 (1977).
12. Б.С. Сорока, Сжигание природного газа с недостатком окислителя и саже образование. II. Сажеобразование при базовых схемах сжигания газового топлива, *Экотехнологии и ресурсосбережение,* 1, 9–19 (2005).

CURRENT STATUS AND CHALLENGES WITHIN FLUIDIZED BED COMBUSTION

M. HUPA*
*Åbo Akademi University, Process Chemistry Center,
Biskopsgatan 8, 20500 Åbo/Turku, Finland*

Abstract. Fluidized-bed technology is rapidly expanding. Today, more than 600 large (20+ MWth) FBC boilers with a total installed thermal capacity of more than 70,000 MWth have been built. Around 75% of this capacity is circulating fluidized beds (CFBC) technology, the rest mostly bubbling fluidized beds (BFBC). The size of the boilers has increased steeply; the largest CFBC units being constructed approach a thermal capacity of 1,000 MWth. At present a major expansion in the FBC capacity is happening in the Far-East, especially in China. Several hundred major boiler projects for coal are underway which may double the global FBC capacity within just 5 years. The boilers in Europe are more and more often designed to burn several fuels simultaneously. Besides coal, wood, biomasses and various waste derived fuels have become important fuels in various recent FBC projects. Creative analysis and laboratory-scale fuel characterization have become very important in the assessment of the feasibility of the various fuels for FBC. Firing of several solid fuels simultaneously has in principle been considered to be an advantage of the FBC technology. However, some fuel mixtures show surprising behaviors and there is a lot of interesting research to be done to more quantitatively establish the influence of the fuel mixture on the key operating parameters, such as boiler efficiency, flue gas emissions, ash behavior and corrosion.

Keywords: combustion, fluidized bed, flue gases, emissions, ash, biofuels

---

*To whom correspondence should be addressed. Mikko Hupa, Åbo Akademi University, Process Chemistry Centre, Biskopsgatan 8, 20500 Åbo/Turku, Finland, mikko.hupa@abo.fi

N. Syred and A. Khalatov (eds.), Advanced Combustion and Aerothermal Technologies, 87–101.
© 2007 *Springer.*

1. Introduction

Fluidized bed combustion technology is becoming increasingly important in the global heat and power industry. New FBC capacity is being installed at a pace of several thousand MW (thermal) every year. The technology has advanced in many ways since its introduction in the early 1980s. Both circulating fluidized beds (CFBC) and bubbling fluidized beds (BFBC) have shown to be competitive depending on the application. Pressurized fluidized bed (PFBC) boilers are currently being demonstrated in several larger-scale projects.

This paper discusses the present state of FBC technology for large-scale heat or power production. Statistical information of commercial FBC projects is summarized and various trends in the evolution of the technology are presented. Most of the data concern boilers in Europe and North America. However, some recent statistical data from China are also included.

The issue of the fuel or fuels being selected in the different projects are discussed in more detail. Fuel mixtures, including biofuels and wastes of various kinds, present a special challenge to the FBC technology and this is elaborated on with examples of recent research.

2. Fluidized Bed Boilers in the World

In this paper, recent statistical data on commercial FBC boiler projects is summarized in several figures. Based on various datasets in the literature, plus information from the three leading Western FBC manufacturing companies, the data describes fairly accurately the status of large FBC boilers (>20 MWth) worldwide.

The total installed capacity of large-scale FBC boilers is around 70,000 MWth and the total number of boilers is around 600. Most of the increase in capacity in recent years has been in CFBC technology, but BFBC boiler capacity has also increased. Figure 1 shows the cumulative thermal capacity and Figure 2 the cumulative number of FBC boilers worldwide, where in both cases the first PFBC installations are already visible. Figures 1 and 2 also show the contribution of fluidized bed gasifiers (FBG) to the overall capacity. Most of these gasifiers feed their product gases to either conventional large-scale utility furnaces or to some process furnaces.

Figure 3 shows how the thermal capacity of individual boilers has changed over the years. The average size of larger-scale FBC boilers has practically doubled within the last 10 years. The largest CFBC boilers are now approaching the full-scale utility size of some 1,000 MWth or 500 MWel in coal firing applications. The first such large CFBC with supercritical steam cycle is being erected in Poland. BFBC boilers are smaller but the largest examples have a thermal power of almost 300 MW.

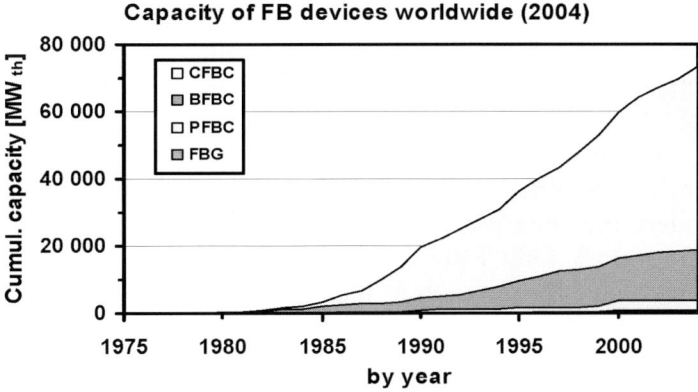

Figure 1. Cumulative thermal capacity of fluidized bed installations worldwide

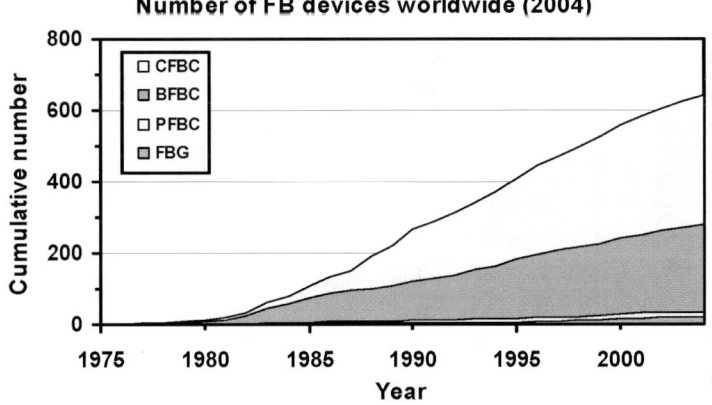

Figure 2. Cumulative number of FBC boilers worldwide

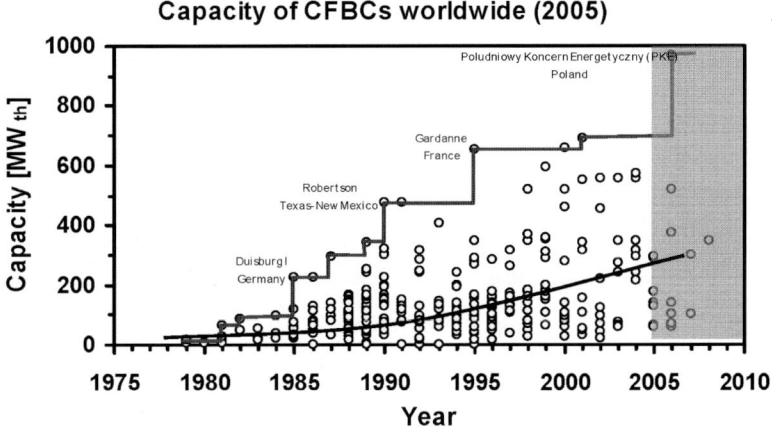

Figure 3. Design fuels in FBC boilers worldwide (*left*: CFBC; *right*: BFBC)

Figure 4 shows the distribution of design fuels for FBC boilers worldwide. It is clearly apparent that CFBC is the dominating technology for larger-scale coal boilers, while BFBC has found its most suitable market in the burning of biofuels of various kinds.

It is also interesting to follow the distribution of the design steam temperatures in the different boilers. The average design steam temperature in CFBC boilers has increased slowly but continuously and the highest steam temperature values for CFBC boilers are already above 550°C. The steam temperatures have been chosen more conservatively in BFBC projects, obviously due to the more demanding flue gas conditions in those units (alkali fly ash, chlorides).

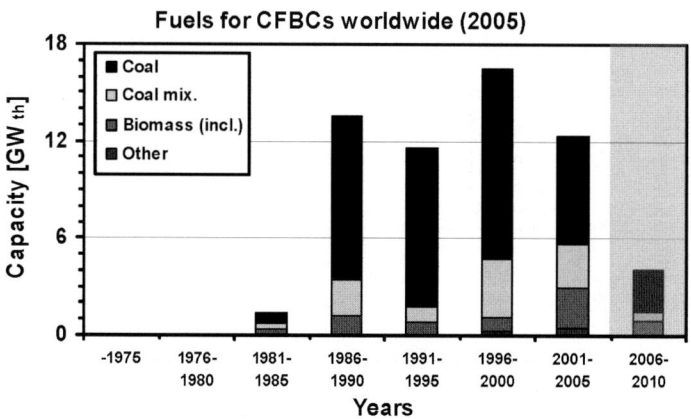

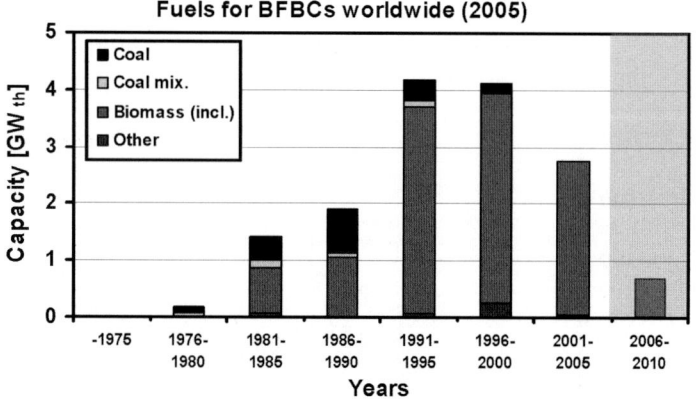

Figure 4. Design fuels in FBC boilers worldwide (*top*: CFBC; *right*: BFBC)

In recent years, many novel solutions have been introduced into boiler design. In particular, cyclone solutions for CFBC systems have seen many alternative designs, including Kvaerner's internal cyclone ("Cymic") and Foster-Wheeler's "rectangular" cooled cyclones ("Pyroflow Compact"). Further major variations can be seen in the location of the heat exchangers. Superheater tubes feature in several new CFBC boilers, placed in the return leg sand bed. This is thought to allow for higher steam temperatures than superheaters in the flue gas flow due to the less corrosive conditions.

The global FBC capacity is going to grow steeply in the future. Figure 5 includes recent information on CFBC projects in China. Currently, several hundred boilers are either at the planning stage or under construction in China, which – when realized – will double global capacity in the near future. Most of this capacity consists of large-scale CFBC boilers for coal.

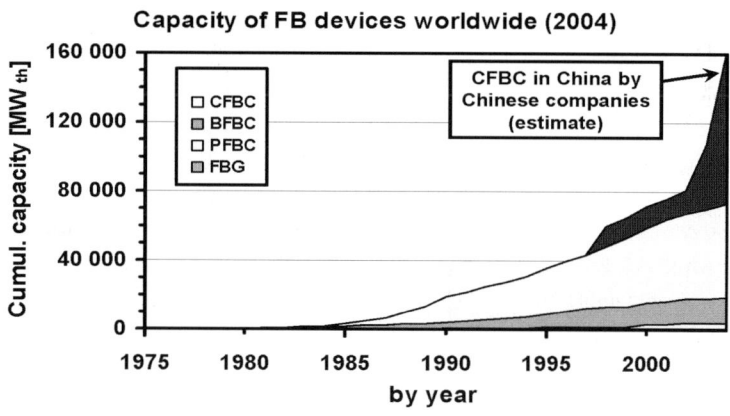

Figure 5. Global FBC capacity including China

3. "New" Fuels and Multifuel Firing in FBC

Today, more and more FBC installations in Europe and North America are being designed for unconventional fuels such as biofuels and waste derived fuels of various kind. In addition, multifuel firing (the use of more than one fuel simultaneously) has become a common feature in many new projects. Table 1 shows examples of some recent CFBC and BFBC projects around the world. The table contains all principal boiler deliveries by two of the major suppliers in 2001–2002 and gives a good overview of the fuel types on the FBC boiler market.

The table shows that of these 23 boiler projects 20 were designed for more than one solid fuel and only three were planned for single fuel firing. Interestingly, in 16 of the 23 projects, wood or some wood-based material was one of the design fuels. Peat was a design fuel in 10 projects, coal in 8, and petroleum coke in 4.

TABLE 1. Examples of major FBC projects 2001–2002 with their design thermal power, superheated steam temperature and fuels (FW = Foster-Wheeler, KP = Kvaerner Power)

CFBC Projects

	MWth	T (°C)	Fuels	
JEA, Jacksonville	689	540	Pet coke, coal	FW
Turow Power Stations	557	568	Brown coal	FW
Alholmens Kraft	550	545	Wood waste, peat, sludge, REF, coal	KP
Thaiheiyo Cement	341	570*	Coal, pet coke, RDF	FW
Indah Kiat Pulp & Paper	317	540	Coal, peat	KP
Shanghai Petrochemical	218	540	Pet coke, coal	FW
Mälarenergi AB	157	540	Wood residues, peat, coal	FW
Sinopec Jingling	156	540	Coal, pet coke	FW
Jämtkraft AB	125	545	Wood wastes, peat, bark	FW
Vattenfall SCA	98	480	Bark, wood residues, paper reject	FW
Norrköping Miljö och Energi	75	470	RDF, sewage sludge	KP
Heizkraftwerk Kehl GmbH	44	500	Demolition wood	FW

BFBC Projects

	MWth	T (°C)	Fuels	
Kymin Voima	269	541	Bark, forest residue, sludge, peat, oil	KP
Jämsänkosken Voima	185	535	Bark, wood residue, sludge, peat, oil	FW
Äänevoima	157	535	Peat, bark, chips, sludge, oil	FW
International Paper Svetogorsk	119	440	Bark, saw dust, sludge, gas	KP
Vamy/Vattenfall	88	525	Bark, sludge, gas, peat, wood residue	FW
Järvi-Suomen Voima	74	482	Bark, plywood, wood dust, peat, oil	KP
Kokkolan Voima	70	482	Wood waste, peat	KP
Visy Pulp and Paper Tumut	55	460	Bark, wood waste, gas	KP
Salmi Voima	45	515	Peat, bark, wood dust, recovered fuel	FW
Turku Energia	40	(204)	Wood residue, chips, bark	FW
Katrinefors Kraftvärme	36	480	Sludge, wood residue, wood waste	FW

Many recent papers have touched on the issue of fuel mixtures in FBC, e.g., the recent review by Reh[1]. Leckner and Åmand[2] reported the number of publications per year dealing with "cocombustion" or "cofiring" as extracted from the data bank of SciFinder in January 2003. It shows a steep increase in the interest in cofiring during the 1990s: the number of publications has increased from 10–20 per year to more than 100 per year. However, these publications have dealt with all combustion technologies, and only a fraction of them have dealt with FBC.

Design and operation of boilers using more than one fuel presents a number of challenges. The overall capacity and efficiency of the boiler is always strongly dependent on the fuel mix, and the supplier has to be able to guarantee the capacity and efficiency within the whole range of the fuel mixture being burned. It is also well known that flue gas emissions are strongly dependent on the fuel. Consequently, it is a major challenge to be able to maintain all emissions below their given limits for all fuel combinations required.

Fuel ashes cause various types of problems, and the interaction between ashes from different fuels is especially poorly understood, with a number of surprises having been reported when fuel mixtures have been used in FBC boilers. Fouling of heat exchanger surfaces, bed sintering and also superheater corrosion are problems that the operator is forced to closely follow when fuel mixtures are used. In addition, the possibilities of utilizing the ash residues from an FBC plant for applications such as the cement industry, etc. are strongly dependent on the fuels being burnt.

To systematically study combustion properties of fuel mixtures is a challenging task. Changes of the fuel mixture influence a number of factors in the furnace process, and it may be very difficult to establish clear relationships between the various furnace behavior factors and the fuel mixture. It is known that factors such as flue gas emissions, fouling tendency or bed sintering tendency are seldom simple linear functions of the fuel mixture. Rather, nonlinear relationships of some kind are much more common.

Below are examples of some of the recent research work into multifuel firing in FBC. The presentation is not a review of the research of fuel mixtures; rather it gives some examples of present activities at Åbo Akademi University. Additional systematic research is required to better understand the interaction between fuels and assist designers and operators when they evaluate the feasibility of different fuel blends in their installations.

4. Flue Gas Emissions

There are very few systematic results in the literature concerning flue gas emission as a function of the fuel mixture in FBC. Chalmers University of

Technology in Sweden has produced some interesting studies based on careful measurements at their 12 MW CFBC boiler. In the early 1990s, they produced the first extensive emission vs fuel mixture data for coal with wood[3], and more recently for coal with sludge[2]. Their experimental setup allows for exceptionally well-controlled tests of fuel mixtures in such a way that most of the furnace conditions, such as the air distribution and the temperature profile in the furnace, can be maintained constant and independent of the fuel mixture. This strongly helps when interpreting the results of the measurements and when trying to identify which effects the various fuels give rise to.

For pure fuels the NO emission was found to be roughly the same, around 60–80 ppm. This was an interesting finding, when taking into account the fact that the nitrogen content in the coal being used was about 1.6% while the nitrogen content for the wood was only roughly a tenth of that, around 0.14%.

An explanation for this finding was found in the measurements of the nitrogen species profiles along the vertical axis of the Chalmers furnace[2,3]. Such profiles clearly show that for wood the NO is rapidly formed in the lower part of the furnace, and the level obtained is maintained throughout the rest of the furnace height. For coal the situation is very different. There is a strong NO formation in the lower part of the furnace resulting in a very high local NO concentration in the bottom of the riser. However, this NO is effectively reduced when moving up in the furnace. This effect is explained by the NO reduction reaction caused by the high amount of char in the fluidized bed during coal combustion. In wood combustion, the char inventory in the bed is an order of magnitude smaller, and no significant NO reduction takes place. This effect could also be qualitatively reproduced by an FBC chemistry model developed at Åbo Akademi. This modeling approach is based on a simplified, one-dimensional (ID) fluidized bed flow description combined with a detailed combustion chemistry model[4–7]. The latter consists of a fuel particle burning model, which also describes the release and conversion of the fuel nitrogen. Furthermore, a comprehensive set of gas phase reactions is included in the model and also some of the key heterogeneous reactions involving NO are taken into account[8]. By adjusting the char inventory in the bed to a typical level for the fuels, the right type of NO profile can be found, as in the Chalmers tests[4].

It is much more difficult to try to model the NO vs fuel mixture curve in Figure 6. A good qualitative explanation was, however, given by Leckner et al. When small amounts of coal were added to the wood, the input of fuel nitrogen increased steeply. However, at that stage the char inventory did not increase in

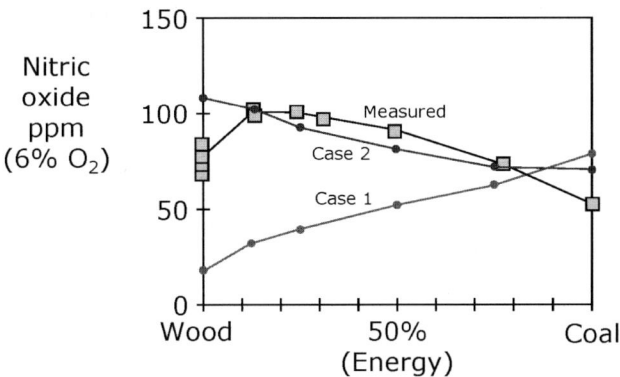

Figure 6. Nitric oxide emission as function of the fuel mixture during combustion of wood and bituminous coal. Experimental results from the Chalmers 12 MW CFBC[2]. Recent modeling attempts at Åbo Akademi University[9]

the same proportion and the net effect was an increase in the overall NO. Finally, when the share of coal increased to some 50%, the char inventory became large enough to compensate for the increased NO formation by the increased NO reduction in the upper furnace. Figure 6 also includes two of the first trials to model the NO vs fuel mixture trend by the type of ID detailed chemistry model described above[4-7]. Unfortunately, the calculated trend curves are very sensitive to some of the input assumptions of the model. For these two curves the difference is in the assumed distribution of fuel in the lower furnace. In case 1, the devolatilization of the fuel particles takes place in the dense bottom bed. In case 2, some of the devolatilization is assumed to also happen at higher elevations in the furnace. This difference, which is influenced by the details of the feeding of the fuels into the furnace, seems to be essential to the NO emission, particularly for the wood rich mixtures. Obviously, much more work is needed for the quantitative interpretation of these types of results.

The European waste combustion directive has become an important factor in the development of technologies for waste combustion (Table 2). That directive defines, besides the SO_2 and NO_x, other pollutants including dioxins and furans, the so-called heavy metals. This directive also necessitates strict emission monitoring.

The heavy metal limits given in the directive – 0.05 mg/m^3n for each of the metals – corresponds to an emission of roughly 0.5 mg/kg of the fuel.

TABLE 2. EU Directive on incineration of waste containing fuels. EDD = European Dirty Dozen, the twelve elements listed in the Directive as "heavy metals"

Acidic gases	
HCl	10 mg/m^3n (on-line)
HF	1.0 mg/m^3n (on-line)
SO_2	50 mg/m^3n (on-line)
NO_x	200 mg/m^3n (on-line)
Dust	10 mg/m^3n (on-line)
EDD	
Hg	0.05 mg/m^3n (measurement twice a year)
Cd + Tl	Total 0.05 mg/m^3n (measurement twice a year)
Sb + As + Co + Cr + Cu + Pb + Mn + Ni + V	Total 0.5 mg/m^3n (measurement twice a year)
Dioxins and furans	0.1 ng/m^3n (measurement twice a year)

The EU Directive can be compared with the typical content of these metals in a recovered fuel. In Figure 7, all 12 EDD elements are analyzed in the recovered wood fuel sample by selective leaching. This implies analysis of the metals in leaching solutions of gradually increasing strength: water, ammonium acetate, hydrochloric acid, etc.

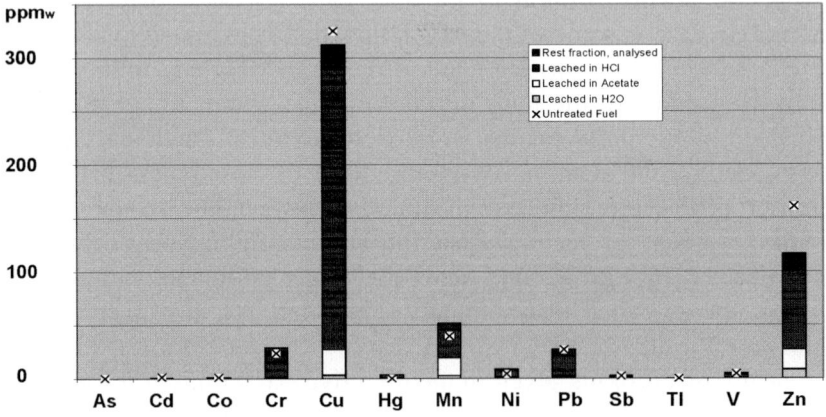

Figure 7. Example of heavy metal contents in a waste recovered fuel (parts per million, mass basis)

This type of selective leaching gives additional information of the chemical state of each of the elements in the fuel and this information can be used to better estimate the behavior of these elements in combustion.

For some of the metals, in this example Cu, Mn, Cr, and Pb capture from release in the flue gases has to be higher then 99%, which sets new demands for the flue gas-cleaning systems.

5. Interaction of Biomass Ashes

Biomass fuels are known to have reactive ashes, which may cause various difficulties in the FBC. Several recent studies have been focused on novel fuel characterization methods or prediction techniques to establish the behavior of these ashes in cofiring[10-16]. The most recent example concerns cofiring of two interesting biomass fuels, rice husks, and eucalyptus bark in a BFBC boiler. The focus of this work was to establish the fouling tendency as a function of the fuel mixture [17].

The ash forming elements in the two fuels are totally different. Rice husk has a very high overall ash content, about 20%. The main constituent in the ash is silica, SiO_2, with small amounts of potassium and chlorine. The ash content of the eucalyptus bark is also high (around 8%), but the main ash forming matters are calcium, potassium, and magnesium. The chloride content in this eucalyptus bark is also very high, 0.6–0.7% in the fuel dry solids.

The ashes produced by the two fuels are very different; not only does the chemical composition differ, but also the physical properties are very different. When burnt, rice husks form coarse, hard silica particles. These silica ash particles maintain the shape of the original plant even when all organic material has been burnt.

This ash has a typical particle size of several hundreds or thousands of micrometers. The burning of eucalyptus bark, however, results in much finer ash. Typically, the ash particles formed are in the micrometer scale, or they are released as vapors which condense later as tiny aerosol particles.

Åbo Akademi University has performed fouling tests of mixtures of rice husk and eucalyptus bark in collaboration with the University of Toronto[17]. Using an electrically heated 9 m tall entrained flow reactor any mixtures of the two fuels could be burned under well-controlled conditions. The fly ash produced could be captured on an air-cooled probe with controlled surface temperature. By this technique clear indications of fouling could be seen with exposure times of only 30–60 min.

Figure 8 shows the deposition rate as a function of the fuel mixture as determined by weighing the probe deposits after exposure. Rice husk ash does not seem to stick on the test probe surface at all. The in situ videos of the probe surface clearly showed how the large rice husk ash particles hit the probe surface and bounce away. On the other hand the bark ash very readily sticks on the surface and forms a gradually growing deposit on the "windward" side of the probe.

The most interesting result from the tests with different fuel mixtures was the finding that no significant buildup of ash was seen until the share of bark was more than 50% (dry solids basis).

This relationship between the deposition rate and the fuel mix is another interesting example of the strongly nonlinear dependencies discussed earlier. In this case the nonlinearity may be an effect caused by variations in physical interaction between the two ashes. The coarse rice husk ash particles are not only nonfouling themselves, but they also seem to be able to keep the tube surface clean from the more sticky bark ash. We have later found some other pairs of fuels also showing a similar strongly nonlinear fouling tendency when burned together.

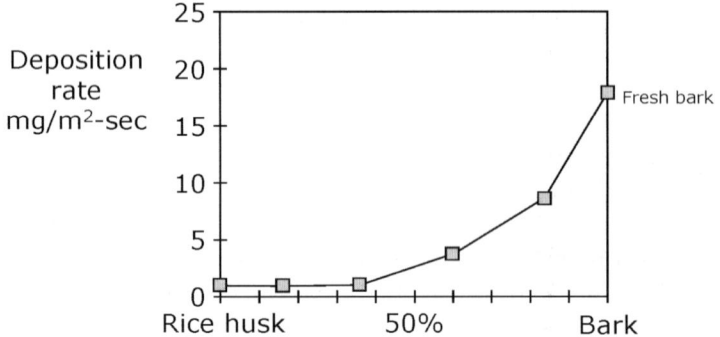

Figure 8. Effect of bark content on fouling rate of the test tube during tests using mixtures of rice husk and bark. Results from the Fouling Test Facility at the University of Toronto[17]

6. Conclusions

Fluidized-bed technology is rapidly expanding. Today, more than 600 large FBC boilers (>20 MWth) with a total installed thermal capacity of more than 70,000 MWth have been built. Around 75% of this capacity is CFBC technology, the rest mostly BFBC. The size of the boilers has increased steeply; the largest CFBC units constructed approach a thermal capacity of 1,000 MWth.

The global FBC capacity is going to grow steeply in the future. Several hundred boilers are either in the planning stage or under construction in China, which – when realized – will double the global capacity in the near future. Most of this capacity consists of large-scale CFBC boilers for coal.

In Europe and North America more and more boilers are being designed to burn several fuels simultaneously. Besides coals, wood, biomasses, and various waste derived fuels have become important fuels in the different recent FBC projects.

In principle, simultaneous firing of several solid fuels has been considered to be an advantage of FBC technology. However, some fuel mixtures show surprising behaviors and there is a lot of interesting research to be done to more quantitatively establish the influence of the fuel mixture on the key operating parameters, such as boiler efficiency, flue gas emissions, ash behavior, and corrosion.

In general, the only way to learn about the firing properties of fuel mixtures is to test the fuel mixture in a pilot plant or full-scale boiler. These kinds of tests are naturally highly important, but also difficult to perform in a way which allows for clear conclusions to be drawn. Without fundamental understanding of the behavior of fuels in mixed firing, the pilot plant results are difficult to extrapolate to other conditions or fuel mixtures. Hence, mixed-fuel firing presents a big opportunity for more fundamental research. Fuel analysis techniques should be improved to give more information on the behavior of the fuels when they are burned in mixtures. Fluidized bed combustion and emission modeling should be extended to give at least some qualitative understanding of mixed fuel firing. Characterization of the formation of fuel ash should make it possible to differentiate between the different ash constituents in such a way that mixed fuel effects could be predicted.

Acknowledgments

I want to thank Professor Bo Leckner at Chalmers University and Professor Honghi Tran at the University of Toronto for long-term collaboration and their great help in preparing this overview.

Åbo Akademi combustion research ("ChemCom" Program) is presently supported by the National Technology Agency (Tekes) and the industrial consortium of the companies Andritz Oy, Foster Wheeler Energia Oy, Kvaerner Power Oy, Oy Metsä-Botnia Ab and Vattenfall Utveckling AB and International Paper. This work is part of the activities of the Åbo Akademi Process

Chemistry Group within the Center of Excellence program by the Academy of Finland.

At Åbo Akademi University I want to acknowledge my close colleagues Rainer Backman, Bengt-Johan Skrifvars and Christian Mueller for their continuous support. Further, Maria Zevenhoven, Patrik Yrjas, Pia Kilpinen, Jukka Konttinen, Sirpa Kallio, and Mischa Theis have been key contributors to the research referred to in this presentation.

The Åbo Akademi FBC boiler database was maintained by Dan Lundmark and Edgardo Coda Zabetta. The help from the boiler companies Alstom, Kvaerner Power and Foster Wheeler to the data bank is warmly acknowledged.

References

1. L. Reh, Development potentials and research needs in circulating fluidized bed combustion, *China Particuology,* **1**(5), 185–200 (2003)
2. B. Leckner and L.-E. Åmand, Co-combustion of sludge with coal or wood, *International Journal of Power and Energy Systems,* **24**(3), 203–3413 (2004)
3. B. Leckner and M. Karlsson, Gaseous emissions from CFB combustion of wood, *Biomass and Bioenergy,* **4**, 379–389 (1993)
4. S. Goel, A. Sarofim, P. Kilpinen, and M. Hupa, Emissions of nitrogen oxides from circulating fluidized bed combustors: modelling results using detailed chemistry, Proceedings: *26th International Symposium on Combustion*, 28 July–2 August 1996, Naples, Italy
5. P. Kilpinen, S. Kallio, and M. Hupa, advanced modeling of nitrogen oxide emissions in circulating fluidized bed combustors: parametric study of coal combustion and nitrogen compounds chemistries, Proceedings: *15th International Conference on Fluidized Bed Combustion*, 9–13 May 1999, Savannah, GA, CD-ROM, Paper No. FBC99-0155 (1999)
6. P. Kilpinen, S. Kallio, J. Konttinen, C. Mueller, A. Jungar, M. Hupa, L.-E. Åmand, and B. Leckner, Towards a quantitative understanding of NO_x and N_2O emission formation in full-scale circulating fluidized bed combustors, Proceedings: *16th International Conference on Fluidized Bed Combustion*, 13–16 May 2001, Reno, NV, CD-ROM, Paper No. FBC01-079 (2001)
7. S. Kallio, P. Kilpinen, J. Konttinen, B. Leckner, and L.-E. Åmand, Sensitivity study of fluid dynamic effects on nitric oxide formation in CFB combustion of wood, Proceedings: *7th International Conference on Circulating Fluidized Beds*, Niagara Falls, Canada, 5–8 May 2002
8. R. Zevenhoven and M. Hupa, The reactivity of chars from coal, peat and wood towards no, with and without co, *Fuel,* **77**(11), 1169–1176 (1998)
9. M. Engblom, Prediction of Nitrogen Oxides from Circulating Fluidized Bed Combustion when Co-firing Wood and Coal, Master's thesis, Åbo Akademi University, Turku, Finland (2003)
10. B.-J. Skrifvars, R. Backman, M. Hupa, G. Sfiris, T. Åbyhammar, and A. Lyngfelt, Ash behavior in a CFB boiler during combustion of coal, peat or wood, *Fuel,* **77** (1–2), 65–70 (1998)

11. B.-J. Skrifvars, M. Öhman, A. Nordin, and M. Hupa, Predicting bed agglomeration tendencies for biomass fuels fired in FBC boilers: A comparison of three different prediction methods, *Energy and Fuels,* **13**(2), 359–363 (1999)

12. M. Zevenhoven-Onderwater, J.-P. Blomquist, B.-J. Skrifvars, R. Backman, and M. Hupa, The prediction of behavior of ashes from five different solid fuels in fluidized bed combustion, *Fuel*, **79**(11), 1353–1361 (2000)

13. M. Öhman, A. Nordin, B.-J. Skrifvars, R. Backman, and M. Hupa, Bed agglomeration characteristics during fluidized bed combustion of biomass fuels, *Energy and Fuels,* **14**(1), 169–178 (2000)

14. B.-J. Skrifvars, M. Zevenhoven, R. Backman, and M. Hupa, Predicting the ash behavior of different fuels in fluidized bed combustion, Proceedings: *16th International Conference on Fluidized Bed Combustion*, 13–16 May 2001, Reno, NV, Paper No. FBC01-113 (2001)

15. C. Mueller, D. Lundmark, B.-J. Skrifvars, R. Backman, M. Zevenhoven, and M. Hupa, CFD-based ash deposition prediction in a BFB firing mixtures of peat and forest residue, Proceedings: *17th International Conference on Fluidized Bed Combustion*, 18–21 May 2003, Jacksonville, FL (2003)

16. M. Zevenhoven, B.-J. Skrifvars, P. Yrjas, R. Backman, C. Mueller, and M. Hupa, Co-firing in FBC a challenge for fuel characterization and modeling, Proceedings: *17th International Conference on Fluidized Bed Combustion*, 18–21 May 2003, Jacksonville, FL (2003)

17. B.-J. Skrifvars, P. Yrjas, T. Laurén, J. Kinni, H. Tran, and M. Hupa, The fouling behavior of rice husk ash in fluidized-bed combustion. Part 2. Pilot-scale and full-scale measurements, *Energy and Fuels,* **19**(4), 1512–1519 (2005)

FLUIDIZED BED COMBUSTION IS THE UNIVERSAL TECHNOLOGY OF FIRING FOSSIL FUELS AND VARIOUS TYPES OF WASTES

V. A. BORODULYA
A.V. Luikov Heat and Mass Transfer Institute of the National Academy of Sciences of Belarus; 15, P.Brovka Str., Minsk, 220072, Belarus, e-mail: bor@itmo.by

Abstract. Since the 1940s when the fluidized – solids technique was introduced by U.S. petroleum companies for catalytic cracking, this method has enjoyed an expansion to numerous gas–solids contacting operations including roasting, drying, calcining and incineration. Nowadays fluidized bed combustion is one of the most promising technologies for the effective and environment friendly combustion of broad range solid fuels and wastes with varying ash, moisture, sulfur and nitrogen contents without auxiliary equipment. In this paper, several examples of FBC installations that have been already realized in Belarus will be also discussed to further illustrate the potential of this technology.

Keywords: fluidized bed, combustion, boilers, low-grade coals, biofuel, waste materials, emissions

1. Introduction

Fluidized bed combustion technology is the only technology which is capable of firing fossil fuels and various types of wastes efficiently without extensive pretreatment in an environmentally sound manner and whilst utilizing the energy of the fules[1-4].

It can be justifiably claimed that fluidized bed combustion is the only really new approach to the combustion of fossil fuels that has emerged for at least six decades. The potential merits of burning coal in a fluidized bed of mineral matter in which the boiler tube surface is immersed were recognized as long ago as 1962, and throughout the 1960s and 1970s efforts were increased to apply them to the combustion and gasification of fuels such as coal. The eventual result has been the development of fluidized bed boilers, gasification

N. Syred and A. Khalatov (eds.), Advanced Combustion and Aerothermal Technologies, 103–112.

plants, furnaces, drying plants and heat recovery systems which utilize coal more economically and with less pollution than at present.

From the 1970s onwards, attention has increasingly turned towards emission control technologies for the reduction of oxides of nitrogen and sulfur, the so-called acid rain precursors. It was possible to reduce NO_x formation and to retain sulfur by the application of fluidized combustion in which coal or generally critical fuels burn at a much reduced combustion temperature, at which calcium sulfate, the reaction product of SO_2 and additive lime, is still stable. It is primarily for environmental reasons that fluidized bed combustion has been become more attractive for the utilization of low-grade coals and the incineration of wastes.

During the last decade, several new factors emerged which influenced the development of combustion for power generation. CO_2 emission control is gaining increasing acceptance as a result of the international greenhouse gas debate. In consequence, the biomass and, in particular, its main representative – wood, will be increasingly burned using FBC technology. In this case the emission results will be extremely good, and it is evident that this fuel type can be burned in a fluidized bed with extremely low emissions of SO_2, and negligible production of polyaromatic hydrocarbons, dioxins and furans.

2. Fundamental of Fluidized Bed Combustion

2.1 FLUIDIZATION

Fluidization is an operation by which a bed of solid particles acquires fluid-like properties by passing a gas or liquid through it. A bed of particles rests on a gas-distribution grid in the bottom of a vessel. Through a perforated bottom, a gas is injected into the vessel. The sequence of events, as the gas flow rate increases, is typically as follows:

- For low flow rates the bed particles rest upon each other. The gas flows in the space between the particles. The velocity is lower than minimum fluidization velocity and we have a static or fixed bed.
- As the gas velocity increases, the gas velocity is greater than the minimum fluidization velocity. The particles can move with respect to each other. The excess gas from minimum fluidization passes the bed in the form of bubbles. Such a bed is called a BFBC.
- When the gas velocity is higher than the terminal velocity of bed particles, some particles will be carried away with the gas, and the suspension of particles then extends itself over the entire space of the vessel above the dense bottom bed. In this case the bed material carried away from the dense bed

has to be removed from the gas leaving the vessel and recirculated in order to maintain the bed. This regime is termed as a suspended or circulating bed.

The states of the fluidized bed combustor are illustrated in Figure 1.

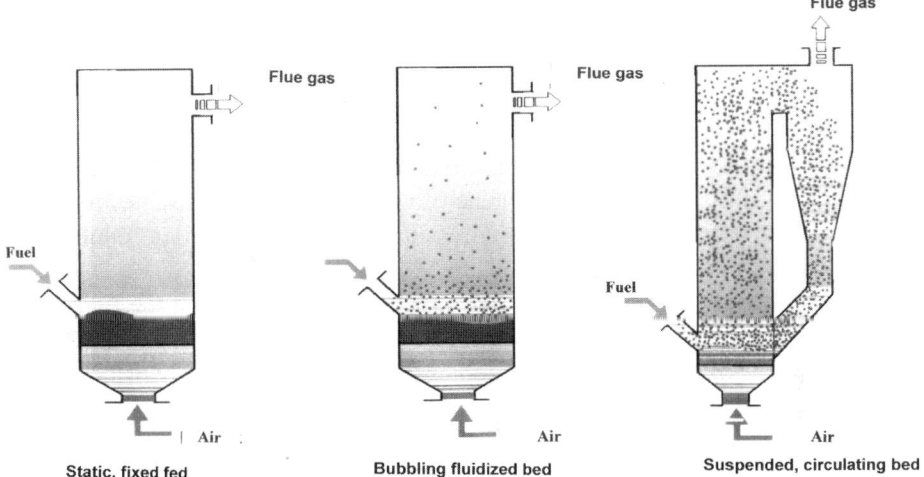

Figure 1. Fluidization states of the FB combustor

2.2 COMBUSTION EFFICIENCY

The fluidizing mechanism offers several advantages over conventional combustion:

- Increased combustion efficiency due to superior combustive turbulence. The fluidized bed provides an intensely turbulent intimate mixture of fuel, hot inert bed particles and air.
- The improved combustion efficiency results in a lower unburnt carbon (ash) loss, and the ash does not have to be recycled back into the furnace for further combustion.
- More stable combustion due to a "thermal flywheel effect". A tremendous amount of heat is absorbed and retained by the large mass of inert particles making up the fluidized bed. The large thermal mass and extreme turbulence greatly reduce the potential for cold and hot spots to occur in the furnace.
- A decrease in the formation of NO_x and possible capture of SO_2 in the bed.

2.3 EMISSIONS

2.3.1 Nitrogen Oxide and Other Emissions

At the relatively low temperatures typical of FBC systems, the major source of nitrogen oxides (90%+) is from the fuel nitrogen rather than nitrogen from the air. A simplified model of the different possible processes for the formation of NO_x and N_2O which takes into account that fuel which is introduced in a fluidized bed combustor/boiler decomposes into volatiles and char fractions is shown schematically in Figure 2.

The emission of NO from the combustion of wood is normally higher than that from the combustion of coal because of a smaller amount of char in the bed during the combustion of wood, which results in a much smaller reduction of the NO formed in this case.

The formation of nitrogen oxides from the nitrogen content of the fuel can be effectively prevented by the recirculation of the flue gas and staging the combustion air, when part of the combustion air is introduced into the free-board. The end results are extremely low NO_x emissions even with fuels rich in nitrogen.

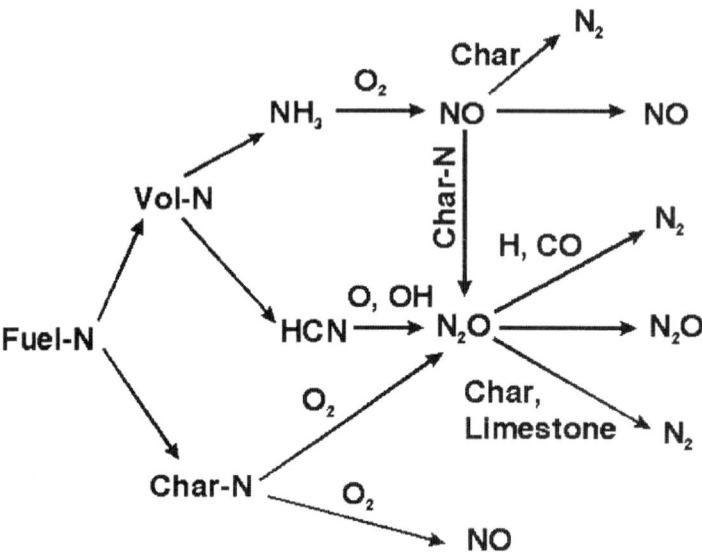

Figure 2. NO and N_2O formation and reduction pathways

2.3.2 *Sulfur capture in FBC*

The use of limestone for in-furnace desulfurization is one of the advantages of fluidized bed combustion. In units operating at atmospheric pressure the limestone or dolomite first calcines, and then the CaO reacts with SO_2 via the global reactions

$$CaCO_3 = CaO + CO_2 \qquad \Delta H = 182.1 \text{ kJ/gmol} \qquad (1)$$

$$CaO + SO_2 + \tfrac{1}{2}O_2 = CaSO_4 \quad \Delta H = -481.4 \text{ kJ/gmol} \qquad (2)$$

Due to the favorable conditions in the fluidized bed, over 90% of the resulting sulfur oxides can be removed at temperatures of 850 870°C.

3. Fluidized Bed Firing Systems

The main advantages of this type of firing are:

- The possibility to burn high inert content fuels as well as fuels and fuel mixtures with widely different characteristics
- High firing efficiency
- Simple fuel preparation and feeding
- Good load changing and partial load behavior
- Low SO_2 content of the flue gas due to the addition of limestone or dolomite to the fuel
- Low NO_x production due to low combustion temperatures (~850°C) and suitable adjustment of the firing
- High heat transfer rates to the heating surfaces located in the combustion chamber

4. Today's Status of FBC Development in Belarus

4.1 BACKGROUND

In countries, like Belarus, where the structure of primary energy resources are limited, there is and will remain a need to use local fuels. In order to substitute for expensive imported liquid fuels or natural gas, especially in the field of energy generation in industry and district heating. This means that low-grade coals, biomass and waste will in many cases be the fuels of choice. In these situations, FBC boilers currently have and hot-gas generators can have a significant role in heat and power production in the industrial and agricultural sector as well as for district heating.

In particular, according to the adopted National program on attaining 25% energy consumption by domestic energy resources till the year 2012, annual consumption of domestic and renewable fuels in structural balance of energy-producing fuel should be increased from 4.2 million tce (16.7%) in 2003 to 6.75 million tce (25%) by 2012.

This should be followed by a further technical development of plants for energy production on the basis of low-grade coals and biomass. These goals are to be achieved first of all through the activity of the national boiler's manu-facturers. In Belarus there are more than 20 plants which manufacture small boilers operating on different kinds of solid fuel. One of them is the Joint-Stock Company "Head Specialized Designing Bureau" (HSDB) placed in Brest.

On the basis of the scientific knowledge acquired by the A.V. Luikov Heat and Mass Transfer Institute of the National Academy of Sciences of Belarus and the engineering experience gained by the Joint-Stock Company "Head Specialized Designing Bureau", a set of steam and water heating boilers with a capacity in the range of 0.4–1.25 MWth, based on the fluidized-bed technology have been developed[6–8].

At first, the boilers have been designed to be fired with anthracite culm. But, as it can be seen from Table 1, two other fuels were used too (Volyn and Silesia coals).

It was confirmed that it is possible to carry out stable and continuous operation of the boilers with all of these fuels, which were supplied in lump sizes of up to 100 mm as received.

The first 0.4 MWth demonstration unit was put into operation in 1993, and has been run throughout the heating season producing 6,620 GJ of heat. It has provided a lot of operating experience with the boiler itself, as well as other components e.g., fuel feeding and dust carryover[6].

TABLE 1. Technical characteristics of small developed boilers based on fluidized-bed technology

N	Parameter	KP-F-0.4 T	KP-F-1.25 T	KW-F-1.25 T
1.	Heat capacity (MW)	0.4	1.25	1.25
2.	Rated steam capacity (t/h)	–	1.85	–
3.	Final steam conditions			
	- Temperature (°C)	120	133–179	–
	-Pressure (MPa)	0.02–0.06	0.2–0.9	
4.	Actuating water pressure (MPa)	0.6	0.6	0.6
5.	Min. water consumption (m³/h)			43.0
6.	Design temperatures (°C)			
	- Water in	80	80	70
	- Water out	–	–	95
7.	Water volume of boiler (m³)		2.5	1.8
8.	Steam volume of boiler (m³)	0.8	0.34	–
9.	Hydraulic resistance of boiler (MPa)	–	–	0.02
1	2	3	4	5
10.	Fuel fired			
	Coal	x	x	
	Peat			x
	Wood			
11.	Low calorific value (GJ/t)			8.12
12.	Heating surface (m²)	13.5	42.3	42.6
13.	Reference fuel consumption (tce/h)	0.170	0.179–187	0.187
14.	Flue gas temperature (°C)	258	250	200
15.	Thermal efficiency (%)*	80–82	80–82	82
16.	Power consumed (Kw)	6	17.6	21
17.	Mass of the boiler plant (t)	3.3	6.7	9.0
18.	Overall dimensions (mm)			
	Length	2,130	8,250	11,380
	Width	1,500	2,380	3,125
	Height	2,900	4,235	5,250
19.	Emission levels for dry gases based on zero excess air (mg/Nm³)			
	CO	375	375	1,100
	NO$_x$	750	750	750

*Based on lower heating value.

TABLE 2. Fuel tested in the steam fluidized bed boilers

Kind of coal	Anthracite culm	Volyn coal	Silesian coal
Ultimate analysis (wt%)			
C	70.5	53.7	65.5
H	1.4	3.6	4.8
O	1.9	5.1	8.7
N	0.8	0.7	0.8
S	1.7	3.1	1.8
Ash	16.7	25.8	9.5
Moisture	7.0	8.0	8.9
Total	100.0	100.0	100.0
Higher heating value (MJ/kg) (Dry)	33.4	21.5	22.5

4.2 DESIGN FEATURES OF THE DEVELOPED FLUIDIZED BED BOILERS

It is known that the bubbling bed principle has specific advantages over that of circulating beds owing to the comparatively simple design and low power consumption when establishing the bed[5].

Typical BFBC boilers which were developed in Belarus have some specific design features:

- The bed combustor is based upon the standard vertical shell-type technology. Combustion is confined to the water-cooled cylindrical section of the continuous welded wall chamber. In the case of a steam boiler, hot gases are exhausting through a fire-tube section which provides an additional heat surface for superheating steam generated in the furnace. Hot gases from start-up burners are used for ignition.

- The gas distributor consists of a serious of "bubble-cap" nozzles fed from an air distributor plate covered by a layer of sand particles and ensures intensive stirring of fuel particles and inert material and their circulation in the bulk of the bed.

- There is a system for maintaining a required fraction composition of the beds inert material and for controlling the fluidized bed height, which allows operation with fuels producing different kind of ash, with relatively large particle sizes and small velocities in the combustion chamber.

- The ash collection and transport system may be used to recirculate fines into the bed in order to increase the combustion efficiency.

- The automatic control system maintains the preset temperature by flue gas recirculation. Periodic fuel supply of wide-fraction fuel ensures reasonably low average fuel rates and allows simplification of the automatic control system.

- The multicyclones have so far been sufficient, in most cases, for cleaning flue gases. There is additional equipment, if necessary, for reducing SO_2 and particulate losses and a system for NO_x control.

4.1.1 Description of 1.25 MW Steam Boiler as a Typical Developed Unit

Based on the design experience and performance of the 0.4 MWth steam boiler, as a prototype, a 1.25 MWth steam boiler has been constructed. The boiler as designed is also capable of burning low-grade coals. Oil is used only as a start-up fuel. The peculiarity of this boiler is the configuration of the combustion chamber. The lower part of the chamber has an inner diameter of 1.0 m, the upper part or freeboard section is 1.2 m in diameter. All combustion surfaces are 100% water cooled for a long maintenance free life. The gas distributor is a

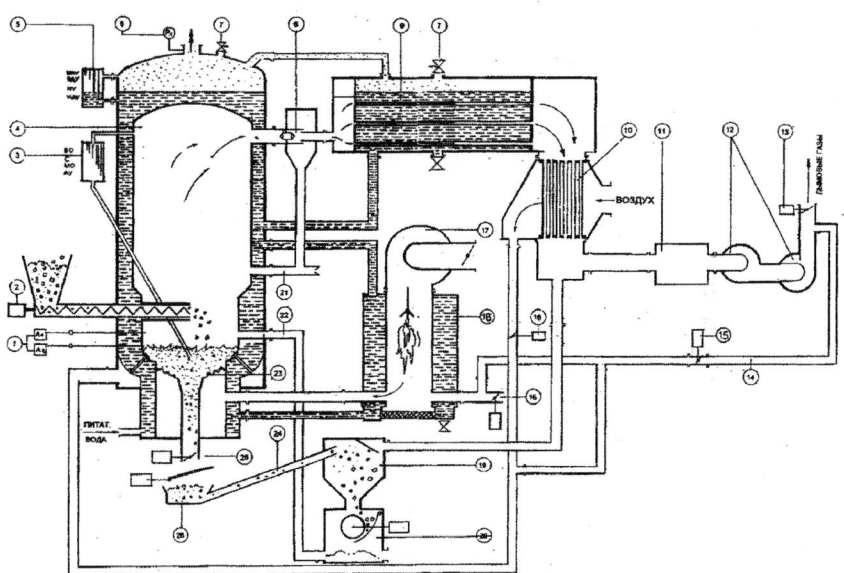

Figure 3. Scheme of 1.25 MWth fluidized bed steam boiler: 1 – bed temperature regulator, 2 – coal screw feeder, 3 – bed level sensor, 4 – furnace chamber, 5 – gauge-glass, 6 – manometer, 7 – safety-valve, 8 – precipitator (cyclone), 9 – superheater, 10 – recuperator, 11 – fly ash filter, 12 – exhaust fan, 13 – exhaust fan vent, 14 – line for recirculation flue gases, 15 – recirculation vent system for maintaining required fractional composition of the bed material, 16 – air vent, 17 – start-up burner, 18 – lighting chamber, 19 – ash hopper, 20 – ash crusher, 21 – ejector, 22 – line for feeding inert material, 23 – gas distributor, 24 – ash draining line, 25 – ash draining system, 26 – ash separator

water cooled hemispherical cap-type of steel construction, which has an additional annular section with an entirely separate air fluidizing line. The boiler is suitable for working pressures of up to 0.9 MPa. A flow scheme of the 1.25 MWth fluidized bed steam boiler plant is shown in Figure 3.

In October, 2000 the boiler successfully passed tests under conditions of full load and was then exported to Poland where it was put into operation. Starting a test program including the firing of Polish coal with high ash (up to 37%) and moisture (12%) content.

5. Conclusions

A set of universal 0.4–1.25 MWth BFBC boilers are now available, with the most appropriate boiler design dependent upon the unit size and the type of the fuel that is to be burnt. The boilers are a proven technology, and are a low-risk alternative to grate-fired boilers. Additional knowledge and operating experience continues to be accumulated, covering a wide range of fuels and waste products.

References

1. Lee Yaverbaum, 1977, Fluidized Bed Combustion of Coal and Waste Materials, Noyes Data Corporation, Park Ridge, New Jersey.
2. Borodulya V.A. and Vinogradov L.M., 1980, Fluidized Bed Combustion of Solid Fuels, Nauka i Technika, Minsk (in Russian).
3. Borodulya V.A., Dikalenko V.I. and Palchonok G.I., Fluidized bed combustion of low-grade coals and wastes: research, modeling and development, in: Proc. Int. school-seminar Superadiabatic combustion and its applications, Minsk, 1995, pp. 121–126.
4. Werther J., Oganda T., Borodulya V.A. and Dikalenko V.I., Devolatilization and combustion characteristic of sewage sludge in a bubbling fluidized bed furnace, in: Proc. Second Int. Conf. on Combustion and Emissions Control, London, 3–5 December 1995, pp. 149–158.
5. Oka S., 2001, Is the Future of BFBC Technology in Distributive Power Generation? Thermal Science, 5(2):33–48.
6. Borodulya V.A., Dobkin S.M. and Telegin E.M., Design and operating experience of the first commercial 0.4–0.5 MW fluidized boiler in Belarus, in: Proc. 3rd European Conf. on Ind. Furnaces and Boilers, Lisbon, Portugal, 18–21 April, 1995.
7. Borodulya V.A. and Dobkin S.M., Today's status of FBC developments in Belarus, in: Proceedings of the 3rd Int. Conf. on Coal Utilization Science and Technology. CUSTNET, Bucharest, 1998.
8. Borodulya V.A. and Dobkin S.M., Development experience of small industrial coal fired fluidized bed boilers, Paper presented at the 11th Symposium of Thermal Engineers from Serbia and Montenegro, 1–4 October Zlatibor, 2003 (unpublished).

SOLID FUEL COMBUSTION TECHNIQUES AND AUGMENTATION

PLASMA-SUPPORTED COAL COMBUSTION MODELING AND FULL-SCALE TRIALS

V. E. MESSERLE, A. B. USTIMENKO*
Combustion Problems Institute, al-Farabi Kazakh National University, 172 Bogenbai batyra str., Almaty, 050012, Kazakhstan

Abstract. Plasma-supported solid fuel combustion is a promising technology for use in thermal power plants (TPPs). Development of this technology comprises two main steps. The first is the execution of a numerical simulation and the second involves full-scale trials of plasma-supported coal combustion through plasma-fuel systems (PFS) mounted on a TPP boiler. For both the numerical simulation and the full-scale trials, the 200 MW boiler at Gusinoozersk TPP (Russia) was selected. The optimization of the combustion of low-rank coals using plasma technology is described, together with the potential of this technology for the general optimization of the coal burning process. Numerical simulation and full-scale trials have enabled technological recommendations for improvement of existing conventional TPP to be made. PFS have been tested for plasma start-up and flame stabilization at 27 TPP boilers in various countries. Steam-productivity of these boilers varied between 75–670 t/h (TPH), with each boiler equipped with different types of pulverized coal burner. During PFS testing, coals of all ranks (brown, bituminous, anthracite and their mixtures) were used, the volatile content of which varied from 4% to 50%; ash – from 15% to 48% and calorific values – from 6,700 to 25,100 kJ/kg. In summary, it is concluded that the developed and industrially tested PFS improve coal combustion efficiency and decrease harmful emission from pulverized coal-fired TPP.

---

*To whom correspondence should be addressed. A.B. Ustimenko. Combustion Problems Institute, al-Farabi Kazakh National University, 172 Bogenbai batyra str., Almaty, 050012, Kazakhstan. Tel.: +73272 615148; Fax:+(73272)675141; e-mail: ust@ntsc.kz

N. Syred and A. Khalatov (eds.), Advanced Combustion and Aerothermal Technologies, 115–129.
© 2007 *Springer.*

Keywords: coal, combustion, thermochemical preparation, plasma-fuel system, simulation, industrial tests

1. Introduction

Plasma-assisted coal combustion is a relatively unexplored area in coal combustion science and only a few references are available on this subject[1,2]. Coal-fired utility boilers face two problems, the first being the necessity to use expensive oil for start-up and the second being the increased commercial pressure, which requires operators to burn a broader range of coals, possibly outside the combustion equipment manufacturer's specifications. Each of these problems results in a negative environmental impact. Oil firing for start-up increases the gaseous and particulate burden of the plant. The firing of poorer quality coals has two disadvantages: reduced flame stability necessitating oil support, which comes with inherent implications for emissions and cost; and reduced combustion efficiency due to increased amounts of carbon in the residual ash, leading to an increase in emissions per MW of power generated. Plasma aided coal combustion represents a new, effective and ecofriendly technology, which is equally applicable to alternative "green" solid fuels.

One of the prospective technologies is Thermo Chemical Plasma Preparation of Coals for Burning (TCPPCB). This technology addresses the above problems in TPP. Development of TCPPCB technology comprises two main steps. The first includes numerical simulations and the second involves full-scale trials of plasma-supported coal combustion in a TPP boiler. For both the numerical study and full-scale trials, the 200 MW boiler at Gusinoozersk TPP (Russia) was selected. Within this concept a portion of pulverized solid fuel (pf) is separated from the main pf flow and undergoes activation by arc plasma in a special chamber – PFS (Figure 1).

The air-plasma flame is a source of heat and additional oxidation. It provides a high-temperature medium enriched with radicals, where the fuel mixture is heated, volatile components of coal are extracted, and carbon is partially gasified. This active blended fuel can ignite the main pf flow supplied into the furnace. This technology provides boiler start-up and stabilization of pf flame and eliminates the necessity for additional highly reactive fuel.

The numerical simulations were performed using the Cinar ICE 'CFD' code[3]. Cinar ICE has been designed to provide computational solutions to industrial problems related to combustion and fluid mechanics. The Cinar code solves equations for mass, momentum and energy conservation. Physical models are employed for devolatilization, volatiles combustion (fast unpremixed combustion), the combustion of char and turbulence ($k-\varepsilon$). Comparison

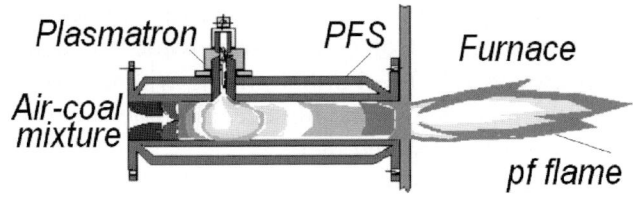

Figure 1. Sketch of PFS

of the calculations with data generally reveals excellent agreement. The maximum discrepancies between measured and calculated furnace temperatures do not exceed 15%.

Numerical simulation and industrial trials have enabled technological recommendations for the improvement of existing conventional TPP to be made. PFS have been tested for plasma start up and flame stabilization at 27 TPP boilers in various countries. Steam productivity of these boilers varied from 75–670 t/h, with each boiler equipped with different types of pulverized coal burner[4]. At PFS testing power coals of all ranks (brown, bituminous, anthracite and their mixtures) were used, the volatile content of which varied from 4% to 50%; ash – from 15% to 48% and calorific values – from 6,700 to 25,100 kJ/kg. In summary, it is concluded that the developed and industrially tested PFS improve coal combustion efficiency and decrease harmful emission from pulverized coal-fired TPP.

2. Numerical Simulation

The PFS is a cylinder with a plasma generator placed on the burner (Figure 2). In PFS, since the primary mixture is deficient in oxygen, the carbon is mainly oxidized to carbon monoxide. As a result, at the exit from the PFS a highly

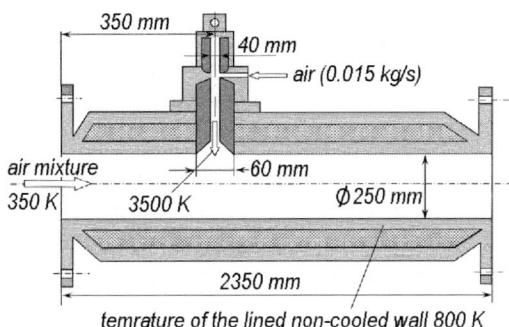

Figure 2. Schematic view of PFS

reactive mixture of combustible gases and partially burned char particles is formed, together with products of combustion, while the temperature of the gaseous mixture is around 1,300 K. Further mixing with the secondary air, upon the introduction of the mixture into the furnace, promotes intensive ignition and complete combustion of the prepared fuel.

The numerical experiments were performed for a cylindrical direct flow burner – PFS shown in Figure 2. The efficiency of plasmatron was around 85%. All parameters of the PFS are presented in Table 1.

"Tugnuiski" bituminous coal (TBC) was used for the experiments. Both its proximate and ultimate analyses, as well as its particle size distribution are presented in Table 2. From the available experimental data of the PFS operation, the measured composition of the gas phase at the exit of the PFS was (volume %): $CO = 28.5$; $H_2 = 8.0$; $CH_4 = 1.5$; $CO_2 = 2.0$; $N_2 = 59.5$; $O_2 = 0.0$; others = 0.5, including NO_x=50 mg/m^3.

The flame from the plasmatron feeds the PFS 0.35 m axially downstream of the PFS inlet plane (Figure 3). For modeling purposes the plasma flame is assumed to be a heat/mass source with an exit temperature of 2,800 K and a mass flow of 54 kg/h.

The thermal and chemical equilibrium approach was selected for cal-culations of the PFS, for which the TERRA code was used[5]. This method included the formation of the libraries containing the values of species concentrations as a function of the mixture fraction and temperature level. Although the combustion is not a thermal equilibrium process, the application of this approach could be justified by the existence of charged species and radicals, which are highly active and probably act as catalysts, increasing the rate of chemical reactions. In addition to this, the high-energy input and maximum temperature level make the chemical reactions fast so that they are probably close to the equilibrium condition.

TABLE 1. Specification of PFS operating parameters

Operating data		Plasmatron	
PFS Length (m)	2.35	Electric power (kW)	100
PFS Inner diameter (m)	0.25	Plasma gas	Air
Primary air		Mass flow (kg/h)	54
Air flow (kg/h)	3,500	Inlet air temperature (K)	298
Velocity (m/s)	20.0	Outlet air temperature (K)	2,800
Temperature (K)	80	Inner diameter (m)	0.04
Coal-dust concentration (kg/kg)	0.50	Outlet velocity (m/s)	118.2

TABLE 2. Specification of TBC

Proximate analysis	mass (%)	Particle size distribution*
Moisture	14.00	160 μm–10%
Volat. Matter	36.27	130 μm–10%
Fixed carbon	44.33	74 μm–20%
Ash	19.40	50 μm–40%
Ultimate analysis	mass (%)	24 μm–20%
Carbon	61.7	
Hydrogen	4.10	Lower calorific value: 23,000 kJ/kg
Nitrogen	1.20	
Sulfur	0.39	Coal feed rate: 1,750 kg/h
Oxygen	13.20	

*Assumed particles size distribution

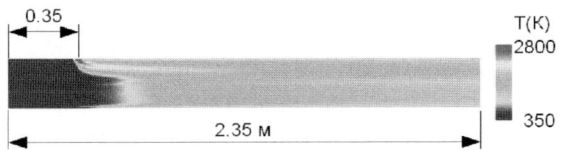

Figure 3. Predicted temperature contours along the PFS axial direction

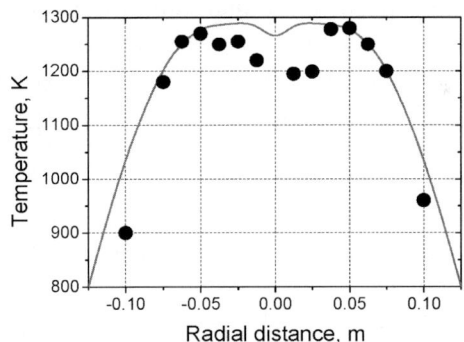

Figure 4. Predicted temperature radial profiles at the exit of PFS. Line is calculation, • is experiment

The numerical results for the radial temperature profiles at the burner exit are presented in Figure 4, while Figure 3 shows the predicted temperature contours along the axial direction of the PFS[6]. The numerical results are only validated with the measured data at the exit of the PFS. The radial temperature profile is shown for an axial location of 2.0 m from plasmatron axis ($x = 2.35$ m). The predicted profile is revealed to be axis-symmetric in accordance with the experimental profile. The measured profile shows a distinctive local minimum temperature at the chamber central line. However, in the case of the predicted temperature profile this minimum is insignificant. This could be the reason for the underpredicted penetration of the plasma jet into the coflowing stream of air–coal mixture. In the practical situation, it may be expected that the plasma jet will separate the air–coal mixture flow into two streams, leaving the central part of the flow with lower fuel concentration. The high-energy concentrated plasma jet, with high initial momentum, may act as a solid body[7] penetrating through the cross-flow, while the coal particles' trajectories are divided into two streams, showing two temperature maxima on both sides of the center line.

The mean calculated species concentrations of the gas phase composition at the exit of the PFS was (volume %): $CO = 26.1$; $H_2 = 12.1$; $CH_4 = 0.1$; $CO_2 = 0.8$; $N_2 = 60.9$; $O_2 = 0.0$.

The values of temperature and species concentration were used as input values to numerically simulate the boiler working in the plasma regime.

The PFS was incorporated in the furnace of a full-scale boiler with a steam productivity of 640 TPH (Gusinoozersk TPP, Russia). The schematic view of the boiler equipped with PFS and its main dimensions are shown in Figures 5 and 6. The furnace is characterized by two symmetrical combustion chambers, each having four tangentially directed main double burners in two layers. Each of the main burners is divided into two sections; the air–fuel mixture supply section and the secondary air supply section. The combustion chambers are joined by a central section. The cooling chamber is above the combustion chambers and the turning chamber is above that again. Fuel consumption of the boiler was 121.3 TPH and the total amount of air in the boiler was 553,128 m^3/h, with a secondary air temperature of 350°C. The coefficient of air surplus at the fire-chamber's outlet was 1.2. Initial data for the main burners are the same as for PFS shown in Table 1.

Four PFS are mounted instead of four lower sections of the main double burners, as shown in Figure 5. During the period of boiler warmup and flame stabilization, the plasmatrons are operating. When the boiler performance is stabilized, the plasmatrons are switched off and PFS work as conventional pf burners. In the case of flame instabilities, the plasmatrons are easily switched on.

The grid for the mathematical simulation is defined by $118 \times 52 \times 68$ grid lines in three directions (x, y, and z).

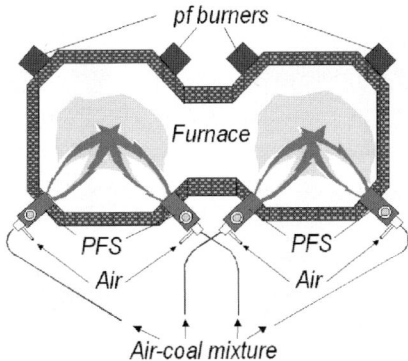

Figure 5. BKZ 640-140 boiler furnace equipped with four PFS (*top view*)

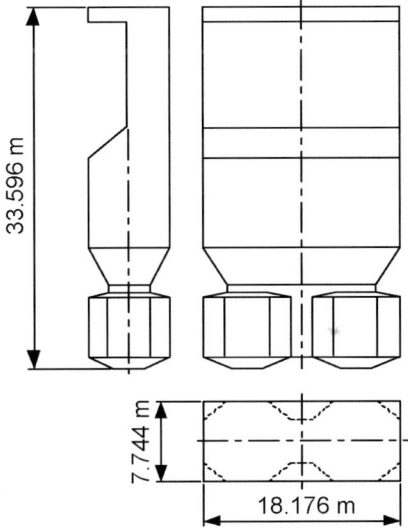

Figure 6. Scheme of the industrial furnace of BKZ 640-140 boiler

The main results from numerical simulations of the furnace velocity vectors, temperature profiles and oxygen concentrations, are presented in Figures 7–9. The velocity vectors field within the single semifurnace can be found on the left of Figure 7. The specific alignment of the burner ports in each corner of the combustion chamber generates the tangential flow of the fluid and particles, increasing the intensity of combustion. It can be seen that the initial momentum of the supply streams deviates from its original angle due to the intensity of tangential momentum created in the central part of the chamber. The velocity vectors for the plasma operational regime can be seen on the right of Figure 7. As was the case for the standard operation regime, the initial momentum of the

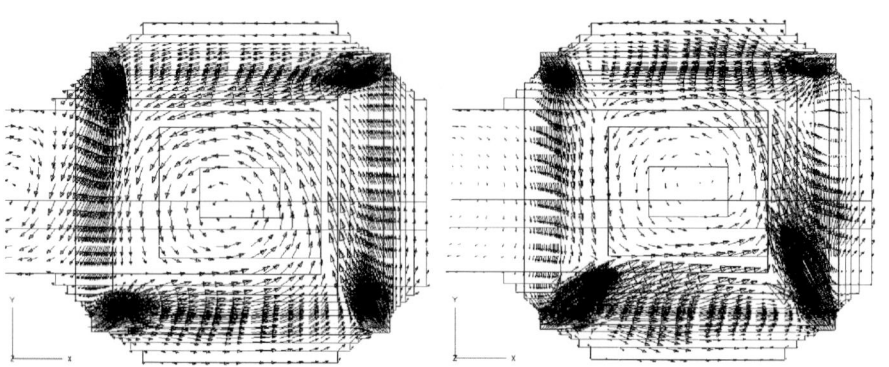

Figure 7. Velocity vectors within the combustion chamber at the level of the burners. Figure on the left is standard operational regime; figure on the right is plasma operational regime

supply streams deviates from its original angle due to the intensity of the tangential momentum generated in the central part of the chamber, which is further increased by the contribution of high velocity jets coming from the PFS. It can be seen that under these conditions coal particles tend to move from the center of the chamber towards the conventional burners. It should be noted that the velocity vector fields are presented in such a way that the standard (on the left) and plasma operational regimes (on the right) can only be compared quantitatively as different scaling factors were applied to the vector quantities. This was necessary, as the velocity profile values at the PFS outlet were significantly higher then the field values.

In Figures 8 and 9, panel 1 presents the predicted values for the center line along the furnace height, while panel 2 gives the values at the exit, along the furnace width. The numerical results represent the boiler performance for the standard operational regime and for operation in the plasma regime. Values of boiler performance in the standard regime at the exit from the furnace were validated. The measured average temperature at the exit is around 1,400 K, and this value agrees with the numerical results, while the averaged concentration of measured oxygen is around 4%. In the plasma regime (Figure 9) the tempe-rature levels along the furnace height are lower, while rapid combustion of pf within the combustion chamber results from the introduction of the mixture of hot combustible gases and unburned char from the PFS.

The zero concentration of oxygen calculated along the center line of the furnace in the conventional regime (Figure 8), almost until $K = 40$, suggests the rapid combustion of coal, followed by an increase in oxygen concentration in lower parts of the cooling chamber due to mixing with surrounding gases. The concentration of oxygen at the outlet from the furnace was around 3.75%, which agrees well with the measured value of 4% [3]. The oxygen concentrations for the plasma-assisted regime (Figure 9) differ significantly (panel 1) due to the modified flow and mixing pattern caused by the increased tangential momentum. Panel 1 suggests that the combustion of pulverized coal has been displaced from the center line of the combustion chamber in the radial direction, probably due to increased centrifugal forces which tend to move the particles closer to the surrounding walls. However, in the upper level of the furnace, after around $K = 50$, the oxygen concentration profiles in both cases look similar. In the plasma operation regime, the concentration of oxygen at the outlet from the furnace (panel 2) is slightly lower, which suggests an increase in overall burnout of coal particles.

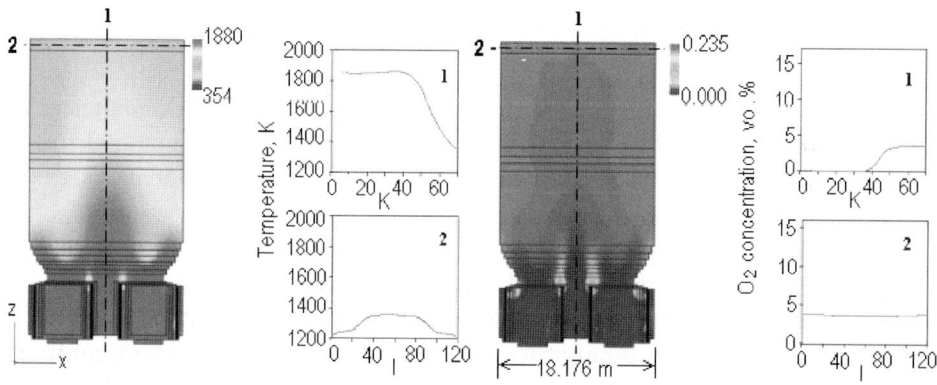

Figure 8. Predicted temperature and oxygen level contours and profiles for the conventional operational regime (center line, exit of the furnace)

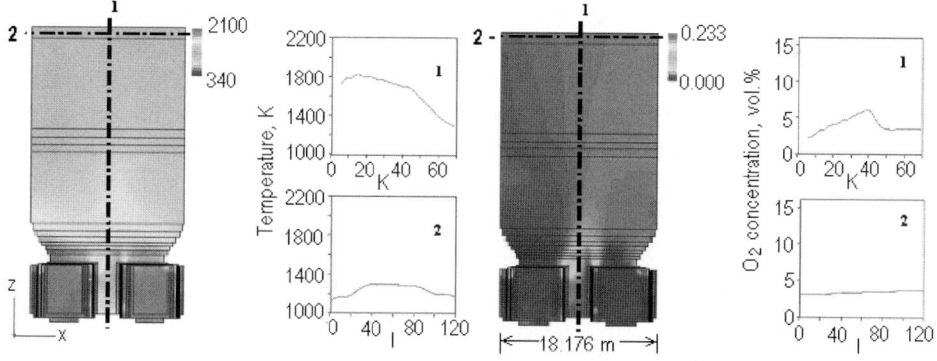

Figure 9. Predicted temperature and oxygen level contours and profiles for the plasma operational regime (center line, exit of the furnace)

The averaged mass flow of volatile matter released from coal particles is presented in Figure 10. The peak mass flows of volatiles of around 0.53 kg/s and 0.9 kg/s are reached within the region of the burners, while zero mass flow

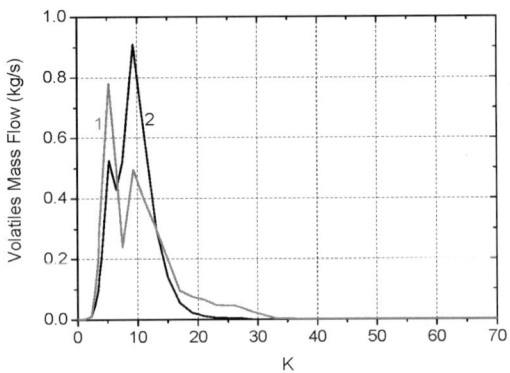

Figure 10. Averaged mass flow of volatiles along the furnace height. 1 is standard operational mode; 2 is plasma operational mode

of volatiles at $K = 23$ indicates the earlier consumption of volatile matter in the plasma-assisted regime than the standard operational regime. The graph also indicates a more intense and rapid devolatilization in the plasma operational mode.

Predictions in the plasma operational regime (Figure 9) show satisfactory agreement with the experimental data. The following data were measured when the boiler power was 120 MW and the excess air factor was 1.24: the concentration of oxygen (O_2) in the exhaust gas was 6.1%; NO_x was 700 mg/nm^3 (1,431.5 ppm) and unburned carbon was 0.8%. The temperature in the body of the flame was 1,270°C and the temperature in the furnace outlet was 1,050°C. The concentration of carbon dioxide (CO_2) in the exhaust gas, calculated through the O_2 concentration, was 14%. It can be seen that the difference between the measured and predicted temperature of the combustion products inside the furnace is no more than 17% and about 6% at the furnace outlet. However, the difference between the measured and predicted concentrations of oxygen in the exhaust gas is about 30%. A possible explanation of this discrepancy is the 60% boiler load factor while measurements were made. There is no need to operate in the plasma-aided regime at a boiler load factor of 100%. But when the load factor decreases plasma is needed for pf flame stabilization.

3. Industrial Tests

The experiments and computer simulations of plasma-aided pf combustion showed the applicability and efficiency of PFS for their use in coal-fired boilers. On the basis of the completed research PFS were incorporated into the furnaces of industrial coal-fired TPP. The PFS have been tested successfully for 27 pulverized coal boilers at 16 TPP (Russia, Kazakhstan, Korea, Ukraine,

Slovakia, Mongolia and China) with steam productivities between 75–670 TPH since 1994[4]. The boilers were fitted with different pf preparation systems (such as direct pf injection and systems with pf hopper). In total, 70 PFS were mounted and tested on the boilers.

Figure 5 presents a scheme of the arrangement of four PFS on the 200 MW power block at Gusinoozersk TPP with the use of high concentration pulverized coal (top view).

Figure 11 presents a scheme of the arrangement of PFS on the BKZ-420 boiler at Ulan-Bator TPP-4 (top view). Twelve burners are placed in three layers on the corners of the boiler. Two PFS were mounted opposite each other on the lowest layer. After 2–3 s from the PFS start, the temperature of both pulverized coal flames achieved 1,100–1,150°C. After 1 h the temperature of the flames, which were 7–8 m in length, achieved 1,260–1,290 C. In accordance with operating instructions, the boiler start-up duration was 4 h. All eight boilers of the TPP were equipped with PFS for fuel oil-free boiler start-up

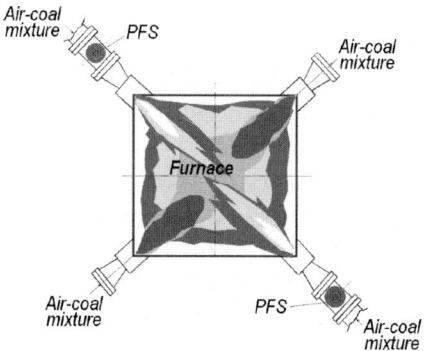

Figure 11. BKZ-420 boiler furnace equipped with two PFS (*top view*)

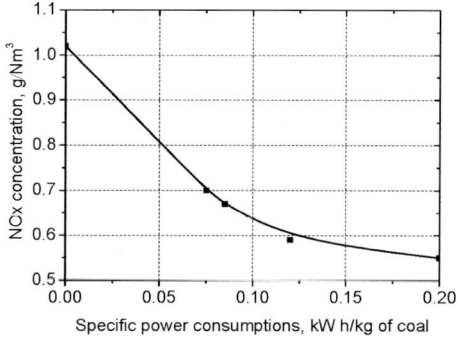

Figure 12. Specific power consumption influence onto reduction of nitrogen oxides concentration at plasma aided pulverized coal combustion

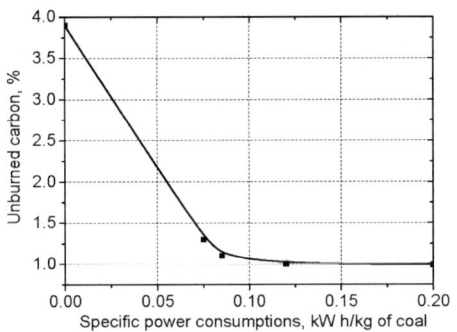

Figure 13. Specific power consumption influence onto reduction of unburned carbon at plasma-aided pulverized coal combustion

Figures 12 and 13 present the results of investigations into both NO_x and unburned carbon reduction for plasma ignition of coal at the outlet of a furnace[8]

When a plasmatron operates in the plasma flame stabilization regime, NO_x concentration is halved and unburned carbon is reduced by a factor of 4. NO_x is reduced due to two-stage pulverized coal combustion. The first stage occurs in the PFS, where ignition and partial gasification of coal in primary air takes place. The second stage occurs in the boiler furnace, where combustion of the products in secondary air takes place. Fuel-nitrogen is released with coal volatiles inside the PFS. It forms molecular nitrogen due to the deficit of oxygen in the air–fuel plasma mixture treated in the PFS. In the second stage, thermal nitrogen oxides are mainly formed. It is well known[9] that fuel-nitrogen is a main source of nitrogen oxide emission. Thermal NO_x can reach 10–15% of all NO_x emissions in a furnace at temperatures above 1,700 C. In the case of unburned carbon (Figure 13), its decrease can be explained by the increase in fuel reactivity caused by the formation of pulverized coal and the explosion of heat generated by its interaction with arc plasma.

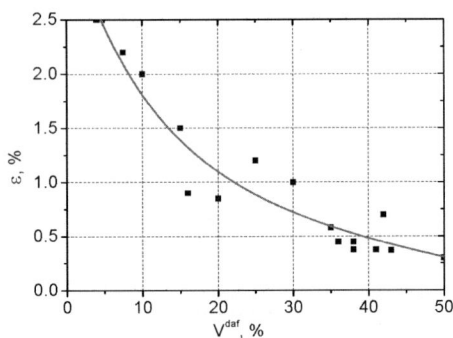

Figure 14. Generalized experimental dependence of specific electric power consumption on a plasmatron (ε) vs coal volatiles content (V^{daf}) from PFS tests at 16 different TPP. $\varepsilon = P/(Q \cdot G)$; here P is plasmatron electric power; Q is coal calorific value; G is the coal consumption through a PFS. Line is polynomial fit

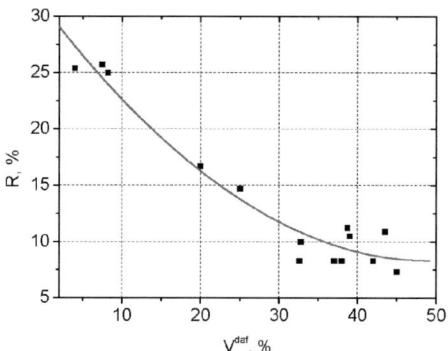

Figure 15. Generalized experimental dependence of PFS specific heat power (R) vs coal volatiles content (V^{daf}) from PFS tests at 14 different TPP (1994–2004). $R = (G_{start}/G_{nominal}) \cdot 100\%$, here G_{start} and $G_{nominal}$ is consumption of coal at plasma boiler's start-up and nominal consumption of the coal for the boiler respectively. Line is polynomial fit

Figures 14 and 15 represent a generalization of the data collected from full-scale plasma-aided boiler start-up and pf flame stabilization.

It can be seen (Figure 14) that the specific power consumption (SPC) for PFS decreases with coal volatile content increase, i.e., when coal reactivity rises. Note power consumption of the plasmatron is within 2.5% of heat power from the pf burner, even for the anthracite coal with minimal reactivity. For example, auxiliary power requirements for a boiler are 8–10%. Thus, if we take the coal volatile yield from the curve in the figure, we can determine the electric power of the plasmatron needed for PFS consumption of a specified coal of known calorific value.

Figure 15 shows that the relative heat power of PFS decreases from 25% to 10% as the content of coal volatiles (fuel reactivity) increases. Note for comparison, that the relative heat power of oil systems for boiler start-up and pf flame stabilization is some 30% in existing TPP. In other words, even for incineration of a low-rank coal of 4–5% volatile content, the relative heat power of the PFS is 5% less than for conventional incineration of the coal. In the case of midband reactivity coal ($V^{daf} = 15$–30%) and high reactivity ($V^{daf} = 30$–45%) the relative heat power of the PFS is only 7–10%. This means that the energy efficiency of PFS is 3–4 times higher than conventional oil systems used in TPP for boiler start-up and pulverized coal flame stabilization. Note that oil system (oil nozzles and metal consuming, environmentally unfavorable oil equipment) operating costs are three times higher than PFS[2]. Thus, if we take the coal volatile yield using the curve from the figure, we can determine the total heat power of PFS needed for boiler start-up of specified nominal coal consumption for the boiler. Then, knowing the coal consumption of the boiler burners we can find a number of PFS for this boiler plasma start-up.

4. Conclusions

– Developed, investigated and industrially tested plasma-fuel systems improve coal combustion efficiency, while decreasing harmful emission from pulverized-coal-fired Thermal Power Plants.
– PFS eliminate the need for expensive gas or oil fuels on start-up.
– PFS improve coal ignition and burnout without the need for such remedies as increasing the mill temperature, augmenting the excess air factor, or finer grinding.
– The application of numerical modeling approaches of pulverized coal preparation for combustion within the plasma burner has shed new light on the possible chemical and physical mechanisms of the coal–plasma interaction.
– Although the combustion process of pulverized coal may not be in thermal equilibrium, the present thermal equilibrium calculations resulted in predictions close to the experimental data.
– Simulation of an industrial boiler in conventional and plasma operational modes reveal that the operation of PFS results in stable ignition and intensive burning of the pf at reduced temperature; conditions which reduce the amount of nitrogen oxide formation.
– Prior to the wider implementation of PFS, additional data relating to further coal types and their blends are ideally required.

Acknowledgments

The authors gratefully acknowledge the European Commission for this work funding through the ISTC project and personally Professor F. Lockwood for his support and coordination of the research project.

References

1. M. Sugimoto, K. Maruta, K. Takeda, O.P. Solonenko, M. Sakashita, M. Nakamura, Stabilization of pulverized coal combustion by plasma assist, *Thin Solid Films*, **407**, 186–191 (2002).
2. M.G. Drouet. La Technologie Plasmas. Potentiel d'Application au Canada, *Revue Generale d'Electricite*, **1**, 51–56 (1986).
3. F.C. Lockwood, T. Mahmud, M.A. Yehia, Simulation of pulverised coal test furnace performance, *Fuel*, **77**(12), 1329 (1998)

4. Z. Jankoski, E.I. Karpenko, F.C. Lockwood, V.E. Messerle, A.B. Ustimenko, *Plasma Supported Solid Fuel Combustion. Numerical Simulation and Full-Scale Trials*. Proceedings of 8th International Conference on Energy for a Clean Environment, Lisbon, Portugal, 2005. – CD of Proceedings – N 18.2; Book of Abstracts – p. 69

5. M. Gorokhovski, E.I. Karpenko, F.C. Lockwood, V.E. Messerle, B.G. Trusov, A.B. Ustimenko, Plasma technologies for solid fuels: experiment and theory, *Journal of the Energy Institute*, **78**(4), 157–171 (2005)

6. Z. Jankoski, F.C. Lockwood, V.E. Messerle, E.I. Karpenko, A.B. Ustimenko, Modelling of the pulverised coal plasma preparation for combustion, *Thermophysics and Aeromechanics*, **11**(3), 461–474 (2004).

7. F.J. Weinberg, Electrical discharge-augmented flames and plasma jets in combustion, in *Advanced Combustion Methods*, edited by F.J. Weinberg. Academic Press, London, 1986.

8. V.E. Messerle, V.S. Peregudov, ignition and stabilisation of combustion of pulverised coal fuels by thermal plasma, thermal plasma and new materials technology, in *Volume 2: Investigation and Design of Thermal Plasma Technology,* edited by O.P. Solonenko, M..F. Zhukov. Cambridge International Science Publishing, London, 1995.

9. D.H. Tike, S.M. Slater, A.F. Sarofim, J.C. Williams, Nitrogen in Coal as a Source of Nitrogen Oxide Emission from Furnace, *Fuel*, **53**, 120–125 (1974).

INVESTIGATION OF PULVERIZED COAL COMBUSTION

A. P. BURDUKOV
Institute of Thermophysics Sibirian Branch,
Russian Academy of Sciences, 1, Acad. Lavrentyev Ave,
630090, Novosibirsk, Russia, e-mail: burdukov@itp.nsc.ru

V. I. POPOV
Institute of Thermophysics Sibirian Branch,
Russian Academy of Sciences, 1, Acad. Lavrentyev Ave,
630090, Novosibirsk, Russia, e-mail: vipopov@itp.nsc.ru

Abstract. This report presents investigation results on micronized coal combustion. Experimental data for the ignition and combustion of coals under different stages of metamorphism was obtained using a large-scale facility with a capacity of up to 1 MW. Proposals for the application of micronized coals in power and industrial boilers are considered.

Keywords: autothermal mode coal, combustion, combustion chamber, ignition, micronized fine grinding, metamorphic stage, primary furnace

1. Introduction

Among the available technologies of coal combustion, the coal-dust one is the most ultimate, and the application of micronized coals is of a particular interest.

The duration of coal-dust fuel combustion depends mainly on the diameter of coal particles and its physical-chemical characteristics. As it was already noted in [1], together with a significant decrease in particle mass and an increase in the reacting surface, fine grinding considerably increases the chemical activity of coal due to a breakdown of chemical bonds and formation of free radicals. Simultaneously, its physical-chemical properties become close to those of fuel oil, i.e., the rate of combustion, flame sizes, volumetric calorific intensity, required volume of furnace for pulverized coal and fuel oil combustion are commensurable. This provides widespread opportunities for micronized coal use by different power technologies: i.e., for the ignition and

N. Syred and A. Khalatov (eds.), Advanced Combustion and Aerothermal Technologies, 131–140

stable combustion of coal-dust flame at power boilers, substitution of fuel oil and gas at gas-and-oil industrial boilers, etc.

2. Physical-Chemical Alterations of Coals at Fine and Ultrafine Grinding

There are typically two processes for fine dispersion of coals: grinding with an increase in specific surface S and the aggregation of fine particles during relatively long grinding, which decreases value S This provides a drastic intensification of coal reactivity in different processes and chemical reactions. The fine dispersion is accompanied by an enlargement of outer and inner surfaces due to the opening of blocked pores and the development of new porosity, caused by microcracks in coals. The volume of micropores and transitional pores multiply, i.e., a fundamental transformation of initial porous structure occurs.

Structural changes aimed at the reduction in structural order are determined by radiography; they cause transformation of many properties. It is obvious in difractograms that lines become wider, scattering background appears, and intensity of reflex maximums becomes lower. All these prove deep violations of the fine structure of coal substance, caused by plastic deformations.

It is important to note that the density decreases significantly from 1.59 to 1.39 g/cm^3, and this proves that structure loosening due to a break of chemical bonds and rearrangement of macromolecule structure.

The intensity and character of alterations in coal composition and properties depend on coalification degree (metamorphism), medium, type and techno-logical parameters of grinding.

Local concentrations of mechanical and heat energy are created in a mill due to mechanical effects, and this leads to a break of chemical bonds.

In coal molecules, the detachment of side chains with formation of free radicals is the most probable. The radical way of decay is well studied by the method of electron paramagnetic resonance (EPR). In coals, free radicals are formed due to a break of C–C bonds, and an unpaired electron belongs both to the first and to the second carbon atoms. High chemical activity of coals, together with oxidation and ignition tendency, is explained by formation of free radicals, including macroradicals.

The generation of products with a low organic mass, which increases reactivity, is a general tendency of organic substance alteration.

The above sequences of activation grinding provide widespread opportunities for application of this process in different fields of coal use.

3. Application of Micronized Coals for Power Technologies

Now it is common throughout the world for power engineering to use micronized coals for different applications. One of these is the substitution of fuel oil by coal for flame kindling and lighting in coal-dust boilers [2–5]. Usually, gas or fuel oil is used for boiler ignition as well as for stabilization of combustion at decreased loads or for the application of low-reaction coals. Thus, more than 50 million tons of fuel oil is annually used for ignition and lighting of boilers and the stabilization of liquid slag output [6]. Simultaneously, the combined combustion of a high-reaction fuel (gas, fuel oil) with coal significantly increases mechanical underburning, decreases boiler efficiency and leads to other negative after-effects.

One of the methods for fuel oil substitution in these processes is the application of fine-dispersed coal powder.

The most favorable combustion conditions are achieved when micronized coal is supplied directly from the mill to the furnace, since increased coal reactivity (developed during grinding) is kept, and the problems connected with the storage of fine-dispersed coal are excluded.

The schemes of micronized coal separation and the system of micronized coal preparation were tested in the beginning of the 1980s at Lumen electric power station in Germany for units of 150 and 350 MW power.

Similar investigations are carried out in the USA, where the share of fuel oil, burnt at coal power plants, makes up about 6% of the main fuel consumption (according to Union Carbide Corporation data). Therefore, the substitution of expensive fuel oil by coal is undoubtedly reasonable. The projects on pulverized coal use by the Reburning systems are developed there for suppression of nitrogen oxides [7]. The boilers, using micronized coal, which can substitute fuel oil in fuel boilers, are also developed there.

The reacting surface of micronized coal particles is higher (approximately by the factor of 3) than in conventional systems, its chemical activity is also increased, which provides characteristics of the coal flame similar to those of dispersed fuel oil. This allows the complete combustion of coal in a smaller furnace volume than required for combustion of a coal-dust with lower underburning, higher boiler efficiency and lower NO_x content.

The main burning characteristics of micronized coal are as follows: a reduction of the flame length by 60%, the doubling of combustion rate, an increase in the amount of heat transferred by radiation in the furnace, the reduction of ash and slag depositions (which miniaturizes the systems of ash removal), and the allowance to make the vapor superheater and waste gas heater more compact.

Another application of micronized coal is its use in oil-gas boilers [8]. The method of micronized coal production is of a particular importance. First mills used for this purpose were the air and vapor ones (the 1950s), however, their power consumptions were five times higher than in mechanical mills.

The American project [8] uses mechanical mills with an obligatory system of particle separation, which prevents penetration of large fractions into the combustion chamber.

In general, micronized coal may be important for the systems of ignition and lighting of coal-fired boilers. A specific requirement is the conversion of oil-gas boilers for the combustion of micronized coals, which will solve the problems of industrial power engineering, dealt with a lack and high price of fuel oil. Naturally, there will be problems relating to flue gas cleaning because there are almost no such systems in boilers, combusting gas and fuel oil.

A special setup is being developed at the Institute of Thermophysics, Siberian Branch of the Russian Academy of Sciences. Investigations into the combustion of different micronized coals have started, and systems for flue gas cleaning are being developed and tested. Simultaneously, the problems related to the application of devices for micronized coal combustion with direct powder supply into a furnace are being considered. These problems are as follows:

1. The study of the mechanism and rate of combustion and gasification of micronized coal under different metamorphic stages to determine the working parameters of technological devices in systems for the reduction of NO_x emissions, boiler ignition and lighting, and fuel substitution in oil-gas boilers.

2. Pilot and industrial testing of micronized coal combustion at the objects of heat and power engineering; testing of the systems for flue gas cleaning.

3. Tests on ignition and lighting of coal-dust flame in power boilers, using micronized coals.

Since the properties of coals, used by power installations, vary widely: by a metamorphic degree, ash content, and moisture content. This should be considered by the technologies of micronized coal combustion.

At the first stage of investigations, the main task was aimed at the determination of conditions for the stable burning of micronized coal suspension ($d_{ave} = 20–40$ μm) in a furnace chamber with a cylindrical muffle vortex primary furnace and different burners (direct-flow and vortex ones).

The types of coals used for the investigations are described in Table 1.

TABLE 1. Description of coal fuels used in studies

Coal	W^a (%)	A^a (%)	S^a (%)	V^a (%)	Q^a (kcal/kg)
Brown coal – B-2 from Nazarovskoe deposit	12.3	11.6	0.28	36.2	4,705
Black coal from Tikhonskoe deposit (Yakutsk region)	5.8	20.95	0.29	22.1	4,710
Black coal from Stary Tyrgan deposit, Krasny Brod siding, TR grade	7.3	5.8	–	11.1	7,000

The experimental setup for the investigation of the mechanic-thermo-chemical processing of the fuel, the micronized coal combustion and gasification is shown in Figure 1. Coal powder with particle size of up to 5 mm is fed into bunker 25, it is supplied into pulverizer 23 through batcher 24 and after grinding it is fed into funnel 26. The micronized coal is then carried by transporting air 28 to the burner. The coal-dust mixture is formed before being supplied into a plasma jet of plasma torch 3. The plasma jet ignites the coal-dust mixture, and it burns under lean conditions, i.e., the ignition and partial gasification of the coal with volatile products occurs. After partial gasification or thermochemical processing of the fuel, the high reacting combustible mixture is fed into muffle primary furnace 5, where it is mixed with secondary air 29, tangentially supplied into vortex chamber 4, and then into primary furnace 5. The main part of secondary air is fed into the primary furnace before the input into the furnace volume.

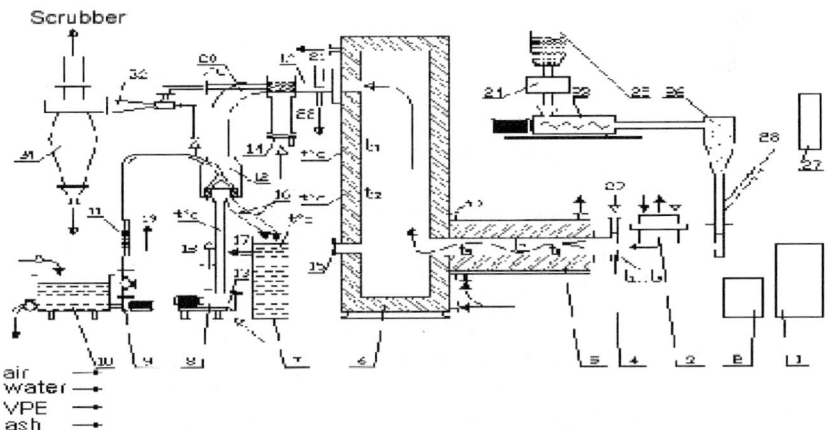

Figure 1. Heat-and-power setup

The combustion process is controlled by eight thermocouples, located along the primary furnace and furnace. The setup is equipped with pipes for the sampling of gaseous combustion and gasification products (21, 22). The furnace pressure is controlled by the additional air supply at furnace outlet 14.

The coal-dust flow is fed into pipe 1 with flow rate G_1. Simultaneously, some part of primary air is fed coaxially to the coal-dust flow through pipe 2 with flow rate G_2 and through pipe 3 with flow rate G_3. The mixed coal-dust flow is supplied to the zone of plasma torch 4, and the ignited coal-dust mixture from the pipe junction of plasma torch 5 is fed into the lined primary furnace.

The direct-flow burner with partial swirl of initial air allows a qualitative estimate of the swirl effect on the process of coal-dust ignition and partial combustion.

The direct-flow burner was used for experiments with brown coal from Nazarovskoe deposit; Tikhonskoe deposit (Yakutsk region), and lean coal of TR grade.

According to experiments, the swirl of initial air intensifies the process, i.e., the stable heat mode in the primary furnace is reached faster, but the autothermal mode is determined by the content of volatiles in the used coal.

Typical diagrams of the temperature variation with time at the primary furnace inlet under the autothermal mode are shown in Figure 3 (curve 1). When the plasma torch is switched on, the temperature increases up to 950°C, and in 5 min of operation at the haul moment it drops to 900°C and stays constant during the whole period of coal supply. The diagram of temperature alteration in time without autothermal thermochemical fuel processing is shown in Figure 3 (curve 2), in this case the coal has a lower output of volatiles; the primary furnace is heated slower, and at plasma torch switch-off the temperature decreases rapidly below the temperature of coal-dust ignition.

The combustion products are fed to scrubber 12, where they are cooled and cleaned by water 19 from solid admixtures.

After the scrubber the cleaned flue gases are thrown out by blower 8 into the atmosphere.

Water is supplied to the scrubber by pump 9, the water flow rate is controlled by a rotameter, and regulated by a valve.

The flow rates of the carrier, initial, secondary and plasma-forming air are controlled by rotameters and manometers. The plasma torch is fed by power source 1 and controlled by control panel 2.

Experiments were carried out with three types of burners: a direct-flow burner with partial swirl of the initial air flow (Figure 2); a cylindrical vortex burner with tangential fuel input, and a cyclone burner with snail input of the coal-dust mixture.

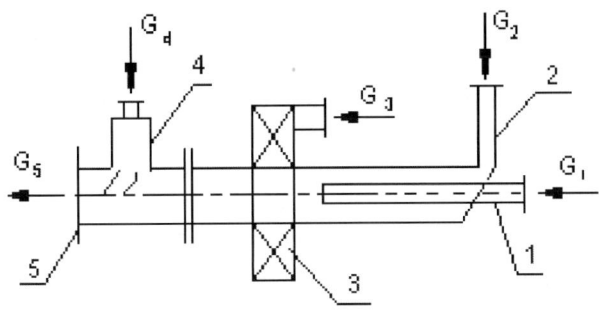

Figure 2. Direct-flow burner

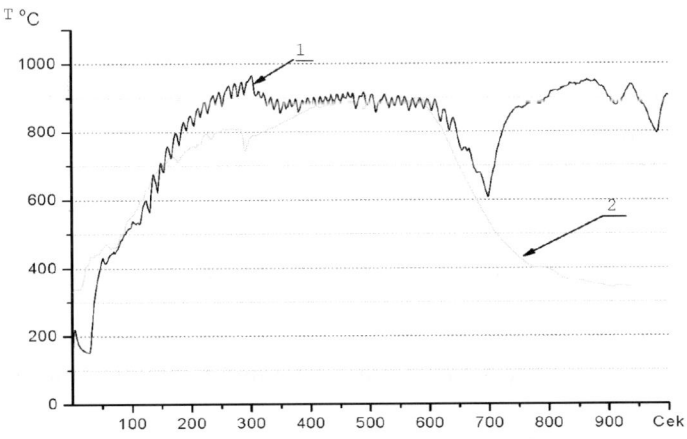

Figure 3. Ignition temperature of pulverized coal

The cylindrical vortex burner with tangential fuel input was tested both with ignition from a plasma torch and with ignition from a propane burner.

Experiments with brown coal and coal of the TR grade were carried out with the cylindrical vortex burner. According to these experiments, the ignition of the coal-dust mixture both from the plasma torch and gas burner in the cylindrical vortex burner is stable for all coal grades.

The process mode is determined by the ratio of fuel and air flow rates in the burner, primary furnace and furnace chamber inlet; the mode can be regulated by these ratios.

Ignition by the cyclone burner with the snail input of coal-dust mixture is accomplished from an additional gas (propane) burner, located coaxially to the snail inlet.

The air and coal-dust flows were fed at a small angle to the axis of the snail inlet. Secondary air was supplied before the burning coal-dust flow into the furnace.

The feature of the snail input is that in comparison with the cylindrical vortex burner, a smaller part of the burning coal-dust flow affects a new coal flow, that allows a reduction of the temperature load at the primary furnace inlet and a finer regulation of the modes of dry slag removal by the initial air flow.

Experiments with brown coal from Nazarovskoe and Tikhonskoe deposits (Yakutia) were carried out at the cyclone burner with the snail fuel input. According to these experiments, both of these coal grades in the snail burner can be converted into the mode of autothermal thermochemical fuel processing before combustion in the furnace.

In all experiments (mill productivity 2.75, 5.5, and 8.0 g/s), the size of coal particles and ash particles in the furnace chamber (Figure 4) is controlled together with concentrations of CO, CO_2, O_2, SO_2, NO along the gas flow in the furnace chamber.

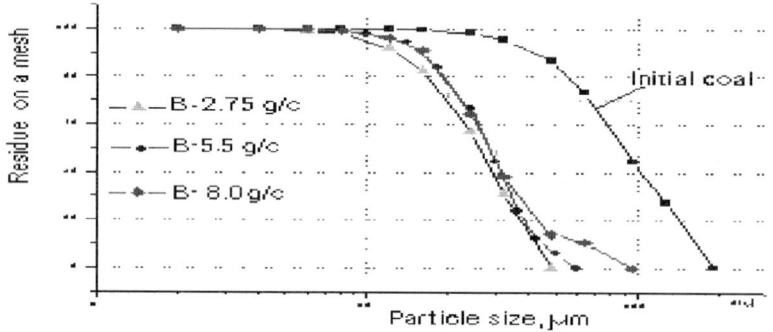

Figure 4. The main experimental parameters were sent to PC in real time

4. Development of the Mathematical Model and Calculations of the Processes of Micronized Coal Combustion

To check technological proposals on optimal schemes of combustion and design solutions, mathematical models of coal combustion were developed for the conditions of fine grinding and vortex thermal processing, including gaseous and plasma activation.

Two mathematical models are used for the investigation of coal-dust combustion with plasma activation. The first one-dimensional (1D) model is called "Plasma-coal" and describes a two-phase chemically active flow in the primary furnace – muffle with a plasma heat source. This model determines the thermal contribution of arc as the difference between electric power produced by arc and heat losses into the chamber wall. The heat losses in the channel are expressed via experimental distribution of the heat flux power into the wall. Thermochemical transformations of coal are described there by a single mechanism with a single set of kinetic reaction parameters. The total kinetic scheme (116 chemical reactions) together with the extraction of initial volatiles considers the reactions of their following transformations, including formation of nitrogen and sulfur oxides as well as reactions of carbon oxidation and gasification. The model is written as a system of ordinary differential equations, including equations of component concentrations, conservation equations for mass, number of particles, momentum, and energy, equations of motion and gas and particle temperature, as well as the state equation of ideal gas.

The results of calculations by the "Plasma-coal" model have allowed the determination of the product concentrations of plasma activation of the air–coal mixture; temperatures and velocities of gas and particles; the initial conditions for modeling the combustion of micronized coal, activated by plasma, in the experimental furnace. The three-dimensional (3D) "Cinar" model was used for the mathematical modeling of the furnace. This model is written as the generalized system of conservation equations in Cartesian coordinates, it includes the conservation equations of mass, momentum and energy for the gas phase and particles. For this purpose the $k-\varepsilon$ model of turbulence was used together with a simplified model of coal particle combustion with consideration of volatile extraction, coke residue burning and gas phase combustion, and the discrete radiation model. The properties of the gas phase are calculated using the Euler system, and the motion of the particles is modeled via the Lagrange approach. The mass transfer from particles to gas is determined by a term in Euler equations of conservation and the effect of turbulence is considered by the model for the turbulent dispersion of particles. A change in mass and temperature of particles is controlled by the equations of energy and mass conservation. "Fuel", "thermal", and "fast" nitrogen oxides are taken into account, when determining NO_x concentrations.

The fields of temperatures, gas velocity, CO_2, O_2, and NO_x concentrations in different furnace cross sections, and the trajectories of coke residual particles in the combustion chamber were determined through calculations of the primary furnace and furnace of the heat-and-power setup.

5. Conclusions

The promising designs of burners for micronized coal combustion were found through a series of experiments. Thes designs allow the combustion of coals at different stages of metamorphism under the autothermal mode, which is of a particular interest for development of the systems for ignition and combustion stabilization of the flame without fuel-oil, and for the conversion of the gas-oil boilers to combustion of the coal-dust fuel.

The ignition and combustion of micronized coals with different stages of metamorphism were modeled for different systems of coal mixture ignition.

Proposals on micronized coal application for the systems of coal-dust flame ignition, for stabilization of coal-dust flame combustion in power boilers and for substitution of fuel-oil by coal in industrial boilers were developed and sent to the power plants.

References

1. A.P. Burdukov, L.I. Pugach and T.S. Yusupov, Prospects for use of micronized coal in power industry, in: *Heat-and-Power Engineering. Physical-Technical and Environmental Problems, New Technologies, Technical-Economical Efficiency*, Research Collection, Novosibirsk State Technical University, Novosibirsk, 1999.
2. A.V. Polubentsov, V.V. Vykhovanets and N.A. Belogorlova production of fuel oil with the help of disintegrator equipment, in: *Disintegrator Technology*, Tallinn, 1989, pp. 99–101.
3. Yu. A. Tolasov, G.S. Khodakov and V.S. Zolotukhin, Peculiarities of the granule composition of coal, grinded by an impact mill, in: *Disintegrator Technology*, Tallinn, 1989, p. 101.
4. C.L. Wilson, Coal – Bridge to the Future, *HarperInformation*, 1980.
5. E. Kh. Verbovitsky, and V.P. Kotlyar, Coal instead of mazut for flame ignition and enlightenment in coal-dust boilers, *Energokhozyaistvo za rubezhom*, 1984, No. 1.
6. E.I. Karpenko, V.E. Messerle, Introduction to Plasma-Energy Technologies of Using of Solid Fuel, *Nauka*, Novosibirsk, 1998.
7. Clean Coal Technology Demonstration Program, *US Department of Energy*, April 1996.
8. P.I. Damiani, R.I. Teachout and D.P. Finn, Micronized coal firing boiler facility and Potential utility applications, *Power-Gen'94 7th International Conference*, December 1994, Book 3.

SOLID FUEL PLASMA GASIFICATION

V. E. MESSERLE, A. B. USTIMENKO[*]
Combustion Problems Institute, al-Farabi Kazakh National University, 172 Bogenbai batyra str., Almaty, 050012, Kazakhstan

Abstract. This paper presents a numerical analysis and experimental investigation of the gasification under steam and air plasma conditions of two very different solid fuels, a low-rank bituminous coal of 40% ash content and a petrocoke of 3% ash content; with an aim of producing synthesis gas. The numerical analysis was fulfilled using the software package TERRA for equilibrium computation. Using the results of the numerical simulation, experiments on plasma steam gasification of the petrocoke and air and steam gasification of the coal were conducted in an original installation. Nominal power of the plasma installation is 100 kWe, and total consumption of the reagents is up to 20 kg/h. A high quality synthesis gas was produced in the experiments on solid fuels plasma gasification. It has been found that a synthesis gas content of about 97.4% vol. can be produced. It was demonstrated that the monitoring of the synthesis gas composition can be ensured by modifying the initial parameters of the experiment. A comparison between the numerical and experimental results showed a satisfactory agreement.

Keywords: coal, gasification, arc plasma, plasma chemical gasifier, synthesis gas

1. Introduction

The world's petroleum reserves are limited. Based upon current global consumption, it has been estimated that these reserves will be depleted in approximately 40–60 years. Coal is worldwide the most abundant energy resource and the

---

[*]To whom correspondence should be addressed. Tel.:+(73272)615148; Fax:+(73272)675141; e-mail: ust@ntsc.kz

N. Syred and A. Khalatov (eds.), Advanced Combustion and Aerothermal Technologies, 141–156.

least expensive fossil fuel. An additional energy resource is petrocoke, derivable as a result of hydrocarbon production by the thermal processing of oil sands; for instance in Canada there are mountains of petrocoke. Petrocoke is a solid fuel consisting of fixed carbon, tar and ash. Direct utilization of petrocoke is difficult because of its hardness and high tar content. In connection with this, the development of solid fuels technologies which would be environment friendly and efficient is of a primary importance. Coal gasification is a well-proven technology that started with the production of synthesis gas ($CO + H_2$) for use in urban areas. Gasification-based technologies can be used to convert carbon-containing resources into a clean synthesis gas, with high value as a fuel for combined cycle power generation[1], or as a raw material for the production of liquid fuels and chemicals. Moreover, they have the advantage of being capable of cogenerating electricity and fuel/chemicals efficiently, economically, and in an environmentally acceptable manner. Environmental performance of these technologies can be tailored to any specific requirement. In addition, due to the high efficiency nature of these technologies, the emissions of CO_2 are relatively low.

Plasma gasification of solid fuels is one of many possible technologies for effective and environment-friendly solid fuel utilization[2]. For instance, in China coal gasification under steam plasma conditions[3] has showed quite high indexes (syngas yield of 75%).

This paper presents the numerical and experimental investigation of solid fuels gasification in an air and water steam medium in an arc-plasma reactor. First, the gasification of air–solid fuel and steam–solid fuel mixtures was investigated numerically with the aid of the software code TERRA[4], validated for thermal equilibrium calculations. These mixtures were then investigated experimentally and the numerical results were verified against operational data obtained in the experiments. Kazakhstan Kuuchekinski bituminous coal (KBC) of 40% of ash content and Canadian petrocoke (CP) of 3% of ash content (Table 1) were used for the investigation. Their dry basis higher calorific values are 16,632 and 47,008 kJ/kg correspondingly. The special combined plasma chemical reactor (CPCR) for coal gasification allows the processes for the thermoimpact on solid fuel to be performed to obtain synthesis gas free from nitrogen and sulfur oxides. The experimental installation is intended to work in range of 30–100 kW electric power generation, with a mass-averaged temperature of 1,800–4,000 K, a coal-dust consumption of 3–12 kg/h and gas-oxidant (air or steam) flow of 0.5–15 kg/h. High quality syngas was produced from the both solid fuels.

2. Numerical Simulation

The software Code TERRA[4] was used for the thermodynamic analysis of the solid fuels plasma gasification process. This software was created for equilibrium computations of high-temperature processes, and in contrast to traditional thermochemical methods of equilibrium computation that use the Gibbs energy, equilibrium constants and Guldberg and Vaage law of acting mass, TERRA is based on the principle of maximizing entropy for isolated thermodynamic systems in equilibrium. TERRA has its own database of thermodynamic properties for more than 3,500 chemical agents over a temperature range of 300–6,000 K. It must be noted before discussing the results, that despite the fact that in principle the plasma reactor is an open, not isolated system and there is an exchange of energy and substance with the external medium; the thermodynamic modeling of solid fuel gasification inside the plasma-chemical reactors is possible. Firstly, for the preparation of the heat and material balance of the plasma gasifier, the actual heat losses are taken into account; and in this case the mass mean temperature in the plasma reactor is determined as for a thermodynamically isolated system. Secondly, the time the reagents stay in the zone of reactions is about 1 s, which is many times longer than the thermodynamic equilibration time in the system at the high process temperature[4]. Thirdly, the CPCR is a flow reactor and the quasistationary process of gasification is provided.

The calculations were performed for the process of KBC and CP gasification in the air and steam medium. The chemical composition of KBC and CP is presented in Table 1. All the calculations were performed for atmosphere pressure $P = 0.1$ MPa and within the temperature interval 400–4,000 K. The temperature is suggested to be kept at the expense of the external heat source, which is an arc in a CPCR.

TABLE 1. Solid fuels chemical analysis, % dry mass basis

Solid fuel	C	O	H	N	S	SiO_2	Al_2O_3	Fe_2O_3	CaO	MgO	K_2O	Na_2O
KBC	48.86	6.56	3.05	0.8	0.73	23.09	13.8	2.15	0.34	0.31	0.16	0.15
CP	75.0	0.88	15.53	0.01	5.63	1.31	0.78	0.6	0.1	0.05	0.07	0.04

For the first set of calculations, the feed was assumed to consist of 100 kg of KBC plus 127.5 kg of air; for the second set the feed was assumed to be 100 kg of KBC plus 62.75 kg of water steam; for the third set the feed was supposed to be 100 kg of CP plus 460 kg of air and for the fourth set the feed was assumed to be 100 kg of CP plus 120 kg of water steam. The gas phase species (Figures 1–3), degree of solid fuel gasification (Figure 4) and specific power consumptions (SPCs) for the processes (Figure 5) were calculated for all the variants.

Figure 1 presents the results of the calculations for the air gasification of KBC. The gas phase (Figure 1 on the left) for a temperature interval of 400–4,000 K mainly consists of nitrogen and syngas ($CO + H_2$) thermodynamically steady to the coproducts of the coal gasification process. Its concentration reaches the maximum of 54.79% at $T = 1,800$ K. The oxidant concentration (H_2O and CO_2) decreases to zero as the temperature increases to 1,400 K. Nitrogen content substances are presented mainly in the form of molecular nitrogen (N_2). Its concentration decreases with temperature because of the syngas appearance in the gas phase, the increase in other Nitorgen substances is due to the species of the coal mineral mass converting into the gas phase at temperatures over 1,600 K. The Carbon monoxide (CO) concentration reaches the maximal value of 35.74% at a temperature of 1,800 K. The concentration of molecular hydrogen (H_2), for a temperature interval from 400 to 1,600 K increases promptly and reaches the maximum of 20.3% at $T = 1,600$ K. At the temperature over 1,600 K the H_2 concentration decreases because of its dissociation and in the gas phase atomic hydrogen (H) appears. The H concentration increases with the temperature and reaches 23.34% at 4,000 K. At a temperature over 1,600 K in the gas phase there is hydrogen cyanide (HCN) and cyanide (CN), the total concentration of which reaches 2% at a temperature 2,800 K. There is no NO_x in the gas phase. It is connected with the oxygen deficit and as a consequence of reducing atmosphere. It is known to gasify carbon to CO, the stoichiometric ratio of C:O must be equal to 0.75, in our case the ratio of C:O is 1.36. Thus, the nitrogen is combined into the oxygen free compounds HCN and CN. Sulfur containing species are presented as H_2S and SiS, their concentrations amount to 0.27% and 0.26% correspondingly.

The mineral components of the coal sublimate into the gas phase (Figure 1, center) along with the temperature of the coal gasification process. They are mainly aluminum (Al) and silicon (Si). In the gas phase they appear at temperatures over 2,200 K. Their sum concentration reaches 5.5% at 4,000 K. The total concentration of all the other species of mineral mass in the gas phase reaches 1%. The Carbon concentration in condensed phase (Figure 1, right) decreases due to its gasification in the temperature interval 800–1,200 K. In the interval of 1,200–1,600 K its concentration is practically steady, because of the absence of free oxygen in the gas phase. In the interval 1,600–1,800 K the carbon concentration decreases sharply due to its participation in the reactions of ferric and silicon oxides reducing to the carbides (Fe_3C and SiC). Within the temperature interval 1,800–2,600 K, all mineral components evolve into gas phase as Al, Si, SiS, Fe, Al_2O, SiC_2, etc. (Figure 1, center).

The gas phase of the products of KBC gasification in the steam medium (Figure 2, left) is presented generally by syngas, of which the maximum concentration is 99% at 1,500 K. At that sum, the concentration of atomic and molecular hydrogen is higher than concentration of carbon monoxide for all the temperature range, and varies from 48% to 59% in volume. With the temperature concentration of carbon monoxide decreasing from 47% at 1,500 K down to 34% at 4,000 K. A light decrease in syngas concentration with a temperature increase is connected with the appearance in the gas phase of the components of mineral mass of the coal. They begin to convert from the condensed phase to the gaseous one at temperatures over 1,600 K (Figure 2, center). At temperatures of over 3,000 K the mineral components of the coal are presented in the gas phase, mainly as the elements Si, Al, Ca, Fe, Na, and compounds AlH, AlOH, SiS, etc. With the temperature the last are decomposed intensively. In this case the ratio of C:O is 0.78. It is practically corresponded to stoichiometric ratio. It enabled 100% of the carbon to be transformed to carbon oxides. Furthermore, at temperatures above 2,000 K the oxygen from mineral mass of coal starts actively to participate in the reactions of hydrogen, sulfur and nitrogen oxidation. It leads to their oxides formation and NO particularly. The NO concentration reaches 350 ppm by $T = 4,000$ K. As to the condensed phase (Figure 2, right), with a temperature increase to 3,000 K, all the components of it are transferred into the gas phase. Note, there are no carbides in the condensed phase except one minor peak of Fe_3C concentration.

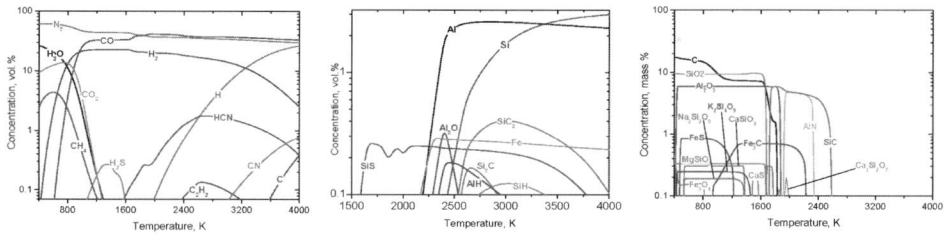

Figure 1. Composition of organic (*left*) and mineral (*center*) components in gas phase and condensed components (*right*) vs temperature at KBC air gasification

Figure 3 presents the gas species concentration vs temperature of the process of CP gasification in air and steam medium. As the mineral mass of CP is very small, we consider the behavior of the inorganic species neither in the gas nor in the condensed phase. Using air gasification (Figure 3, left) in a temperature interval of 1,000–2,600 K, the gas phase mainly consists of molecular nitrogen (N_2) and syngas ($CO + H_2$). When the concentration of hydrogen is 26.9% ($T = 1,400$ K), the concentration of carbon monoxide is 22.9%, and the concentration of nitrogen is 47.4%. Under these conditions the sulfur is presented as hydrogen sulfide (H₂S), with a concentration of less than 0.7%. The oxidant concentration (H_2O and CO_2) in this temperature interval does not exceed 2.1%. At temperatures over 2,000 K in the gas phase, the atomic hydrogen (H) is present. Its concentration reaches 35.9% by 4,000 K. At temperatures exceeding 2,200 K, the hydrogen sulfide dissociates into atomic sulfur (S) and sulfur hydride (SH). The total maximum concentration of them is about 0.5%. In the gas phase at temperatures above 3,000 K, hydroxyl (OH) and atomic oxygen (O) appears; their total concentration is about 1.5%. At temperatures exceeding 3,200 K, in the gas phase nitrogen oxide (NO) is noticed, its concentration reaching 1,500 ppm.

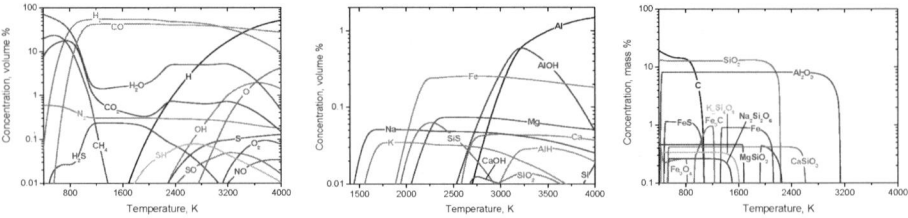

Figure 2. Composition of organic (*left*) and mineral (*center*) components in gas phase and condensed components (*right*) vs temperature at KBC steam gasification

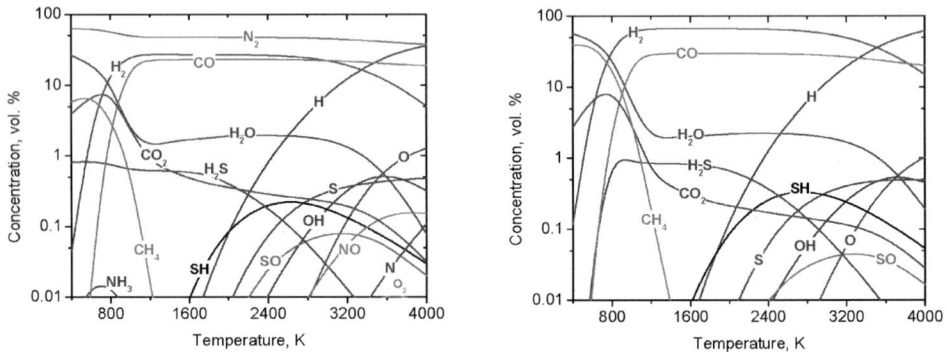

Figure 3. Concentrations of gas phase species (organic mass) variation with temperature of the process of CP air (*left*) and steam (*right*) gasification

Using CP steam gasification in a temperature interval of 1,200–2,800 K, the gas phase (Figure 3, right) mainly consists of syngas. When the concentration of hydrogen was 67% ($T = 1,400$ K), the concentration of carbon monoxide was

29.8% and sulfur was presented as sulfur hydrogen (0.8%). The oxidants concentration (H_2O and CO_2) does not exceed 2.5% in that temperature range. At temperatures greater than 2,000 K, atomic hydrogen appears in gas phase. Its concentration reaches 62.2% at a temperature of 4,000 K. At temperatures greater than 2,200 K, hydrogen sulfide dissociates into atomic sulfur and sulfur hydride, with a total maximum concentration reaching 0.5%. At temperatures greater than 3,000 K, in the gas phase hydroxyl and atomic oxygen appears. Their total concentration was less than 1.5%. In contrast to the gasification in air medium, in the steam gasification process nitrogen oxides are not formed.

Note that the concentration of syngas obtained in the process of steam gasification is significantly higher that that in the process of air gasification for the both solid fuels.

The degree of solid fuel carbon gasification X_C is determined from the carbon content in the condensed phase. Specifically, X_C is calculated according to the following expression:

$$X_c = \frac{C_{bas} - C_{fin}}{C_{bas}} \cdot 100\% ,$$

where C_{bas} and C_{fin} is the initial carbon concentration in the solid fuel and final carbon concentration in the solid residue (condensed phase) at the current corresponding process temperature.

The degree of carbon gasification (Figure 4) increases with the temperature in all variants and reaches 100%. Thus, the carbon from the solid fuels completely transforms into the gas phase mainly in the form of CO (Figure 1–3). At KBC gasification in air medium, in the temperature interval of 1,200–1,600 K, the carbon gasification degree is practically permanent because of free oxygen absence and carbides formation. At temperatures exceeded 1,600 K, the gasification degree increases sharply due to decomposition of ferric and silicon carbides and carbon oxidation in reduction reactions of oxygen containing the compounds Al_2O_3, $CaSiO_3$, $MgSiO_3$, etc. and completely transforms to gas phase by the temperature 2,100 K. At KBC gasification in steam medium, the carbon gasification degree reaches 100% by 1,200 K. In contrast to the KBC gasification in air medium there is no "plateau" in temperature dependence of gasification degree. This is explained by the different levels of temperature necessary for the complete carbon gasification in water steam ($T \sim 1,200$ K) and the formation of silicon, ferric and aluminum carbides in the condensed phase ($T = 1,400–2,400$ K) (Figures 1 and 2). This means that when the temperature of carbide formation in the condensed phase is reached, the carbon has completely transferred into the gas phase as CO. The carbon gasification degree

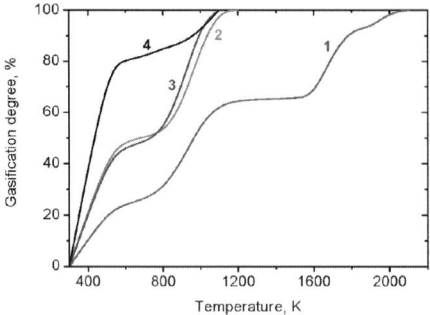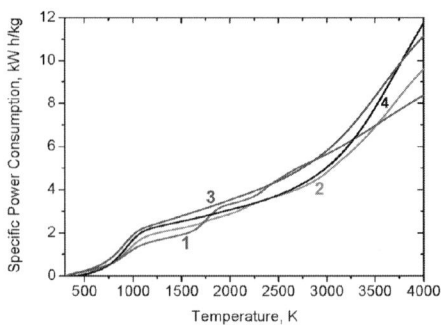

Figure 4. Carbon gasification degree (*left*) and specific power consumptions for the process of plasma gasification (*right*) related to 1 kg of syngas versus temperature. 1, 2 is KBC gasification in air and steam medium and 3, 4 is CP gasification in air and steam medium correspondingly

increases with the temperature in both variants of CP gasification, and reaches 100% at close and at relatively low values of temperature, which is 1,100 K.

The SPCs for the gasification processes were calculated by the difference of working medium (solid fuel and oxidant) enthalpy at initial (T = 298 K) and final (current temperature of the process) conditions. The calculation results for both processes are presented in Figure 4. At temperatures above 1,000 K, the SPC for producing syngas increases smoothly with the gasification processes temperature. Note that for both processes of CP gasification, in the temperature interval of 1,400–2,600 K, where syngas concentration is maximal and practically permanent, the SPC related to 1 kg of syngas are noticeably different. For example, at a temperature of 1,400 K the SPC for the process of CP gasification in steam medium is less than that in air medium, and amounts to 2.43 and 2.64 kWh/kg of syngas correspondingly. The SPC for syngas production for KBC gasification in both air and steam medium, has minor fluctuations in the temperature range from 1,600 to 2,200 K (curve 1), in which the reducing endothermic reactions with carbon run, monotonously increase with temperature. In the temperature range of 1,700–3,200 K, the SPC for KBC gasification in steam medium are less than in air medium. At the temperatures that the coal gasification degree reaches 100%, which are 1,200 and 2,000 K for KBC gasification in steam and air medium correspondingly, the SPC for KBC gasification in air medium (3.32 kW h/kg), are 1.7 times higher than for KBC gasification in steam medium (1.95 kW h/kg). The syngas concentration for both the KBC and CP gasification in steam medium is higher essentially (see Figures 1–3) and reaches 97% against 50% at gasification in air medium for CP gasification and 97% against 61% for KBC gasification. At the same time, the syngas at the solid fuels gasification in steam medium is of a higher quality. For example, the hydrogen concentration at CP and KBC gasification in steam medium reaches 67% and 54.8% against 27.1% and 22.6% at the solid fuels gasification in air medium correspondingly.

Thus numerical investigation showed that from both energy and ecology point of view, for solid fuel gasification the superior process is its gasification in steam medium with use of an internal heat source. This source can be a plasma one, which allows the attainment of the temperature required for the process of gasification without the additional fuel incineration, and the target product syngas dilution with inert products of combustion. Solid fuel plasma gasification under steam conditions allows the converting of the organic mass of the fuel into high-calorific syngas, which is free from nitrogen and sulfur oxides. This conversion can be written as the following overall reaction: $C + H_2O = CO + H_2$. The plasma source energy compensates for the endothermic effect of this reaction, which is equal to 130,500 J/mol.

3. Experimental

3.1 EXPERIMENTAL SETUP

The gasification experiments were conducted in a tube-type setup, which is schematically shown in Figure 5. The principal components are: a plasma generator-reactor, *1*; slag trap, *2*; syngas and slag separator chamber, *3*; syngas cooling chambers, *4, 5, 6*; pulverized fuel feeding system, *8*; and a steam feeding system, *9*. The zones of heat release due to the arc plasma, and of heat absorption by the solid fuel and gas streams, are combined in the same plasma generator-reactor chamber, *1*. This consists of a cylindrical water-cooled jacket, with a top cover carrying a graphite rod electrode, of diameter 0.04 m, and the inlet pipes for pulverized solid fuel (pf) and plasma-forming gas. It is lined with graphite, of thickness 0.02 m, and bounded at its bottom side by a graphite orifice. The inner diameter of the reactor (i.e., of the graphite lining) is 0.15 m and its height is 0.3 m. The direct current arc is established (as sketched) between the rod and the graphite lining electrodes. It is localized at the ring electrode band of 0.07 m by an enveloping electromagnetic coil, *12*, made of water-cooled copper pipe. The plasmotron power is variable between 30 and 100 kW.

The slag trap, *2*, is a water-cooled cylinder of height 0.56 m, containing a slag catcher basket. It is lined with graphite to give an inner diameter of 0.15 m. The syngas cooling occurs in chambers *4, 5, 6*, which are water-cooled stainless steel cylinders, some of which are also lined with graphite to the same diameter.

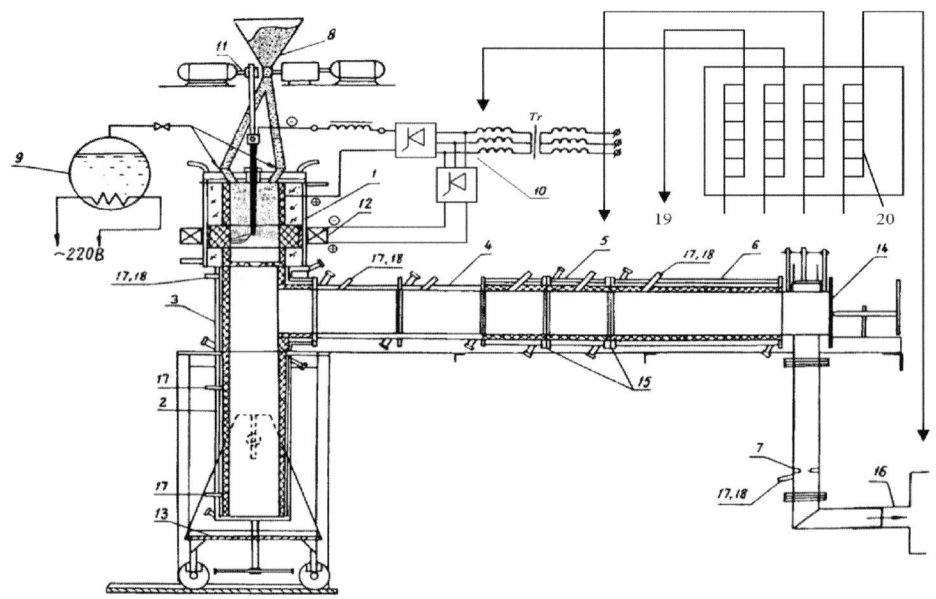

Figure 5. Experimental setup for pf plasma gasification. 1 – DC arc plasma reactor, 2 – slag trap, 3 – syngas and slag separator chamber, 4,5,6 – syngas cooling chambers, 7, 20 – flowmeters, 8 – pf feeding system, 9 – steam feeding system, 10 – power supply unit, 11 –motorised electrode position system, 12 – electromagnetic coil, 13 – handcart-elevator, 14 – waste gas output section with safety valve, 15 – sections for sulfur clearing, 16 –gas extraction system, 17 – pipes for thermocouples, 18 – pipes for gas sampling, 19 – the installation units cooling system, *Tr* – transformer

The pf supply system consists of a screw feeder, *8*. The steam feeding system consists of a hot-water boiler mounted on a weighbridge, *9*, and a steam reheater. The two streams are merged and partially mix just prior to admission to the reactor through pipes in its top cover. The pulverized fuel admission rate is determined by weighing, to within an error of 1.5%. The steam rate is varied in the range of 0–10 kg/h to within an error of 3%. On completion of an experimental run, all components of the apparatus are thoroughly cleaned of the condensed phase, which is weighed to determine the mass of the solid residue from the fuel conversion process. The yield of gaseous products is measured with a Prowirl 77F flowmeter, *7*, at the outlet of the syngas cooling chamber, *6*.

The gas to be analyzed is withdrawn from the syngas cooling chambers through the pipes, *18*, by a quartz probe and directly admitted to the gas analysis system. The gas is subject to chromatographic analysis with the help of a SRI 8610C gas analyzer. The carrier gas is helium or argon, and the detectors are thermal conductivity sensors. The analysis of light gases H_2, CO, O_2, N_2, and CH_4 are provided in columns filled with a molecular sieve CaX, and the analysis of CO_2 in a silica gel column. A method of absolute calibration is used for experimental data handling. The composition of the solid residue was investigated by chemical and x-ray phase analysis. To determine the carbon gasification degree the absorption-weight method is used. According to this method carbon is determined through measuring the carbon dioxide that is formed as a result of the solid residue sample combustion in oxygen and the following absorption of it by ascarit (KOH or NaOH covering asbestos).

An overall calorimetric balance is performed on the setup using the coolant water flows and temperatures. The water temperatures are measured by thermocouples (Chromel-Copel) to within an absolute error of 0.2°C. The overall error in the calorimetric balance is typically 6–10%.

The temperatures of the reactor wall and of the gases exiting from the graphite diaphragm are measured by standard tungsten–rhodium thermocouples. The thermocouple junction exposed to the exiting gases is 10^{-3} m in diameter; the radiation error does not exceed 50°C. The gas temperature in the reactor is not measured. Instead, the mass-averaged temperature of reagents is calculated from the reactor heat.

The mass and heat balance is an important indicator for gasification experiments. The equations of material and heat balance of the experimental setup take the following form:

$$G_2 + G_3 + G_4 + G_5 = G_6 + G_1 + G_7, \text{kg/h, and}$$

$$W_0 + W_1 = W_2 + W_3 + W_4 + W_5 + W_6, \text{kW,}$$

where G_2, G_3, G_4, and G_5 are the mass flow rates of pf, steam or air, carrier gas for fuel pulverization, and electrode graphite respectively. G_6, G_1, and G_7 are the mass flow rates of slag (solid residue), effluent gases, and pulverized coke (fume and fine particles) being removed respectively. W_0 is the heat output of the arc, W_1 is the heat supplied with steam at $T = 405$ K; the heat losses to the cooling water in the unit assemblies are: W_2 from the reactor; W_3 from the gas and slag separation chamber; W_4 from the syngas cooling chamber; W_5 from the slag catcher; and W_6 from the heat carry-over in the effluent gas stream.

The arc electric power is determined from a wattmeter. The heat input to the steam is calculated as:

$$W_1 = G_3 H_1, \text{kW,}$$

where $H_1 = H_{405\,K} + \Delta H_{steam} = 0.05 + 0.63 = 0.68$ kWh/kg of steam and ΔH_{steam} is the heat of vaporization.

The heat loss with the effluent gases is determined from the effluent gas temperature (T_g), flow rate (G_1), and composition obtained through the gas analysis. The measured temperature, pressure, and composition of gases are fed into the TERRA code[4], and the specific enthalpy of the effluent gases is computed for the specified values of the parameters mentioned above. The gas mixture heat output is calculated as follows:

$$W_6 = H_6 G_1, \text{ kW,}$$

where $H_6 = \int\limits_{300}^{T_g} C_p \cdot dT$ is the specific enthalpy of the gas mixture.

Thus, except for W_6, all the components of the heat balance equation are measured during the experiment.

3.2 EXPERIMENTAL RESULTS

The arc is initiated by vaporizing a wire brought into contact with the rod and ring electrodes. The flow of steam or air and finally grained fuel is then initiated. The steam(air)/pulverized fuel mixture entering the arc zone is heated to high temperatures by the arc rotating in an electromagnetic field to produce a two-phase plasma flow where the solid fuel gasification process occurs. The solid residue so produced descends through the orifice plate into the slag catcher, 2 (Figure 5). The gaseous products exit the gas and slag separation chamber, 3, and flow into the cooling chambers, 4, 5, 6. Then gaseous products are exhausted to ventilation.

The duration of the experiments varied from 0.5 to 1 h. The dust of KBC and CP (see Table 1) was used in the experiments. The sieve analysis of the pf using the shaker "Tyler" RX-812 and a set of calibrated sieves with mesh sizes of 43–1,000 μm revealed that the mean sizes of the KBC and CP dust particles were 75 and 105 μm, correspondingly.

Prior to conducting an experiment, an estimate of the steam requirement was made from the water gas reaction. This estimate ignored the oxygen content of the inorganic material of the coal. The conditions of the experiments and their results are shown in Table 2. To calculate the SPC for the process, reduced to one kilogram of syngas output, the following formula is used:

$$Q_{SP}^{SYNG} = (W_0 + W_1 - W_2)/(G_1 - G_4), \text{ kW·h/kg.}$$

The thermal efficiency of the reactor was determined as 76% for all experiments. As a result of KBC plasma gasification in steam and air medium and CP gasification under steam plasma conditions, the concentrations of gas species, the carbon gasification degree X_C and the mass-averaged temperatures T_{AV} in the reactor were revealed.

TABLE 2. Main indexes of the solid fuels plasma gasification

N	Solid fuel	Consumption (kg/h)				W_0 (kW)	Q_{SP}^{SYNG} (kW h/kg)	T_{AV} (K)	CO	H_2	N_2	X_C (%)
		Fuel	Steam	Air	G_1					Volume %		
1.	KBC	8.0	–	8.0	12.3	33	2.1	2,100	27.4	15.9	55.3	89.6
2.	KBC	4.0	–	5.1	7.4	30	3.1	2,850	38.1	18.2	43.7	95.8
3.	KBC	4.0	1.9	–	4.2	25	4.8	3,500	41.5	55.8	2.7	94.2
4.	KBC	6.5	3.0	1.9	8.5	52.8	4.7	3,550	38.6	51.4	9.8	92.0
5.	CP	2.5	3.5	–	5.1	60	9.4	3,800	33.9	65.3	0.8	76.3
6.	CP	2.5	3.0	–	4.9	60	9.6	3,850	36.2	63.1	0.7	78.6

From the table it follows that the KBC gasification degree in air increases from 89.6% to 95.8% as the SPC rises from 2.1 to 3.1 kWh/kg of syngas. At these points, the syngas yield increased from 33.3% to 56.3%. As a consequence of gasification, the products mass consumption G_1 decreases from 12.3 to 7.4 kg/h, the mass-averaged temperatures in the reactor increases from 2,100 to 2,850 K and the corresponding intensification of the process. For KBC gasification under plasma-steam conditions, the SPC are noticeably higher (4.7 and 4.8 kW h/kg), keeping the gasification degree at the high level of 92.0–94.2%. At these conditions, the typical syngas yield is appreciably higher (90.0–97.3%). Note that contrary to the gasification under air-plasma conditions, for the process of plasma gasification in steam medium the concentration of hydrogen in the syngas is higher than the concentration of carbon dioxide. The hydrogen concentration produced for the gasification in steam is 3–4 times higher than that produced in the air medium. This effect is connected with the enriching of syngas with hydrogen due to the reaction of water steam decomposition by the solid fuel's carbon.

For the CP, in view of its high content of tar its direct use for redox is impossible. Whereas the CP reactivity is extremely low, its utilization in the conventional processes of combustion or gasification is very problematic. In connection with this, the experiments for CP gasification under plasma-steam conditions were initiated. To produce syngas possessing good qualities the process was carried out under a relatively high SPC (9.4–9.6 kW h/kg of syngas). Under these conditions, the mass-averaged temperature varied from 3,800 to 3,850 K, and the CP gasification degree varied from 76.3% to 78.6%. Note, the high content of hydrogen in the syngas (63.1–65.3%) at syngas yields of between 99.2% and 99.3%. The obtained ratio of H_2:CO $\approx$ 2:1 corresponds to the optimal content of a gas for synthetic liquid fuel (methanol) synthesis: $CO + 2H_2 = CH_3OH^5$.

The difference between the CP and KBC ash content is more than ten times (3–40%). But the yields of syngas vary insignificantly, reaching practically limiting values (90–99.3%). Note that the H_2 concentration in syngas is substantially higher than that of CO.

It must also be note that the SPC relating to syngas varies in a broad interval from 2.1 to 9.6 kWh/kg. Evidently, for syngas production using the water steam method, the utilization of the heat of the gasification products and the use of the exit gases as a carrier instead of air is required for pf feeding, The SPC would be minimized and consequently the efficiency of syngas production would be higher. An alternative way of reducing the SPC is based on the application of the plasma-autothermal principle of solid fuel gasification in a two-stage gasifier. During the first gasification stage, 30% of the fuel is utilized under air plasma conditions and the following combustion of the gasification products to CO_2 is realized, with the evolution of 33,173 kJ/kg of carbon. This heat, together with heat of the electrical arc is enough to compensate for the endothermic effect (10,875 kJ/kg of carbon) of the water gas reaction ($H_2O + C = C + H_2$). The rest (70%) of the fuel is used in the process of gasification in steam medium. An examination of experiment 3 (Table 2) with a SPC of 8 kWh/kg of carbon, but for the same initial data with plasma-autothermal gasification of the same fuel the SPC would be 0.8 kWh/kg of carbon. This is equal to the arc generation of heat W_0 decreasing from 25 to 2.5 kW.

3.3 COMPARISON OF EXPERIMENTAL AND CALCULATED DATA

In order to estimate the reliability of the TERRA code, comparisons between the calculated results and the experimental data were made. The comparisons are presented in Table 3, which demonstrates that a reasonably satisfactory agreement was achieved.

TABLE 3. Comparison between the calculated results and the experimental data

Method	Content (volume %)			X_C
	CO	H_2	N_2	(%)
Experiment # 2 of Table 2	27.4	15.9	55.3	95.8
Calculation # 2	36.6	16.4	36.6	100
Experiment # 3 of Table 2	41.5	55.8	2.7	94.2
Calculation # 3	38.8	54.1	0.3	100
Experiment # 6 of Table 2	36.2	63.1	0.7	78.6
Calculation # 6	21.3	76.1	–	100

The differences between the calculated and experimental degree of pf gasification is not more than 21.4%, and for the yield of the syngas it is not more than 22.4%. Note that the breakdown of the syngas in the calculations and experiments is similar.

The discrepancies can be explained by deviations from the thermodynamic equilibrium due to the limited residence times in the plasma reactor. It is also impossible to determine all the species of the gas phase at chromatography fulfillment.

4. Conclusions

The completion of a study of two essentially different composition and quality solid fuels undergoing gasification showed that it is possible to produce a high-quality syngas using both a steam and air plasma. It has been found that a syngas with the $H_2 + CO$ content between 43.3% for KBC gasification in air medium and 97.4% for CP gasification in steam plasma conditions can be produced.

The syngas obtained from the solid fuels is a high-quality power gas, which can be used for synthetic liquid fuel (methanol) synthesis: $CO + 2H_2 = CH_3OH$[5]. Syngas of this quality has high-potential as a reducing agent for iron ore direct reduction and can serve as a substitute of metallurgical coke. Also plasma steam gasification is an alternative method for hydrogen production, through water steam decomposition by carbon of a solid fuel.

Further experiments are planned for the gasification of solid fuels under steam plasma conditions, with the aim of increasing the gasification degree and reducing the SPCs of the process.

Acknowledgments

The authors gratefully acknowledge the European Commission for this work funding through the ISTC project, and personally Professor F. Lockwood for his support and coordination of the research project.

References

1. T.L. Wright, *Coal Gasification – Back to the Future? – Perspective on Atmospheric vs Pressurized Coal Gasification* (Proceedings of 28th International Technical Conference on Coal Utilization and Fuel systems, Clearwater, Florida, USA, Published by US Department of Energy & Coal Technology Association of USA, 2003 – CD p. 241–248).

2. A. Ustimenko, F. Lockwood, E. Karpenko, V. Messerle, *Plasma Complex Processing of Power Coals.* (Proceeding of 6th International Conference on Technologies and Combustion for a Clean Environment "Clean Air", Porto, Portugal, 2001, V. III, pp. 1473–1480).

3. Jieshan Qiu, Xiaojun He, Tianjun Sun, Zongbin Zhao, Ying Zhou, Shuhong Guo, Jialiang Zhang, Tengcai Ma, Coal gasification in steam and air medium under plasma conditions: a preliminary study, *Fuel Processing Technology*, (85), 969–982 (2004).

4. M. Gorokhovski, E.I. Karpenko, F.C. Lockwood, V.E. Messerle, B.G. Trusov, A.B. Ustimenko, Plasma technologies for solid fuels: experiment and theory, *Journal of the Energy Institute*, **78**(4), 157–171 (2005).

5. G.N. Kruzhilin. *Plasma Gasification of Coal*. Vestnik (Bulletin) of Academy of Science of the USSR, **12**, 69–79 (1980).

ENVIRONMENTAL AND RELATED STUDIES

SOME ADVANCES IN TWO-PHASE THERMOFLUID RESEARCH FOR ENVIRONMENTAL PROTECTION AND POLLUTION REDUCTION

P. J. BOWEN
*Cardiff School of Engineering, Newport Road,
Cardiff, CF243TA, UK*

Abstract. Two-phase thermofluid processes can be utilized and optimized to contribute to Environmental Protection and Pollution reduction. This paper describes recent advances in understanding and optimizing some of these processes, which are employed within a broad range of engineering applications including combustion engines, power generators and large-scale uncontrolled hazards such as fires or explosions. It describes advances in understanding of novel fuel injection and atomization processes, the development of direct numerical simulation (DNS) techniques for free-boundary two-phase problems and the potential of future modeling approaches. Progress in understanding two-phase combustion systems includes burning rate, ignition characteristics, explosions involving high-flashpoint fuels, flame interaction with water vapor and water mists, and the generation of ultrafine particulate matter. Advances in both experimental and modeling approaches are discussed.

Keywords: thermofluid, two-phase, hazards, power generation

1. Introduction

This paper discusses a range of research advances and developments over the last 10 years at Cardiff School of Engineering in collaboration with several industrial sponsors and government sponsorship in relation to two-phase ther-mofluid problems and processes pertinent to environmental protection and

N. Syred and A. Khalatov (eds.), Advanced Combustion and Aerothermal Technologies, 159–178.
© 2007 *Springer.*

pollutant reduction. The industrial sectors encompassed include the automotive sector, petrochemical, environmental protection, and power generation as well as supporting research of a more fundamental nature.

Two-phase problems naturally arise in many industrial processes, from utilization of liquid fuels for power generation, to spray cooling, or accidental releases of toxic or combustible material. With the ever increasing attention on pollutant reduction and process efficiency gains as we move towards a more sustainable industrial environment, these problems are very likely to receive more attention, with the requirement for further understanding, research and development.

2. Developments in Atomization Technology

This section describes a range of development in atomization and spray technology that have been brought about due to improved diagnostic and/or modeling capability. These technological advances have applications in a broad range of energy-related industrial sectors.

2.1 SUBCOOLED AND SUPERHEATED ATOMIZATION

2.1.1 Subcooled Releases

Release and atomization of subcooled fuel through simple orifices is an area that has received significant attention for diesel injection in compression-ignition, internal combustion engines. However, two-phase problems involving the same elementary physics are of concern when considering accidental large-scale releases of high-flashpoint liquid fuels such as diesel, gas oil or aviation kerosine in industrial hazard consequence modeling. In these cases, the characteristic releases typically are of significantly lower release pressures (typically sub-20 bar) and larger, less-defined orifices (in the order of several millimeters or significantly larger).

These problems were considered in the early 1990s (Bowen and Shirvill, 1994) in relation to area classification codes for offshore oil/gas exploration and production. More recently, this problem has received greater attention due to the involvement of high-flashpoint fuels in high-profile incidents involving transport, namely the Ladbroke Grove (near Paddington, UK) train crash and fireball (1999), the 9/11 Twin Towers attack in New York (2001), and may have played a role in the recent oil refinery explosions and fires in Buncefield (near Hemel Hemstread), UK (2005) currently under investigation.

Figure 1. (a and b) Fatal incidents involving fireballs of liquid fuels. (a) HSE simulation of Ladbroke Grove (London) Train Crash (1999), (b) 9/11 terrorist attack

The nature of the hazard is based upon the quality of spray produced from these nonstandard atomization conditions, and whilst these releases are characteristically of extremely poor quality, nevertheless methods of quantification are required to improve/guide risk assessments and future legislation. For this category of accidental fuel release, the characteristic jet breakup conditions are typically around the first–second wind transition criteria associated with aerodynamic of liquid jets (Bowen and Shirvill, 1994). Hence, associated with these releases is an unbroken jet core of variable length, beyond which the breakup commences. However, sufficiently far downstream where the spray has had sufficient time to develop, meaningful measurements may be taken from which empirical correlations may be derived. In the most recent studies undertaken at Cardiff, the correlation proposed for spray quality from these poorly atomized jets takes the form:

$$\text{SMD}/d = F(We, Re, L/d) = 64.73 We^{-0.552} Re^{-0.014} (L/d)^{0.114} \tag{1}$$

Of interest is the influence of the dimensional variables embedded within this correlation. The exponent of the pressure differential term is circa –0.54, and that corresponding to the release diameter is circa +0.34. These exponents are consistent with a range of correlations derived by other researchers for diesel-type injectors at far higher pressures (several hundred bar) and orifice sizes ($O(10^2)$ μm); it appears that the similarity scaling approach is applicable over a couple of orders of magnitude of the primary variables. The correlation also provides physically "sensible" exponents for the primary liquid fluid parameters. Whilst SMD is a useful parameter for heat/mass transfer problems, it is the range of small droplets below a critical droplet size range – the volume/mass undersize function – which is most useful in hazard analysis. To this end, the following empricial correlation has been derived from phase Doppler anemometry (PDA) data, which allows the mass of fuel contained in droplets below a certain size range (v(D)) to be quantified, given by:

$$1 - v(D) = e^{-0.3(D/SMD)^{1.4}} \tag{2}$$

2.1.2 Superheated (flashing) Releases

Release of superheated liquids – often referred to as "flashing jets" has been of interest to the petrochemical "risk and hazard" communities for some time. This attention has become more focused recently with the proposed large-scale expansion, transportation and utilization of liquified fuels such as liquified natural gas (LNG) and liquified petroleum gas (LPG), particularly in the USA and UK, where imported LNG is an integral part of the national strategy to meet energy demand.

Due the potential efficiency of flashing atomization, the automotive sector has also considered utilizing this technology for improvements in fuel preparation and mixing in IC engines. Furthermore, the technology is most widely used commercially in the form of domestic propellents, for deodorant, etc.. However, again irrespective of the industrial application, the physics of the underlying processes are very similar.

Depending upon the thermodynamic storage conditions, accidental releases of liquified fuels can atomize extremely efficiently producing high-volumes of very fine mists (Figure 2). One of the primary issues for large-scale hazard pre-dictions, is to predict the quantity of fuel which rains out in the near vicinity of the release, and conversely that which remains airborne as very fine aerosol or completely evaporates before touchdown. Hazardous vapor or aerosol that remains within the hazardous plume can travel very large downstream distances, and will be subject to atmospheric dispersion conditions, governing plume entrainment, dispersion and hazardous concentrations.

However, the source term of the release is critical in successfully predicting downstream hazardous conditions. Recently, a simple empirical model (Figure 3) has been proposed as a framework upon which future development for flashing releases of toxic material may be based (Witlox et al., 2005). The model takes account of the various modes of atomization that occur as the critical control parameters vary (here represented by the dimensionless Jakob and Weber

Figure 2. Highly efficient flashing jet release

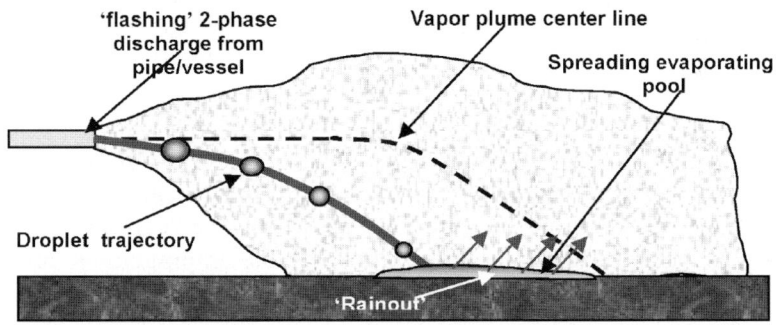

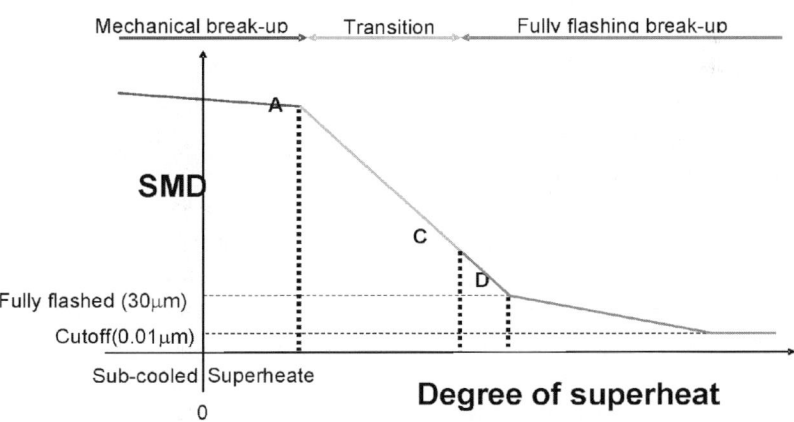

Figure 3. (a and b) Source Term Submodel Concept for Superheated releases of liquified fuels for atmospheric dispersion modeling (Witlox et al., 2005) (a) Source Term Submodel Concept, (b) trilinear Model for Sauter mean diameter as a function of superheat

numbers). Notice the model shows distinct points of transition, which correspond to several changes in atomization modes with release conditions, and is represented by several correlations within the model (Witlox et al., 2005).

2.2 TRANSIENT ATOMIZATION

In the mid-1990s, the automotive industry took considerable steps forward in "clean" gasoline engine technology by introducing the Gasoline Direct Injection (G-DI) engine concept. Transient two-phase fluid diagnostics then became critical in fully understanding and optimizing the state of the pre-ignition fuel charge, and hence some of the basic principles upon which G-DI improvements are based. This provided the challenge for researchers to develop and apply novel techniques to interrogate the two-phase G-DI sprays, which are injected over the timescale of milliseconds, in the form of droplets of size typically less than 20 μm, and at speeds of the order of 100 m/s.

Figure 4 shows a G-DI spray characterization facility designed and commissioned at Cardiff School of Engineering (Comer, 1999), which offers considerable optical access and facilitates controlled optical studies of G-DI sprays at elevated temperatures and pressures (up to 15 bar and 430 K with maximum optical access) representative of in-cylinder conditions. A carefully designed annular air curtain minimizes fouling on the quartz windows, whilst having minimal effect on the quiescent air within the region of injection. The facility has been recently modified allowing the impingement process to be studied. A "dummy" piston which is afforded six degrees of freedom is placed into the rig, and thus can be orientated in any way relative to the injector.

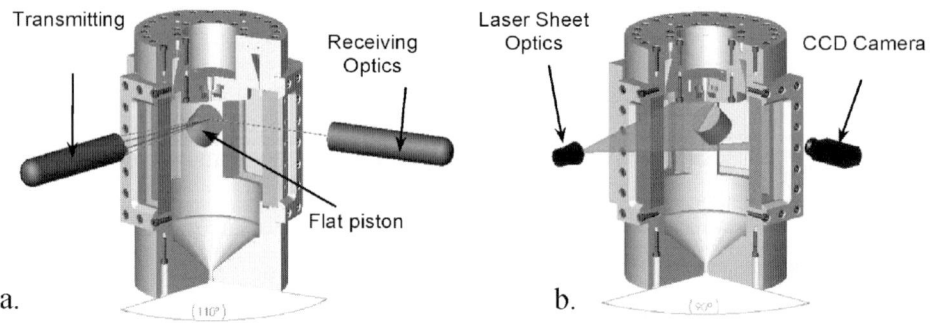

Figure 4. Model of the HT-HP rig setup for (a) PDA and (b) planar imaging diagnostics

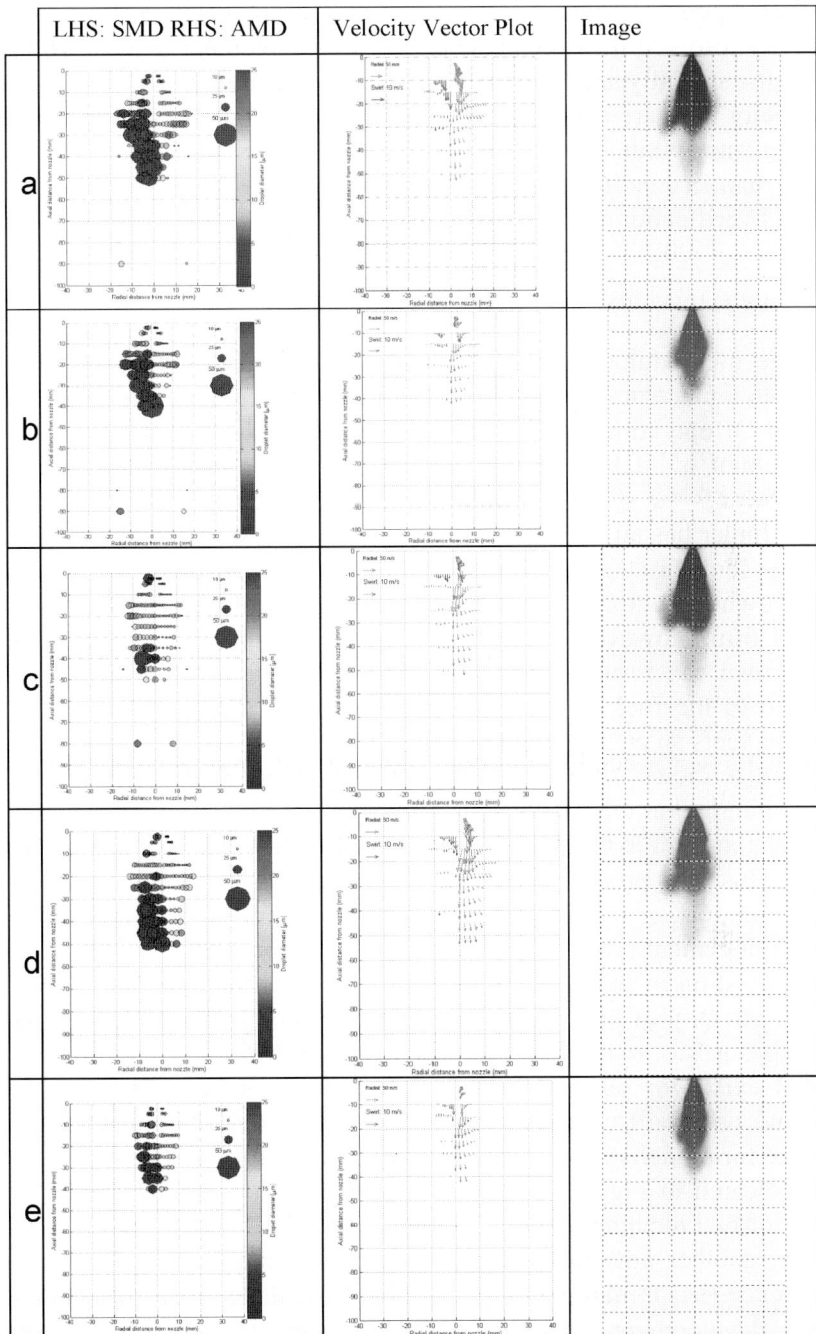

Figure 5. Comparison of droplet diameter, velocity vectors and high-speed images at 1.25–1.50 ms after start of injection. (a) ρ_a = 1.2, T_a = atm, T_{inj} = atm (b) ρ_a = 4.76, T_a = atm, T_{inj} = atm (c) ρ_a = 1.2, T_a = atm, T_{inj} = 358 (d) ρ_a = 1.15, T_a = 423, T_{inj} =303 (e) ρ_a = 4.77, T_a =423, T_{inj} = 303.

Developments in phase-resolved PDA measuring techniques have facilitated temporally resolved measurements of spray characteristics, as shown in Figure 5. Here, the spatially and temporally resolved measurements of fuel droplet size and velocity field are compared with high-speed images of a G-DI spray. Such measurements within the facility shown in Figure 4 facilitate the range of thermodynamic conditions presented in Figure 5, which offer the opportunity for model verification and development.

2.3 SPRAY IMPINGEMENT AND FUEL FILMS

The same methodology can be applied to investigate fuel sprays impinging upon in-cylinder surfaces, utilized for fuel efficiency in so-called stratified charge G-DI operation under low-load conditions. In this operational mode, a small amount of fuel is injected late in the compression stroke. The fuel is guided towards the spark plug where a fuel rich region forms, whereas the rest of the cylinder has an overall lean mixture; air-to-fuel ratios of 30–40:1 can be achieved. Some G-DI strategies utilize the impingement of the fuel spray on the piston crown to control the fuel transportation and mixing, inevitably resulting in a fuel film. However, it has been shown that fuel that is deposited on the piston crown can result in a 60% increase in unburnt hydrocarbon emissions compared to the fuel being completely vaporized (Stanglmaier, 1999) which clearly presents an environmental issue.

Figure 6 shows a time-resolved, two-dimensional data set of a gasoline spray impinging on a heated piston surface. The G-DI spray after impingement upon a piston head in Cardiffs G-DI simulation facility (Figure 4) has also been studied. This data proves invaluable in verification studies of two-phase fluid dynamic models, or indeed as source data in empirical submodel development of the impingement process.

Optimization of the G-DI stratified charge operating mode depends in part upon the quantity of fuel that remains on the piston surface after impingement. Although surface evaporation assists in minimizing this effect, clearly the potential exists for residual fuel to contribute to unburnt hydrocarbon emissions, and potentially degrade the piston surface due to mini pool fires. The challenge for modelers is not only to predict the characteristics of the secondary spray indicated in Figure 6, but also to quantify through the engine cycle the quantity of fuel remaining on the piston crown.

Here, another advanced laser diagnostic technique, namely laser-induced fluorescence (LIF), may play a role in the difficult problem of quantifying the transient development of the thickness of the fuel film on the piston during fuel injection. The LIF technique relies on the principle that upon excitation by laser radiation, the intensity of the fluorescent signal (filtered at 400 nm) from a dopant (3-pentanone) mixed with a nonfluorescent base (iso-octane), is proportional to the film thickness (Kay et al., 2006). A Nd:YAG laser is used as the excitation source (utilizing the fourth harmonic at wavelength 266 nm) and an intensified "HiSense" CCD camera records the results in the form of the fluorescent images.

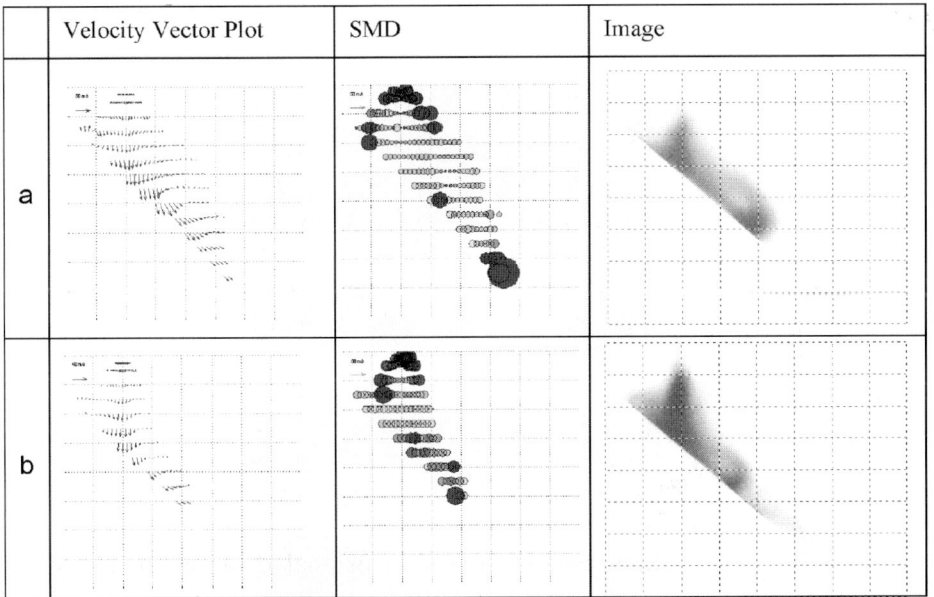

Figure 6. Spatially resolved data for impinging G-DI spray 2.25–2.50 ms after start of injection (a) $\rho_a = 1.2$, $T_a = $ atm, $T_{piston} = 423$ K (b) $\rho_a = 4.76$, $T_a = $ atm, $T_{piston} = 423$ K

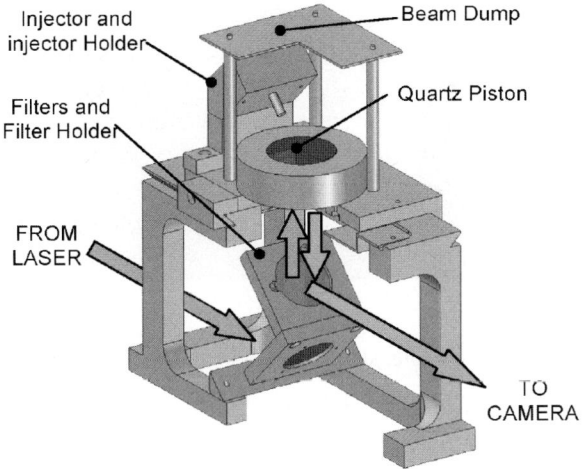

Figure 7. Shows a schematic of a dedicated rig developed to undertake such studies utilizing an optical quartz piston

Figure 7 shows the optical setup. The bottom mirror is chosen to reflect the excitation light with a wavelength of 266 nm only, hence any residual laser light with wavelengths 532 and 1,064 nm are not reflected. For the second optic a dichroic filter was ordered such that it transmitted the laser light at 266 nm and reflected the fluorescent light in the range of 400 ± 20 nm. To ensure that no laser light reflections are collected by the digital camera an additional band pass filter (Schott BG4 glass) is positioned in a "Cokin" filter holder. The filter glass effectively blocks light at 266 and 532 nm.

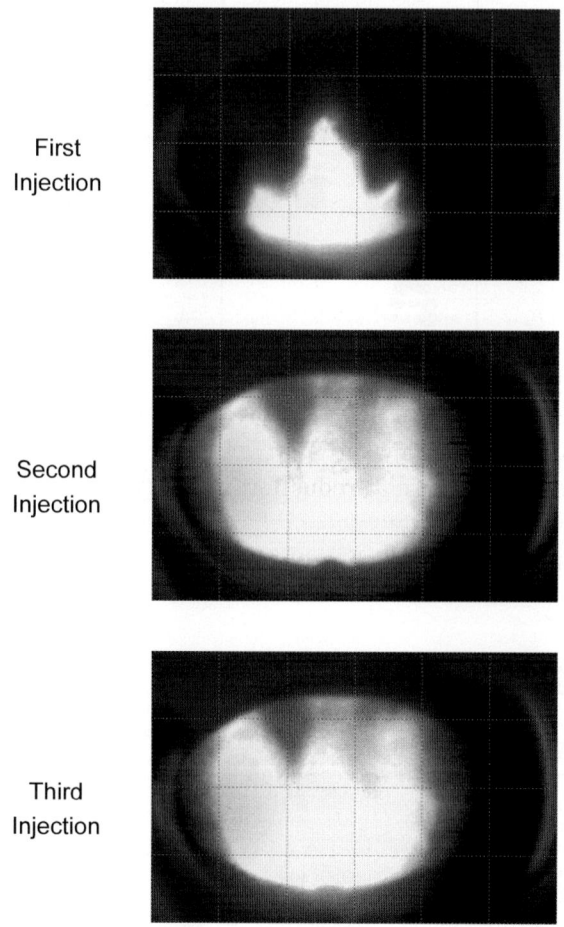

Figure 8. LIF signal 1 ms after start of injection for three consecutive G-DI sprays

Figure 8 shows the accumulation of fuel (via the fluorescent radiation at 400 nm) after multiple injections at 1 ms after start of injection. This initial feasibility study has been undertaken at atmospheric conditions, hence significant challenges remain in terms of quantifying the LIF signal through careful calibration and error elimination, and applying under engine-like conditions.

Again hazard analysis provides the large-scale version of this two-phase flow problem when we consider the accidental release of hazardous liquid which impinges upon an object in the near-vicinity of the release. In the case of high-flashpoint fuel releases for example, secondary atomization may present a case whereby a release which is considered nonhazardous based upon the primary spray characteristics, could translate into a hazardous scenario postimpingement (Marakgos and Bowen, 2002).

Experimental studies of the large-scale impingement problem are very challenging for a range of optical and practical reasons. However, with recent developments in optical diagnostic techniques (e g , PDA signal processing technology for dense sprays), measurements providing quantitative insights into secondary spray characteristics are becoming possible (Marakgos and Bowen, 2002), providing opportunities for hazard model development and appraisal. Figure 9 shows the spray quality and mass of secondary airborne spray measured before and after impingement of release of a low pressure, large orifice liquid jet. Note the salient features from a hazard perspective of significant quantity of secondary spray (up to 60% of the primary mass flow rate from the release) and the substantial reduction in spray quality post impingement. Both these characteristics exacerbate the hazard problem.

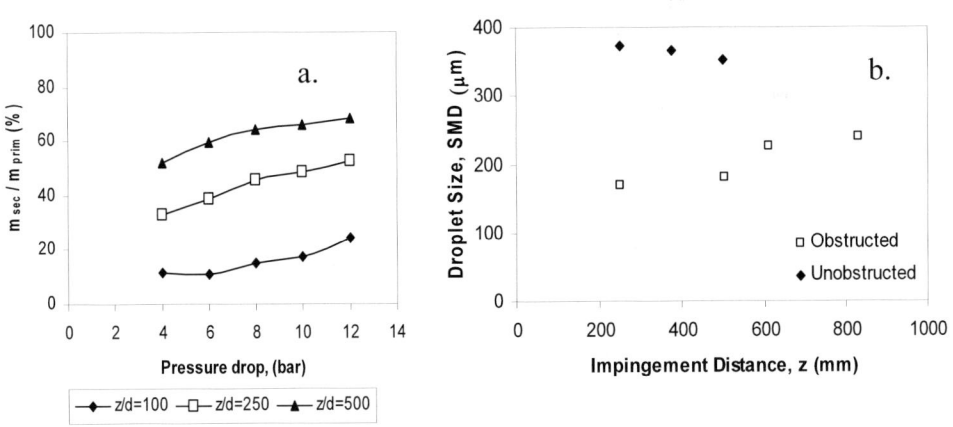

Figure 9. Postimpingement characteristics of large-scale liquid releases. (a) Secondary mass fraction, (b) quality of secondary spray

2.4 ADVANCES IN MODELING ATOMIZATION/SPRAY PROCESSES

Most computational fluid dynamics codes modeling spray or atomization-related problems utilize Euler–Lagrangian models, with considerable use of submodeling for the droplet processes. Figure 10 shows a typical prediction of a spray from one such commercially available model.

However, progress is now also being made in utilizing direct numerical simulation albeit for simple, benchmark two-phase problems. For example, highly computationally efficient spectral element methods have been developed to solve the benchmark atomization problem of droplet deformation in a

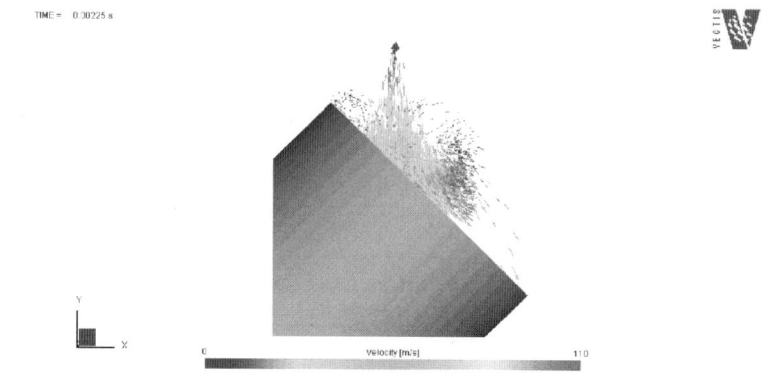

Figure 10. Typical output from two-phase contemporary CFD Model (fuel injection spray)

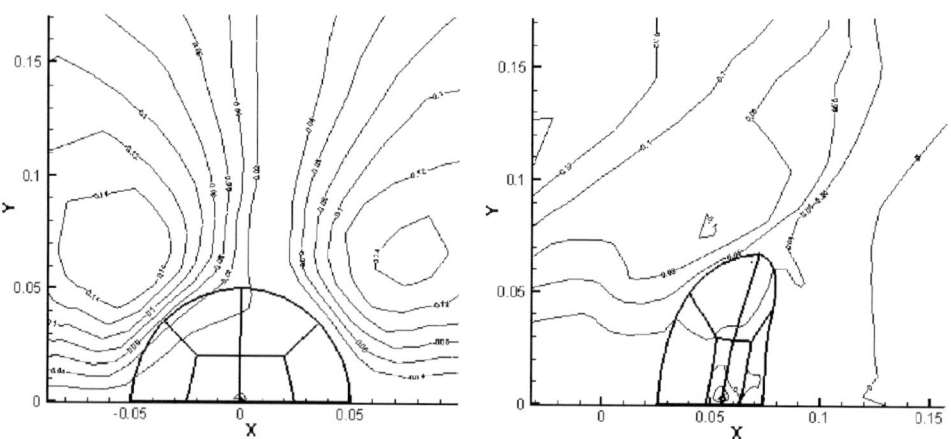

Figure 11. Translation and deformation of a Newtonian liquid droplet prior to breakup in an airstream

gaseous airstream (the "critical Weber number" droplet breakup problem), without submodel utilization (Bowen et al., 2006). Here, the arbitrary Lagrangian–Eulerian (ALE) formulation is utilized to account for the movement of the mesh. Spectral element approximations are used to ensure a high degree of spatial accuracy, and the computational domain is decomposed into two regions, one of which remains fixed in time while the other, located in the vicinity of the droplet, is allowed to deform within the ALE framework. Transfinite mapping techniques are used to map the physical elements onto the computational element, with edges of elements on the free surface described using an iso-parametric mapping. Surface tension is treated implicitly and naturally within the weak formulation of the problem. Figure 11 gives an example of a Newtonian droplet deforming prior to breakup for We = 15, Re = 1,000, density and viscosity ratios of 550 and 16, respectively.

3. Two-Phase Combustion Developments

3.1 DEVELOPMENTS IN IGNITION AND FLAMMABILITY CHARACTERISTICS OF HIGH-FLASHPOINT FUEL RELEASES

On the basis of the empirical findings for two-phase flow characteristics referenced in the preceding section, it was predicted that the large-scale, high-flashpoint fuel release problem introduced earlier could result in flammable mixtures, ignitable with a low-energy electrical discharge even for release pressures below 10-bar in the case of an impinging spray. Whilst 10-bar fuel releases generally produce very poorly atomized sprays, under suitable conditions, sufficient fine mist can be produced in the secondary spray to generate readily flammable mixtures.

This hypothesis was substantiated through a simple demonstration study, shown pictorially in Figure 12. In this demonstration, high flashpoint Shell "Gas Oil" (flash point 68°C) was discharged through a 1 mm circular orifice under a pressure differential of less than 8 bar. An impingement plate was positioned 100 m downstream of and perpendicular to the jet, and a relatively low energy (<200 mJ) igniter repeatedly discharged within the region of high air entrainment near the edge of the impingement plate. The secondary mist was readily ignited, generating a "flash fire" which rapidly consumed the airborne fuel cloud. Minimal overpressure was generated due to the open environment of the test, as to be expected.

Hence, this demonstration consolidates the proposition that high-flashpoint liquid fuels can, even under low release pressures, generate readily ignitable, flammable mixtures which give rise to flash fires or potentially explosions in areas of sufficient confinement and/or congestion.

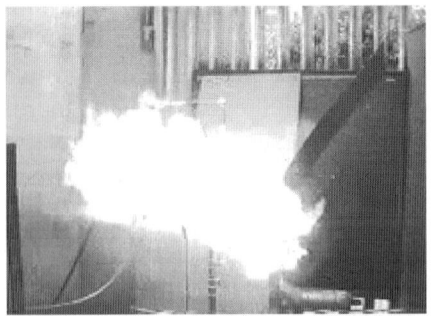

Figure 12. (a and b) Flash fire generated by low energy ignition of high flashpoint (impinging) releases. (a) Preignition secondary mist; (b) postignition flash fire

3.2 DEVELOPMENTS IN DISPERSED-PHASE FUEL BURNING RATES

3.2.1 Cloud Chamber Technology

Laminar burning rates of two-phase fuel mixtures are influential in a range of industrial processes, from engine combustion (internal combustion or gas turbine) to explosion quantification. Generating reliable data for burning rates of dispersed-phase systems has proven a very difficult proposition despite many attempts over the last 50 years, primarily due to the inherent problems in generating uniform, homogeneous two-phase mixtures where the primary two-phase parameters can be readily varied. However, more recently thermodynamic generation of experimental fuel clouds within the laboratory environment has shown promise in generating suitable experimental conditions to facilitate two-phase laminar burning rate studies. The principle of mist generation within the laboratory was reported by Wilson (1897) over 100 years ago, and with the considerable advances in control systems and precision manufacturing techniques since, these underlying thermodynamic principles have been utilized to generate a cloud generation system within an optical combustor (Figure 13).

Considerable effort has been devoted to measuring the thermodynamic control parameters (Crayford, 2004), as indicated in Figure 14. Preexpansion thermodynamic control parameters (pressure, temperature, and humidity) also have to be controlled, as well as ensuring thermodynamic homogeneity through-out the rig. Of particular note in Figure 14 is the relatively long stable period (from 400 ms onwards) of constant, quasi monodisperse droplet size, as measured by a laser diffraction technique, with vignetting errors minimized (Crayford, 2004), as well as stable thermodynamic variables. The stable time

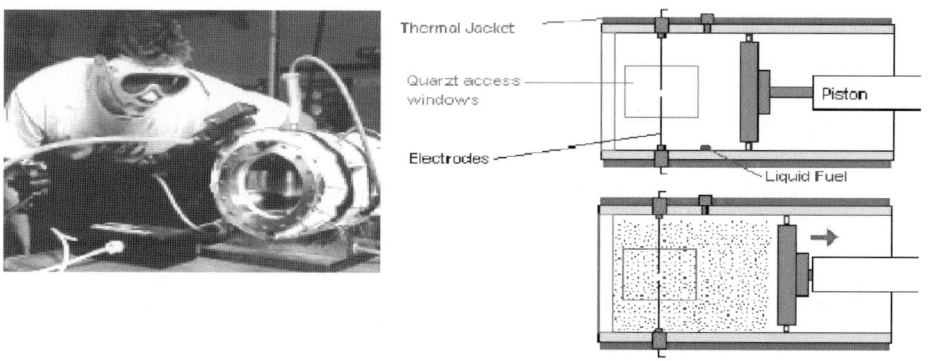

Figure 13. (a and b) Cardiff's optical cloud chamber and basic operational principle respectively

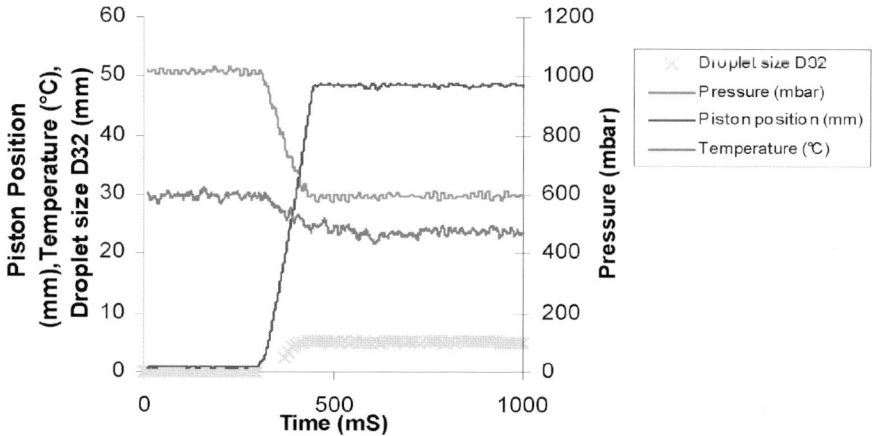

Figure 14. Transient thermodynamic and droplet size measurements during cloud chamber operation

window is more than sufficient to ignite and undertake burning rate measurements through Schlieren photography as indicated in Figure 15.

Notice the undulating, rugged appearance of the flamefront in Figure 15 under homogeneous, quiescent initial conditions. Droplet presence also appears to lead prematurely to highly wrinkled flames (final image) compared with the corresponding, single-phase vapor case.

Although a systematic parametric study varying at least primary variables of equivalence ratio and mean droplet size has yet to be undertaken, no evidence of the controversial enhanced theoretical burning rates has yet been observed in

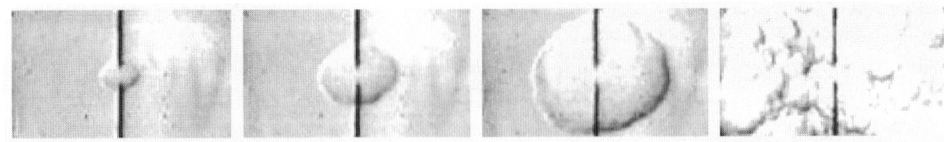

Figure 15. Flame propagation of droplet vapor air flame with natural two-phase "wrinkling" compared with the equivalent vapor case

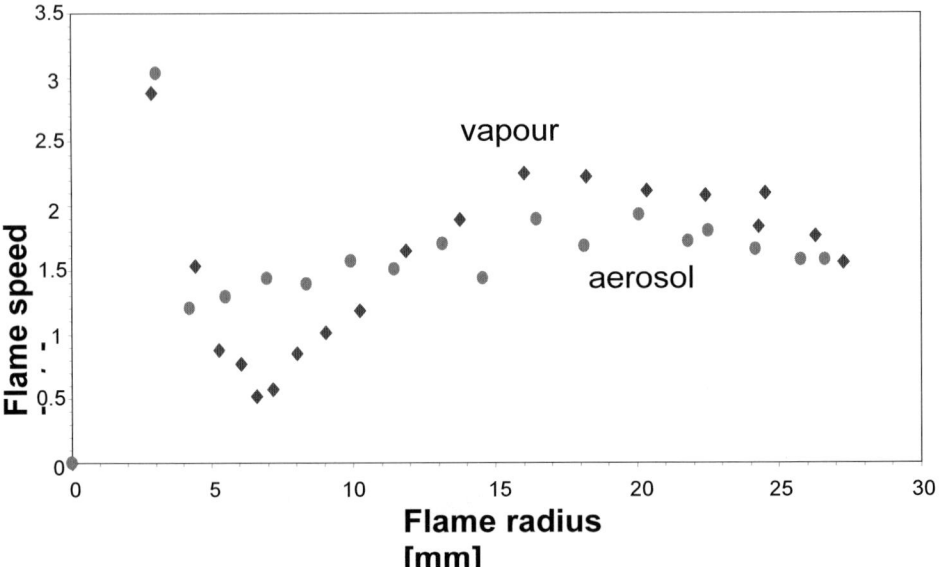

Figure 16. Preliminary measurements of fuel vapor and aerosol flamespeeds in the cloud combustor

the limited experimental programs undertaken on this facility to date, as indicated in the representative Figure 16.

3.2.2 Facility Development for Burning Rates of Alternative Fuels

New optical facilities are being commissioned at Cardiff to facilitate complementary burning rate data to the closed bomb approach offered by the cloud chamber, with the emphasis on higher initial temperatures and pressures, and alternative fuels and fuel mixtures. This will have relevance to the various power generation sectors (stationary and propulsion), and contribute to the challenges of environmental protection and pollutant reduction.

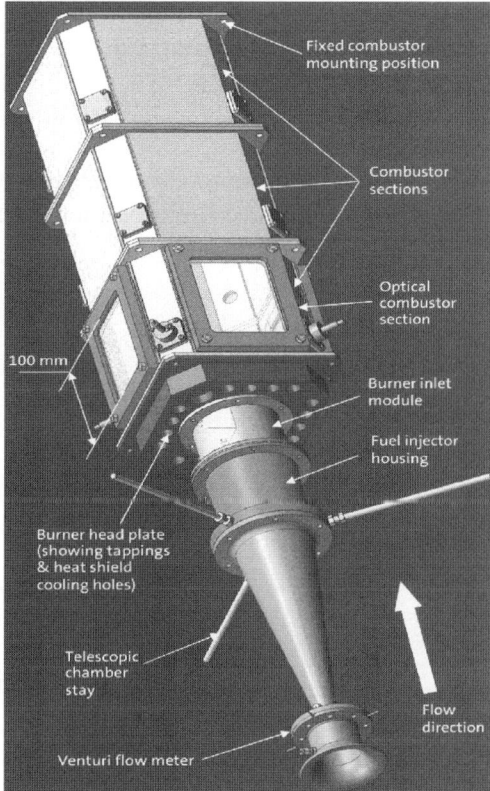

Figure 17. (a and b) High pressure/temperature optical facility for stationary Burner methods to measure burning rate of alternative fuels

3.3 PARTICULATE MATTER POLLUTANT

A lifecycle analysis of particulate matter (PM), with emphasis on ultra-fine PM has been undertaken in collaboration with industrial automotive collaboration to investigate characteristics of combustion-generated PM from both idealized (cloud chamber) and practical systems (Pooley et al., 2003). Diagnostic methodologies from environmental science have been applied to characterize particulate matter, utilizing thermophoretic sampling techniques, followed by electron microscopy. Figure 18(a) shows a vapor-shadowing technique utilized to generate three-dimensional information (and hence surface area estimates) for particulate matter from homogenous fully confined combustor studies. Figure 17(b) demonstrates the importance of defining the appropriate hazard index for particulate matter; with current legislation based on mass undersize indices, and current medical evidence indicating number count of ultra-fine PM is most hazardous to health, Figure 18b shows that conflicting conclusions results from using either of these indices on the same dataset.

3.4   FUEL/WATER MIST EXPLOSION PROTECTION SYSTEMS

As part of an environmental protection program supported by a petrochemical multinational, some of the aforementioned facilities and techniques have been

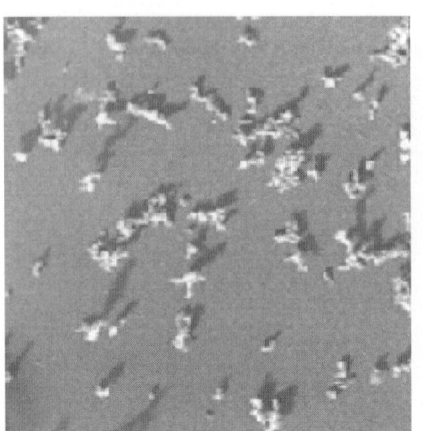
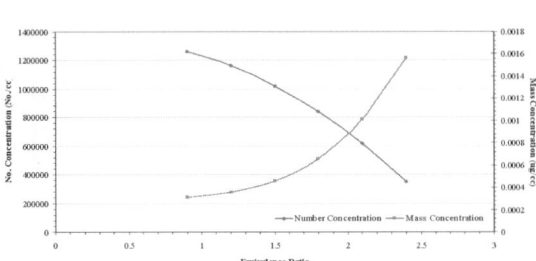

Figure 18. Particulate matter analysis from homogeneous combustor experiments. (a) 3D PM shadow imaging, (b) PM number/mass concentration

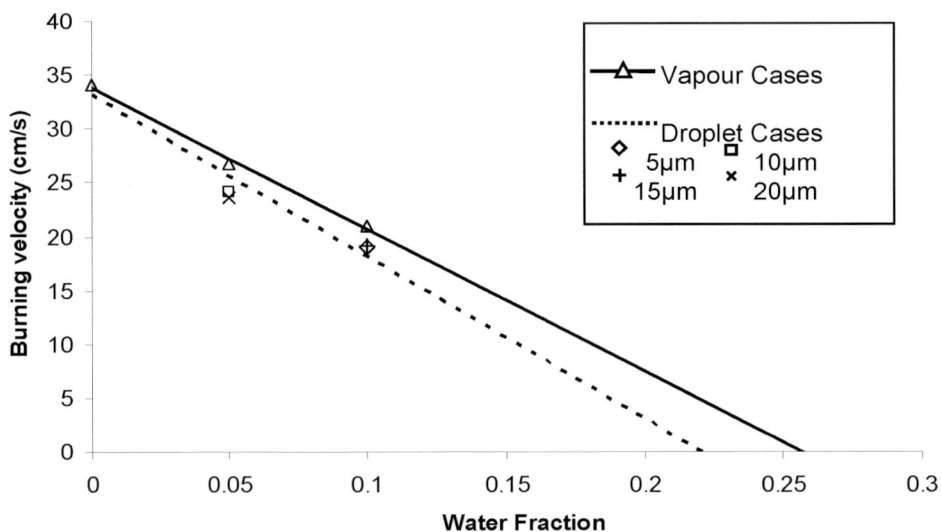

Figure 19. Influence of water vapor and water mist on laminar burning rate and predicted quench conditions

utilized to appraise the effectiveness of superheated water releases as explosion protection devices in place of Halon-based systems. Suitably superheated water releases (e.g., Figure 2) have been shown to generate large quantities of ultra-fine water mist, which is required for suppressing or quenching explosions.

The cloud chamber has also been utilized (Crayford, 2004) to investigate the laminar effect of water vapor and water mist on burning rate characteristics.

Figure 19 indicates the reduction in laminar burning rate as a function of water fraction and constituent mist droplet size, for methane–air mixtures due to the effect of water vapor and water mists (of size 20 μm or less). Extrapolation of the trendlines gives an indication of the water fraction required to quench laminar methane–air explosions, hence offering design guidance for environmentally benign explosion suppression systems.

4. Summary

A range of recent advances in two-phase thermofluid research extending both experimental and modeling capability in the areas of environmental protection and pollutant reduction have been presented. The advantages of cross-fertilization of ideas and methodologies across industries have been emphasized. With the current focus on energy and environmental issues, attention is very likely to increase in these areas, and with the recent improvements in research capability, a period of significant progress is anticipated.

References

Bowen P.J. and Shirvill L.C., 1994, Combustion hazards posed by the pressurised atomisation of High-Flashpoint Liquids, *J. Loss Prevention Process Ind.*, **7**(3): 233–241.

Bowen P.J., Phillips T.N. and Zheng Y., 2006, 'The Prediction of Droplet Deformation Using a Spectral ALE Method, *Submitted to Jnl. Comp. Phys.*

Comer M.A., 1999, Characterisation *of a Transient Gasoline Direct Injection Spray*, Ph.D. thesis, Cardiff University, Cardiff, UK.

Crayford A.P., 2004, *Mitigation of Explosions by Ultrafine WaterMists*, Ph.D. thesis, Cardiff University, Cardiff, UK.

Kay P.J., Bowen P.J., Gold M.R. and Sapsford S.M., 2006, Development of a 2D Quantitative LIF technique towards Measurement of Transient Liquid Fuel Films, *paper ID ICLASS06-181 accepted for presentation at International Conference on Liquid Atomisation and Spray Systems (ICLASS), 27 Aug.–1 Sept., Kyoto (Japan).*

Marakgos A. and Bowen P.J., 2002, Combustion hazards due to impingement of pressurised releases of high flashpoint liquid fuels, *Proceedings of the Combustion Institute,* 29: 305–311.

Pooley F.D., Bowen P.J. and Jones D.A., 2003, Collection and Characterisation of Ultrafine Combustion-Related Particles, *Session 2, paper 2.3, 7th Annual UK Review Meeting on Outdoor and Indoor Air Pollution Research, MRC Institute for Environment and Health, University of Leicester, 1–2 April.*

Stanglmaier, R.H., Li, J. and Matthews, R.D., 1999, The effect of in-cylinder wall wetting location on the HC emissions from SI engines. *SAE publications*, 1999-01-0502.

Wilson C., T., R., 1897, Condensation of water vapour in the presence of dust-free air and other gases, *Proceedings of the Royal Society*, London, pp. 265–307.

Witlox, H.W.M., Harper M., Bowen, P.J. and. Cleary V.M., 2005, Flashing liquid jets and two-phase droplet dispersion, I. Overview (literature survey, experiments and model implementation/ validation), *Proceedings of the 'Mary Kay O'Connor' Symposium, Houston, October 2005.*

CFD MODELING OF GAS RELEASE AND DISPERSION:

PREDICTION OF FLAMMABLE GAS CLOUDS

V. M. AGRANAT*, A. V. TCHOUVELEV, Z. CHENG
A.V. Tchouvelev & Associates Inc., 6591 Spinnaker Circle, Mississauga, Ontario, L5W 1R2, Canada

S. V. ZHUBRIN
Flowsolve Limited, 2nd Floor, 40 High Street, Wimbledon Village, London SW19 5AU, UK

Abstract. Advanced computational fluid dynamics (CFD) models of gas release and dispersion (GRAD) have been developed, tested, validated and applied to the modeling of various industrial real-life indoor and outdoor flammable gas (hydrogen, methane, etc.) release scenarios with complex geometries. The user-friendly GRAD CFD modeling tool has been designed as a customized module based on the commercial general-purpose CFD software, PHOENICS. Advanced CFD models available include the following: the dynamic boundary conditions, describing the transient gas release from a pressurized vessel, the calibrated outlet boundary conditions, the advanced turbulence models, the real gas law properties applied at high-pressure releases, the special output features and the adaptive grid refinement tools. One of the advanced turbulent models is the multifluid model (MFM) of turbulence, which enables to predict the stochastic properties of flammable gas clouds. The predictions of transient three-dimensional (3D) distributions of flammable gas concentrations have been validated using the comparisons with available experimental data. The validation matrix contains the enclosed and nonenclosed geometries, the subsonic and sonic release flow rates and the releases of various gases, e.g., hydrogen, helium, etc. GRAD CFD software is recommended for safety and environmental protection analyses. For example, it was applied to the hydrogen safety assessments including the analyses of hydrogen releases from pressure relief devices and the determination of clearance distances for venting of hydrogen storages. In particular, the dynamic behaviors of flammable gas clouds (with the gas

---

* To whom correspondence should be addressed. Vladimir Agranat, 6591 Spinnaker Circle, Mississauga, Ontario, L5W 1R2, Canada. e-mail: info@tchouvelev.org and acfda@sympatico.ca

N. Syred and A. Khalatov (eds.), Advanced Combustion and Aerothermal Technologies, 179–195.
© 2007 *Springer.*

concentrations between the lower flammability level (LFL) and the upper flammability level (UFL)) can be accurately predicted with the GRAD CFD modeling tool. Some examples of hydrogen cloud predictions are presented in the paper. CFD modeling of flammable gas clouds could be considered as a cost-effective and reliable tool for environmental assessments and design optimizations of combustion devices. The paper details the model features and provides currently available testing, validation and application cases relevant to the predictions of flammable gas dispersion scenarios. The significance of the results is discussed together with further steps required to extend and improve the models.

Keywords: computational fluid dynamics, numerical modeling tool, flammable gas cloud, gas release and dispersion, environmental protection and safety analyses, clearance distance

1. Introduction

In many industries, there are serious safety concerns related to the use of flammable gases in indoor and outdoor environments. It is very important to develop reliable methods of analyses of flammable gas release and dispersion (GRAD) in real-life complex geometry cases. Computational fluid dynamics (CFD) is considered as one of the promising cost-effective approaches in such analyses. The objective of this paper is to describe the advanced GRAD CFD models, which have been recently developed, tested, validated and applied to the modeling of various industrial indoor and outdoor scenarios of releases of flammable gases (hydrogen, methane, natural gas, etc.) in domains with complex geometries.

There are many general-purpose commercial CFD software packages capable of modeling and analyses of fluid flows and heat/mass transfer processes, e.g., the PHOENICS software[1]. However, none of these packages is properly customized for GRAD modeling and analyses of spatial and temporal behaviors of flammable gas clouds. In particular, any direct practical application of these codes to GRAD modeling requires a high level of user's expertise in CFD field due to the complexities of physical processes involved and mathematical models analyzed. Moreover, in the GRAD modeling, proper nonstandard settings are needed for transient boundary conditions, real gas properties, special numerical grid refinements and proper turbulence models. As a result, there is a practical need for developing a user-friendly and validated GRAD CFD modeling tool, which is capable of predicting the behaviors of flammable gas clouds.

Over the last 3 years, significant efforts have been undertaken by Stuart Energy Systems Corporation (SESC) and A.V. Tchouvelev & Associates Inc. in order to develop, test and validate a GRAD CFD modeling tool. Some results of this work have been recently published[2–8]. This paper reviews the previously published results, describes the modeling approach in more detail and provides currently available validation and application cases relevant to the predictions of flammable gas dispersion scenarios.

2. GRAD CFD Modeling Tool Capabilities

The GRAD CFD modeling tool has been designed as a customized module based on the commercial general-purpose CFD software, PHOENICS[1]. The modeling approach, the general governing equations and the additional sub-models are described in this section. Also, the similarity theory is described.

2.1 MODELING APPROACH

PHOENICS CFD software was selected as the flexible framework for performing GRAD CFD analyses, in which pragmatic flammable GRAD models were incorporated for practically affordable predictions using the PHOENICS solvers. PHOENICS is a well-recognized general-purpose CFD package that has been validated and successfully used around the world for more than 20 years. Its main features and capabilities have been described by its developers, CHAM Limited, in references item[1] and on the CHAM's web site, www. cham.co.uk. One of the key features of PHOENICS is its easy programmability, i.e., it enables a user to add user-defined submodels without a direct use of programming languages such as FORTRAN or C. This feature was used to incorporate the nonstandard advanced GRAD submodels described in Section 2.3.

There are three major stages in GRAD modeling: (1) steady state before-the-release run aimed at preparing the initial 3D distributions of pressure and velocity in the computational domain; (2) transient during-the-release run made to describe the spatial and temporal behaviors of flammable gas cloud during the gas release; and (3) transient after-the-release run aimed at predicting the dispersion of the released gas to acceptable levels within the computational domain. First, the modeling is performed under steady-state conditions without any flammable gas leak. The velocity and pressure profiles obtained from the steady-state calculations are then used as the initial conditions for the during-the-release transient simulations, which are performed with a flammable gas leak at the specified rate and time increments. After-the-release transient simulations predict the flammable gas dispersion in the computational domain below the required values of volume concentrations of flammable gas (usually below the Lower Flammability Level (LFL)). It should be noted that both the during-the-release and the after-the-release transient simulations allow for: (1) inclusion of the transient behavior of all calculated variables (pressure, gas density, velocity, and flammable gas concentration); (2) simulation of the movement of flammable gas clouds with time; and (3) evaluation of the safety by analyzing the iso-surfaces of the flammable gas concentration. The flammable gas convection, diffusion, buoyancy, and transience are modeled based on the general 3D conservation equations and the details of various release and dispersion scenarios are introduced via the proper initial and boundary conditions. One of the advantages of PHOENICS is that it contains various turbulence models and enables to select a proper model suitable for a particular practical case. In particular, the unique to PHOENICS turbulence models such as the LVEL model and the multifluid model (MFM) of turbulence were tested in the GRAD modeling.

2.2 GOVERNING EQUATIONS

The transient processes of flammable gas convection, diffusion and buoyancy are governed by the general conservation equations, i.e., the momentum equations, the continuity equation and the flammable gas mass conservation equation. These governing equations are well described in the PHOENICS documentation[1] and could be expressed as:

$$\frac{\partial(\rho u_i)}{\partial t} + \mathrm{div}(\rho U u_i - \rho v_{\mathrm{eff}} \mathrm{grad} u_i) = -\frac{\partial P}{\partial x_i} + \rho f_i, \quad i = 1,2,3 \tag{1}$$

$$\frac{\partial \rho}{\partial t} + \mathrm{div}(\rho U) = 0, \tag{2}$$

$$\frac{\partial(\rho C)}{\partial t} + \mathrm{div}(\rho U C - \rho D_{\mathrm{eff}} \mathrm{grad} C) = C'', \tag{3}$$

where x_1, x_2, and x_3 denote the Cartesian coordinates; u_1, u_2, and u_3 are the velocity components; U is the velocity vector; f_i ($i = 1,2,3$) is the body force component (per unit mass) in the x_i-direction; P is the gas mixture pressure; C is the mass concentration of flammable gas; C'' is the flammable gas source; D_{eff} is the effective flammable gas diffusion coefficient in air; v_{eff} is the effective kinematic viscosity of gas mixture and ρ is the gas mixture density, which is dependent on the flammable gas mass concentration, C, or the flammable gas volumetric concentration, α:

$$\rho = \frac{P}{[CR_{\text{gas}} + (1-C)R_{\text{air}}]T}, \quad \alpha = \frac{CR_{\text{gas}}}{CR_{\text{gas}} + (1-C)R_{\text{air}}} \qquad (4)$$

Here, T is the absolute temperature; and R_{gas} and R_{air} are the gas constants of flammable gas and air, respectively.

The volumetric buoyancy force, acting on the fluid particles in the x_3-direction (vertical direction), is represented by the term, ρf_3, in Eq. (1). Its significance is proportional to the difference between the local transient gas mixture density and the reference density of air under the ambient pressure and temperature. According to the first Eq. (4), the gas mixture density is calculated as an inverse-linear function of the local mass concentration of flammable gas, C, with the coefficients dependent on the gas constants of air and flammable gas and the local pressure and temperature. As a result, the significance of the buoyancy force depends on the transient 3D flammable gas mass concentration distribution.

The local values of effective viscosity and diffusion coefficient, v_{eff} and D_{eff}, include both laminar and turbulent components and are calculated according to the following equations:

$$v_{\text{eff}} = v_l + v_t, D_{\text{eff}} = v_l / \text{Pr}_l + v_t / \text{Pr}_t \qquad (5)$$

Here, subscripts l and t are applied to the laminar and turbulent properties, respectively; and Pr is the Prandtl/Schmidt number.

The laminar kinematic viscosity of gas mixture can be approximated by:

$$v_l = [\alpha \rho_{\text{gas}} v_{\text{gas}} + (1-\alpha)\rho_{\text{air}} v_{\text{air}}]/\rho \qquad (6)$$

Here, v_{gas} and v_{air} are the laminar kinematic viscosities of flammable gas and air, respectively; and ρ_{gas} and ρ_{air} are the densities of flammable gas and air, respectively.

A proper turbulence model was used in each particular practical case in order to calculate the local values of turbulent kinematic viscosity, v_t. Among models used were the LVEL model, the k–ε model, the modifications of k–ε model and the MFM.

2.3 ADVANCED MODEL FEATURES

A few advanced CFD submodels were developed as a part of GRAD CFD module. These submodels simulate the following features: the dynamic boundary conditions, describing the transient gas release from a pressurized vessel; the calibrated outlet boundary conditions; the real gas law properties applied at high-pressure releases; the advanced turbulence models; the adaptive grid refinement tools; and the special output features.

2.3.1 Dynamic Boundary Conditions

In general, the transient (dynamic) boundary conditions should be applied at the flammable gas release location in order to properly describe the released gas mass flow rate, which depends on time. Depending on the pressure in the gas storage tank, the regime of release could be subsonic or sonic (choked). Assuming the ideal gas law equation of state and a critical temperature at the leak orifice and solving the first-order ordinary differential equation for density, $\rho(t)$, the transient mass flow rate at the sonic regime of release could be approximated as[6]

$$\dot{m}(t) = -V\frac{d\rho}{dt} = \rho(t)u(t)A \approx \dot{m}_0 e^{-\frac{C_d A}{V}t\sqrt{\gamma(\frac{2}{\gamma+1})^{\frac{\gamma+1}{\gamma-1}}RT}},$$

$$\dot{m}_0 = C_d A\sqrt{\rho_0 P_0 \gamma(\frac{2}{\gamma+1})^{\frac{\gamma+1}{\gamma-1}}}$$

(7)

where $u(t)$ is the flammable gas velocity at the leak orifice; V is the tank volume; $\dot{m}_0$, ρ_0, and P_0 are the flammable gas mass flow rate, the gas density in the tank and the gas pressure in the tank, respectively, at $t = 0$; A is the leak orifice cross-sectional area; C_d is the discharge coefficient; and γ is the ratio of specific heats for flammable gas: $\gamma = C_p/C_v$, with C_p and C_v being the specific heat at constant pressure and constant volume, respectively. For example, for hydrogen, $\gamma = 1.41$ and the initial hydrogen mass release rate corresponding to the tank with a pressure of 400 bars and a ¼" leak orifice is about $\dot{m}_0 = 0.753$ kg/s, based on the second Eq. (7) with $C_d = 0.95$. It should be noted that the choked release lasts until the ratio of the pressure in the tank over the ambient pressure, namely, P_0/P_{atm} is greater than or equal to $(\frac{\gamma+1}{2})^{\frac{\gamma}{\gamma-1}}$ (it is about 1.90 for hydrogen).

2.3.2 Real Gas Law Properties

Under high pressure, flammable gases display gas properties different from the ideal gas law predictions. For example, at ambient temperature of 293.15 K and a pressure of 400 bars, the hydrogen density is about 25% lower than that predicted by the ideal gas law. In order to account for real gas law behavior, the

GRAD CFD module was provided with additional submodels[6]. In particular, for hydrogen release and dispersion modeling the Abel–Nobel equation of state (AN-EOS) was used to calculate the hydrogen compressibility, z_{H_2}, in terms of empirical hydrogen codensity, d_{H2}:

$$z_{H_2} = \frac{P}{\rho_{H_2} R_{H_2} T} = (1 - \frac{\rho_{H_2}}{d_{H_2}})^{-1}, \tag{8}$$

where ρ_{H_2}, P, T, and R_{H2} are the compressed hydrogen density, pressure, temperature and gas constant, respectively. It should be noted that the hydrogen compressibility, z_{H_2}, is equal to 1 for the ideal gas law. The hydrogen gas constant, R_{H2}, is 4,124 J/(kgK). The hydrogen codensity, d_{H2}, is about 0.0645 mol/cm^3, or 129 kg/m^3. Equation (8) can be simplified as:

$$z_{H_2} = 1 + \frac{P}{d_{H_2} R_{H_2} T} \tag{9}$$

The AN-EOS accounts for the finite volume occupied by the gas molecules, but it neglects the effects of intermolecular attraction or cohesion forces. It accurately predicts the high-pressure hydrogen density behavior[6].

2.3.3 Turbulence Model Settings

The turbulence models tested for GRAD modeling cases were as follows: LVEL model, k–ε model, k–ε RNG model, k–ε MMK model and MFM. It was found that the LVEL model performs better in releases of flammable gas in congested spaces (indoor environment containing the solid blockages) and the k–ε RNG model performs better for jet releases in open space. The details on sensitivity runs related to the turbulence model selection are described in previous papers[2–8]. MFM enables to predict the stochastic properties of flammable gas clouds by way of computing the probability density functions, which record for what proportion of time the fluid at a point in space is in a given state of motion, temperature and composition. However, the MFM approach needs to be further developed for GRAD CFD modeling, with the aim of finding a proper set of model constants and/or functions, which are suitable for the prediction of turbulent flammable gas dispersion in both indoor and outdoor environment.

2.3.4 Local Adaptive Grid Refinement

The local adaptive grid refinement (LAGR) techniques are needed in GRAD CFD modeling in order to accurately capture the flammable cloud behaviors near the release location and in the locations with significant gradients of flammable gas concentration while considering large domains of practical interest. This refinement should be based on the local features of flammable gas mass concentration as a key unknown variable. The iterative technique of LAGR was developed, implemented into the PHOENIS CFD software, tested and validated for the two GRAD CFD module validation cases, namely, the hydrogen release within a hallway, and the helium release within a garage with a car. The results of LAGR modeling were more accurate than the fixed grid solutions obtained with the standard grid refinement tools (see details in Sections 3.2 and 3.3). However, additional development work and testing are needed in order to use LAGR on regular basis for GRAD modeling.

2.3.5 Special Output Features

The dynamics and extents of flammable gas cloud, containing the gas volume concentrations between LFL and UFL, are of major interest in any GRAD modeling. The total volume of space occupied by this cloud and the total mass of flammable gas in the cloud are listed as the special output features. GRAD CFD module calculates these special output quantities as functions of time based on the transient 3D distributions of gas concentrations and gas mixture density.

2.4 SIMILARITY THEORY

The solutions of GRAD governing equations under the prescribed boundary conditions and properties depend on the following dimensionless parameters: the Reynolds number (Re), the Schmidt number (Sc), the Mach number (Ma), the Richardson number (Ri), and the density ratio (k_ρ), which are defined as follows to represent the turbulence, diffusion, compressibility, buoyancy, and density difference effects, respectively:

$$Re=\frac{U_{gas}L}{v_{gas}}, \quad Sc=\frac{v_{gas}}{D_{gas}}, \quad Ma=\frac{U_{gas}}{W}, \quad Ri=\frac{(\rho_{air}-\rho_{gas})gL}{\rho_{gas}U_{gas}^2}, \quad k_\rho=\frac{\rho_{air}}{\rho_{gas}} . \qquad (10)$$

Here U_{gas} is the flammable gas release velocity at the orifice; L is the orifice size; v_{gas} is the laminar kinematic viscosity of the released gas (1.05×10^{-4} m^2/s for hydrogen and 1.15×10^{-4} m^2/s for helium); D_{gas} is the laminar diffusion coefficient of the released gas in the air (6.1×10^{-5} m^2/s for hydrogen, and 5.7×10^{-5} m^2/s for helium); ρ_{air} is the reference density, i.e., the air density, which is 1.209 kg/m3 at 1 atm and 20°C; W is the gas sonic speed, which is equal to $W_{H_2} = \sqrt{\gamma_{H_2} R_{H_2} T} = 1305.61$ m/s for hydrogen and

$W_{He} = \sqrt{\gamma_{He} R_{He} T} = 1005.35$ m/s for helium; and k_{ρ} is the parameter characterizing the variable gas mixture density: $\rho = \rho_{air}(1+(k_{\rho}-1)C)^{-1}$ or $\rho = \rho_{air}(1+(k_{\rho}^{-1}-1)\alpha)$. It should be noted that L is the leak orifice diameter for the circular orifice. If the leak orifice is not circular, a hydraulic diameter, which is defined as $L = \dfrac{4 \times \text{cross-sectional area}}{\text{wetted perimeter}}$, is used for the scaling length. For a rectangular leak hole with sizes of a and b, the hydraulic diameter is defined as $L = \dfrac{2ab}{a+b}$.

In order to validate the CFD modeling results for hydrogen release and dispersion, proper experimental data on hydrogen release and dispersion are required. For reasons of safety, helium was often used in validation experiments as an alternative for hydrogen. However, helium and hydrogen differ in their buoyancy, turbulence, diffusion and density. This can be clearly seen from the following comparison of the dimensionless parameters (10) for these gases and the estimation of the distortions in flows of the two gases:

$$\alpha_{Re} = \frac{Re_{He}}{Re_{H_2}} = 0.91, \; \alpha_{Sc} = \frac{Sc_{He}}{Sc_{H_2}} = 1.17, \; \alpha_{Ma} = \frac{Ma_{He}}{Ma_{H_2}} = 1.30,$$

$$\alpha_{Ri} = \frac{Ri_{He}}{Ri_{H_2}} = 0.47, \; \alpha_{k_\rho} = \frac{k_{\rho,He}}{k_{\rho,H_2}} = 0.50$$

The large distortions result in significant differences in hydrogen and helium release processes: helium is less "turbulent" and "buoyant" but more "compressible" than hydrogen. The hydrogen buoyancy and turbulence effects would be underestimated if helium were used for validation of hydrogen modeling. The choked (sonic) release velocity would be smaller and, as a result, the compressibility would be overestimated as well. Therefore, hydrogen, though combustible, has to be used for the validation of CFD modeling of hydrogen releases and dispersion. Some validation results are reported in the following section.

3. GRAD CFD Software Validation

The GRAD CFD modeling software needs to be validated before it can be widely applied to industrial projects. The predictions of transient 3D distributions of flammable gas concentrations with the GRAD CFD module were validated using the comparisons with available experimental data on GRAD.

3.1 VALIDATION MATRIX

The validation matrix contains the enclosed and nonenclosed geometries, the subsonic and sonic release flow rates and the releases of various gases, i.e., hydrogen, helium, etc. The validation matrix and some validation cases are described in this paper. Seven validation scenarios were selected to cover different industrial release environments and leak types. Table 1 shows the validation matrix, classified by the experiment conditions, such as leak types, release directions and domain types, etc. Seven scenarios covered the leaks from small subsonic releases to large choked releases. The validation work on the wide range of the Reynolds numbers ($50 < Re < 10^7$), the Mach numbers ($0 \leq Ma \leq 1$) and the Richardson numbers ($10^{-5} < Ri < 10^4$) helped validate and calibrate the CFD models and find the suitable settings for the coefficients used in the boundary conditions and the turbulence models for the GRAD modeling.

TABLE 1. GRAD CFD module validation scenarios

Case No.	Case name	Description of experiment				CFD Model	Data source reference
		Domain	Leak direction	Leak type	Experimental data		
1	Helium jet	Open	Vertical	Subsonic, helium release	Steady-state, velocities, concentrations and turbulence intensities	Incompressible, steady state	Reference[11]
2	H$_2$ jet		Horizontal	Subsonic, H$_2$ release	Transient, concentrations	Incompressible, transient	Reference[13]
3	INERIS Jet			Choked, H$_2$ release	Steady-state, concentrations	Compressible, steady state	Reference[14]
4	Hallway end	Semi-enclosed	Vertical	Subsonic, H$_2$ release	Transient, concentrations	Incompressible, transient and steady state	Reference[9]
5	Hallway middle			Subsonic, helium release	Transient, concentrations		
6	Garage with a car			Subsonic, H$_2$ and helium releases	Transient, concentrations		Reference[10]
7	H$_2$ vessel	Enclosed		Subsonic, H$_2$ release and dispersion	Transient, concentrations during dispersion	Incompressible, transient	Reference[15]

3.2   HYDROGEN SUBSONIC RELASE IN A HALLWAY

An example of GRAD CFD validation work was described in detail in the earlier paper[2]. This work was conducted by SESC using the experimental and numerical data[9] published by Dr. M.R. Swain et al. Below is a brief description of this validation work.

A hydrogen release benchmark problem with a simple geometry was used for CFD model validation in this case. In particular, in this scenario (see Figure 1), the hydrogen was released at the rate of 2 SCFM (standard cubic feet per min) from the floor at the left end of a hallway with the dimension of $114 \times 29 \times 48$ in. ($2.9 \times 0.74 \times 1.22$ m^3). At the right end of the hallway, there were a roof vent and a lower door vent for the gas ventilation. Four sensors were placed in the domain to record the local hydrogen concentration variations with time. Figure 1 shows the geometry and the numerical results obtained, i.e., the 3% hydrogen volume concentration iso-surface at 1 min after the start of hydrogen release. The initial grid used was a coarse grid of $36 \times 10 \times 18$ cells. It was found that the concentration differences between the predictions and the measurements were about 20% for sensors 1 and 2 and 10% for sensors 3 and 4.

LAGR was applied to the modeling of hydrogen release in a hallway as illustrated by Figure 1. Table 2, comparing the predicted and the measured hydrogen volume concentrations, confirms that LAGR improves the accuracy of the simulations, both quantitatively and qualitatively. In fact, the simulation on the initial coarse grid of $36 \times 10 \times 18$ fails to predict the increase of concentration at the position of Sensor 4 relative to that of the Sensor 1. This flow feature, however, is realistically captured on the grid with LAGR.

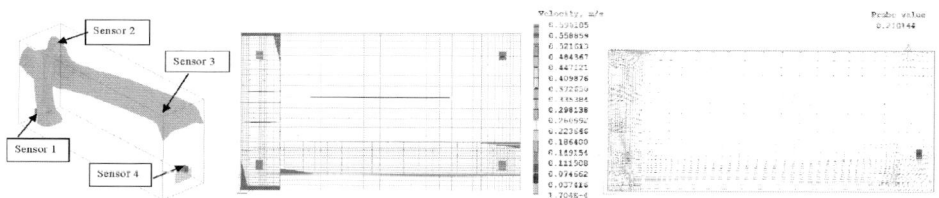

Figure 1. 2-SCFM hydrogen release: hydrogen sensors and predicted 3% hydrogen volume concentration iso-surface at 1 min (*left*), embedded locally refined adapted grid (*middle*) and velocity distribution on adapted grid (*right*)

TABLE 2. Steady-state results for hydrogen release in a hallway (k–ε MMK turbulence model)

Simulations/experiment	Sensor 1 (%)	Sensor 2(%)	Sensor 3(%)	Sensor 4(%)
Experimental observations	1.35	4.90	4.95	1.80
Initial coarse grid, $36 \times 10 \times 18$	1.54	5.58	5.67	1.42
Adaptive refined, $36 \times 20 \times 23$	1.34	5.68	5.77	1.70

3.3 HELIUM SUBSONIC RELEASE IN A GARAGE WITH A CAR

Another GRAD CFD module validation work was conducted using the experimental and numerical data published by Dr. M.R. Swain et al.[10] on the helium subsonic release in a garage with a car. Figure 2 shows the geometry of the case considered. The four small cubes mark the locations of four helium sensors in the domain.

Figure 2. Geometry and helium sensors for helium subsonic release in a garage

LAGR was also applied to this modeling case. Table 3 shows that LAGR helps reduce the predicted concentrations at the locations of sensors 1 and 4 significantly. The predicted results are in accord with the CFD simulations reported elsewhere.

TABLE 3. Steady-state results for helium release in a garage with a car (LVEL turbulence model)

Simulations	Sensor 1 (%)	Sensor 2 (%)	Sensor 3 (%)	Sensor 4 (%)
Swain's CFD results	0.5	2.55	2.55	1.0
Initial coarse grid, 32 × 16 × 16	1.92	2.53	2.52	1.94
Adaptive refined, 39 × 26 × 24	0.98	2.66	2.62	1.08
Adaptive refined, 58 × 26 × 27	0.79	2.70	2.67	1.01

3.4 HELIUM TURBULENT SUBSONIC JET

Another example of GRAD CFD module validation work was described in the reference paper[5]. Below is the brief description of the major findings. In this validation work, a vertical helium jet reported by Panchapakesan and Lumley[11] was simulated using the GRAD CFD module. The real geometry was simplified by a 2D axisymmetric computational domain to save the computational resources. The mixed gas was assumed to have incompressible gas properties so the inverse linear function was used to calculate the mixture density dependent on the local helium mass concentration and the helium and air densities. The k–ε

RNG turbulence model was used while solving the governing equations to predict the velocity and mass/volumetric concentration profiles. The numerical results showed a good agreement with experimental data in both radial and axial directions with the errors less than 10%. The simulation results were also compared with other published helium experimental data obtained by Keagy and Weller, Way and Libby, Aihara et al. and the correlations made by Chen and Rodi[12] for velocity and concentration. The satisfactory agreement (within 10%) between the experimental and numerical data in the three jet regions proved that the GRAD CFD model is robust, accurate and reliable, and that the CFD technique can be used as an alternative to the experiments with similar helium jets. It also indicated that the CFD model can accurately predict similar hydrogen releases and dispersion if the model is properly calibrated with hydrogen coefficients when applying to hydrogen jets.

4. GRAD CFD Software Applications

CFD modeling of flammable gas clouds could be considered as a cost-effective and reliable tool for environmental assessments and design optimizations of combustion devices. In particular, the GRAD CFD software is recommended for safety and environmental protection analyses. The transient behaviors of flammable gas clouds can be accurately predicted with this modeling tool. For example, it was applied to the hydrogen safety assessments including the analyses of hydrogen releases from pressure relief devices (PRD) and the determination of clearance distances for venting of hydrogen storages[2–8]. An example of hydrogen cloud predictions is presented below.

4.1 RELEASE IN A HYDROGEN GENERATOR ROOM

This section discusses one of the potential hydrogen release scenarios – a hydrogen release into the electrolytic hydrogen generator room during self-purging start-up procedure[3]. At start-up, to ensure only high-purity gas is directed for compression, hydrogen is being vented for 10 min. After 10 min, a regulator redirects hydrogen flow from vent to process. The point of potential release is the vent pipe at the roof of the hydrogen generator. The outlet pipe size is 2 and the constant release flow rate is 0.0035 Nm^3/s. First, the CFD modeling was performed under steady-state conditions without any hydrogen leak. The velocity profiles obtained from the steady state were then used as the initial conditions for the during-the-release simulations, which were performed with a hydrogen leak at the specified rate and time increments. After-the-release simulations predicted the hydrogen dispersion in the room below 10% of the LFL.

4.1.1 Before-the-Release Simulation

The existence of a louver and an exhaust fan (flow rate of 1 m³/s) creates a steady-state 3D airflow in the generator room. This flow was simulated first, before trying to simulate the transient 3D behavior of hydrogen cloud introduced by the hydrogen release. Figure 3 shows the steady-state air velocities created by the louver and the exhaust fan.

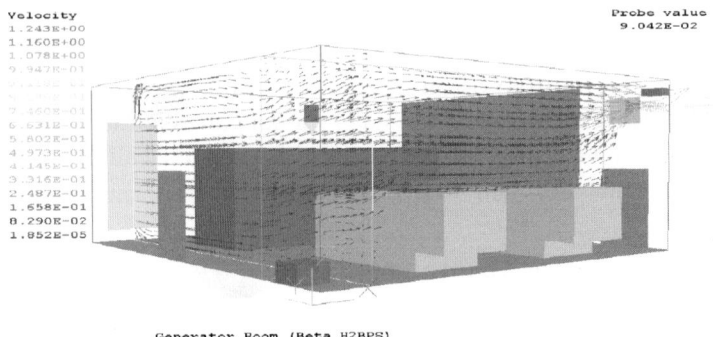

Figure 3. Air velocities at *x*- and *y*-planes before the hydrogen release in the hydrogen generator room

4.1.2 During-the-Release Simulation: Release from Hydrogen Vent Line

The hydrogen release scenario considers the worst case scenario when, for whatever reason, during the hydrogen generator start-up self-purging procedure the hydrogen vent line on the roof of the generator comes off, thus causing all hydrogen being produced during the self-purging procedure (10 min) to leak into the hydrogen generator room. It is also assumed that all hydrogen sensors intended to shut down the generator during the self-purging procedure are disabled. Room ventilation is provided by the louver and the exhaust fan during the release. CFD predictions of 3D hydrogen concentration distribution are shown in Figure 4, which shows the hydrogen LFL (4% vol.) iso-surface at the end of the release (10 min). It is seen that the size of the cloud is very small in comparison to the size of the room.

4.1.3 Size of Flammable Gas Cloud

The size of the flammable cloud was calculated, using the advanced GRAD CFD settings. Three global quantities, DOMV, V4H and V2H were defined as the volume (in m³) of the whole domain (DOMV), the volume of the hydrogen

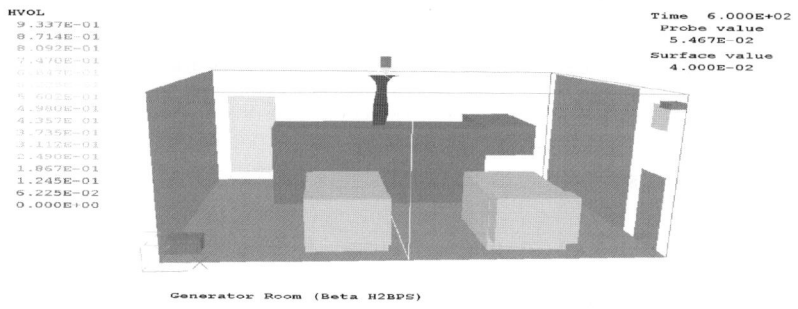

HVOL
9.337E-01
8.714E-01
8.092E-01
7.470E-01

5.605E-01
4.980E-01
4.357E-01
3.735E-01
3.112E-01
2.490E-01
1.867E-01
1.245E-01
6.225E-02
0.000E+00

Time 6.000E+02
Probe value
5.467E-02
Surface value
4.000E-02

Generator Room (Beta H2BPS)

Figure 4. End of 10 min release from the hydrogen vent line: LFL hydrogen cloud

cloud with more than 4% volume concentration (V4H) and the volume of the hydrogen cloud with more than 2% volume concentration (V2H), respectively. The printout from the global calculations file written after the CFD run was as follows: DOMV = 229.95, V4H = 8.072×10^{-2}, V2H = 6.225. It could be seen that the 4% hydrogen cloud volume (V4H), which is about 0.081 m^3, is much smaller than the volume of cloud with 2% volume concentration (V2H), which is about 6.225 m^3. Both clouds are much smaller in volume than the whole domain volume (DOMV), which is about 230 m^3. These findings are significant for understanding of safety of the system considered.

5. Conclusions

Advanced GRAD CFD models have been developed, tested, validated and applied to the modeling of various industrial real-life indoor and outdoor flammable gas (hydrogen, methane, etc.) release scenarios with complex geometries. The models developed include the following options: the dynamic boundary conditions, describing the transient gas release from a pressurized vessel, the calibrated outlet boundary conditions, the advanced turbulence models, the real gas law properties applied at high-pressure releases, the special output features and the adaptive grid refinement tools. The user-friendly GRAD CFD modeling tool has been designed as a customized module based on the commercial general-purpose CFD software, PHOENICS. The predictions of transient 3D distributions of flammable gas concentrations have been validated using the comparisons with available experimental data. GRAD CFD software is recommended for safety and environmental protection analyses. In particular, the dynamic behaviors of flammable gas clouds can be accurately predicted with this modeling tool for environmental assessments and design optimizations of combustion devices.

Acknowledgments

The authors would like to thank Natural Resources Canada (NRCan) and Natural Sciences and Engineering Research Council of Canada (NSERC) for their contributions to funding of this work.

References

1. PHOENICS Hard-copy Documentation (Version 3.5), Concentration, Heat and Momentum Limited, London, UK (September 2002).
2. V. Agranat, Z. Cheng, and A. Tchouvelev, CFD modeling of hydrogen releases and dispersion in hydrogen energy station, *Proceeding of the 15th World Hydrogen Energy Conference*, Yokohama, Japan (June 2004).
3. A. Tchouvelev, G. Howard, and V. Agranat, Comparison of standards requirements with CFD simulations for determining sizes of hazardous locations in hydrogen energy station, *Proceeding of the 15th World Hydrogen Energy Conference*, Yokohama, Japan (June 2004).
4. G.W. Howard, A.V. Tchouvelev, Z. Cheng, and V.M. Agranat, Defining hazardous zones – electrical classification distances, *Proceeding of the 1st International Conference on Hydrogen Safety*, Pisa, Italy (September 2005).
5. Z. Cheng, V.M. Agranat, and A.V. Tchouvelev, Vertical turbulent buoyant helium jet – CFD modeling and validation, *Proceeding of the 1st International Conference on Hydrogen Safety*, Pisa, Italy (September 2005).
6. Z. Cheng, V.M. Agranat, A.V. Tchouvelev, W. Houf, and S.V. Zhubrin, PRD hydrogen release and dispersion, a comparison of CFD results obtained from using ideal and real gas law properties, *Proceeding of the 1st International Conference on Hydrogen Safety*, Pisa, Italy (September 2005).
7. A.V. Tchouvelev, P. Benard, V. Agranat, and Z. Cheng, Determination of clearance distances for venting of hydrogen storage, *Proceeding of the 1st International Conference on Hydrogen Safety*, Pisa, Italy (September 2005).
8. Z. Cheng, V.M. Agranat, A.V. Tchouvelev, and S.V. Zhubrin, Effectiveness of small barriers as means to reduce clearance distances, *Proceeding of the 2nd European Hydrogen Energy Conference*, Zaragoza, Spain (November 2005).
9. M.R. Swain, E.S. Grilliot, and M.N. Swain, Risks incurred by hydrogen escaping from containers and conduits. NREL/CP-570-25315, *Proceedings of the 1998 US DOE Hydrogen Program Review*.
10. M.R. Swain, J.A. Schriber, and M.N. Swain, Addendum to hydrogen vehicle safety report: residential garage safety assessment. Part II: risks in indoor vehicle storage final report.

11. N.R. Panchapakesan, and J.L. Lumley, Turbulence measurements in axisymmetric jets of air and helium. Part 2. Helium jet, *J. Fluid Mech.*, **246**, 225–247 (1993).
12. C.J. Chen and W. Rodi, Vertical Turbulent Buoyant Jets – A review of Experimental Data, in: *The Science and Application of Heat and Mass Transfer*. Pergamon Press, 1980.
13. M.R. Swain, Hydrogen Properties Testing and Verification (June 17, 2004).
14. E. Ruffin, Y. Mouilleau, and J. Chaineaux, Large scale characterisation of the concentration field to supercritical jets of hydrogen and methane, *J. Loss Prev. Process Industry*, **9**(4), 279–284 (1996).
15. Y.N. Shebeko, V.D. Keller, O.Y. Yeremenko, I.M. Smolin, M.A. Serkin, and A.Y. Korolchenko, Regularities of formation and combustion of local hydrogen–air mixtures in a large volume, *Chemical Industry*, **21**, 24 (728)–27 (731) (1988) (in Russian).

TURBULENT COMBUSTION AND THERMAL RADIATION

IN A MASSIVE FIRE

A. YU. SNEGIREV*
*Saint-Petersburg State Polytechnic University,
Polytechnicheskaya 29, St.-Petersburg, 195251, Russia*

S. A. ISAEV
*Saint-Petersburg State Academy of Civil Aviation, Pilotov 38,
St.-Petersburg, 196210, Russia*

Abstract. This study was motivated by an actual large-scale fire of combustibles in open storage, where the fire growth and flame dynamics were greatly affected by the cross-wind. The objectives include development of the model and computer code for studying buoyant turbulent diffusion flames of large fires exposed to cross-winds in the open atmosphere, computer simulations of coherent flow structures in the wind-blown fire plume and of radiative heat fluxes both inside the flame and incident to remote targets. In the developed model, the large eddy simulation (LES) technique is applied for modeling the turbulent flame and thermal plume. In combustion modeling, the presumed probability density function approach is used, within the framework of fixture fraction formulation. Thermal radiation transfer is modeled using the Monte Carlo method, adopting either the weighted sum of gray gases (WSGG) approach or the gray media assumption. Coherent flow vortical structures in the fire plume and the radiative impact of the flame are analyzed through computer simulations and compared with the available experimental data. In special cases, the predicted radiative heat fluxes incident to both targets engulfed in fire and remote surfaces have been found to be in reasonable agreement with the empirical correlations and simple engineering approaches.

Keywords: turbulent diffusion flames, thermal radiation, fire modeling

* To whom correspondence should be addressed. Alexander Yu. Snegirev, Saint Petersburg State Polytechnic University, Polytechnicheskaya 29, St.-Petersburg, 195251, Russia. e-mail: a.snegirev@as1810. spb.edu

N. Syred and A. Khalatov (eds.), Advanced Combustion and Aerothermal Technologies, 197-209.

1. Introduction

The vast amounts of combustibles on industrial sites and in large-scale storage increase the risk of large-scale fire development. As well as the potential for material damage and the direct hazard to occupants posed by industrial fires, they can also result in the dispersion of toxic pollutants in the atmosphere. Over the years, the practical significance of *controlled* combustion in industrial applications has fed and inspired intensive long-standing research that has produced the background for studies of *uncontrolled* combustion occurring in fires and explosions. The use of mathematical modeling is of particular importance to the investigation of the dynamics and consequences of large-scale fires. The fundamental challenge in fire modeling is the need for adequate simultaneous consideration of three physical phenomena – turbulence, combustion, and thermal radiation. All three are all closely related, directly influencing each other as well as governing the dynamics of turbulent radiating flames[1-4].

This study was motivated by an actual large-scale fire of combustibles in open storage, where the fire growth and flame dynamics were greatly affected by the cross-wind. Intense heat radiation caused the ignition of stacks adjacent to the fire origin, damage of surrounding buildings and vehicles, and prevented fire-fighters from getting close to the fire. The accident investigation highlighted the deficiency of existing regulations and codes for determining safe separating distances (in particular, the effects of cross-wind are currently not taken into account). Hence, the role of computer modeling has increased accordingly. It is worth noting that worldwide, computer modeling is increasingly becoming the key component of performance based design in fire and explosion safety along with (and sometimes instead of) the prescriptive codes.

The objectives of this study include the development of the model and computer code for studying buoyant turbulent diffusion flames of large fires exposed to cross-winds in the open atmosphere, computer simulations of coherent flow structures in the wind-blown fire plume and of radiative heat fluxes both inside the flame and incident to remote targets.

A vast amount of experience has been gained from modeling turbulent radiating flames, mainly based on *jet flames* used in industrial combustors and jet engines. *Buoyant flames* developing in fires are known to be less ordered and more susceptible to large-scale hydrodynamic instability. This results in the formation of large-scale coherent vortical structures (flame "puffing") and makes it difficult to use traditional RANS turbulence models effectively. In this study, the approach currently making its way into engineering practice – large eddy simulation (LES) – is applied. In combustion modeling, we use the presumed probability density function approach, within the framework of fixture fraction formulation. Thermal radiation transfer is modeled using the Monte Carlo method adopting either the weighted sum of gray gases (WSGG) approach or the gray media assumption. Coherent flow vortical structures in the fire plume and the radiative impact of the flame are analyzed through computer simulations and compared with the available experimental data. In special cases, the predicted radiative heat fluxes incident to both targets engulfed in fire and remote surfaces have been found to be in reasonable agreement with the empirical correlations and simple engineering approaches.

Model description, numerical procedures, and validation studies can be found elsewhere[5–9]. Here, the results of computer modeling of turbulent flow, combustion and thermal radiation in a massive open fire are presented.

2. Modeling results

Snegriev and Yu[5] previously presented RANS simulations of pool fires exposed to cross-winds. Although the use of the RANS model appeared to be useful, finer details of the flow structure required use of the LES approach. In LES, local isotropy of subgrid scale (SGS) fluctuations is assumed in modeling SGS stresses and fluxes, which implies that the smallest resolved flow structures should correspond to the inertial subrange of the energy spectrum. The latter requirement is met, for example, in large-scale free flows with very high Reynolds numbers, and this is why for a long time the LES methodology has been applied with particular success in large-scale environmental flows. High Reynolds number flows also evolve in industrial applications such as jet turbulent diffusion flames in combustors, in which the use of LES is gradually becoming a part of engineering practice, as demonstrated, for example, by Mare and Jones[10] among many others. In those forced flows, *far field* region (tens of nozzle diameters downstream) is of interest, where turbulence is fully developed. This is not the case in buoyant flows occurring in fires, where *near-field* flow adjacent to the fuel bed and containing the flame must be predicted. In this region, transition to fully developed and isotropic turbulence may not be complete. In addition, the inertial subrange of subgrid-scale fluctuations is not established. Hence, the proper resolution of large-scale hydrodynamic instability (producing large vortical structures) is very important. Furthermore, unlike laboratory jet flows, detailed quantitative data of large fires are scarce and often not available for model validation. In this study, the means of validation of the numerical solutions include qualitative analysis of the flow: major vortical structures observed in the experiments should be resolved in computations. Accordingly, the LES predictions of transient coherent flow vortical structures are demonstrated in this work.

A range of approaches are available to model buoyant turbulent diffusion flames in cross-winds. Integral models with simple entrainment relationships are used in engineering evaluations[11]. Most published CFD studies[5,12] have been undertaken using RANS turbulence models (e.g.,[5,12]). Recent developments in computer technologies have allowed direct numerical simulations of wind-blown nonreactive isothermal jets in incompressible media[13] to be undertaken. However, LES studies[14,15] are still rare especially when the flow is modeled jointly with turbulent combustion *and* thermal radiation.

Snegriev et al.[6] conducted an LES study of large open wind-blown flames representing steady burning of a rectangular area with horizontal dimensions of 40×15 m^2 (fuel surface area $A_{\text{fuel}} = 600$ m^2). Figure 1 presents a prototype of a single pallet stack of combustible materials as investigated in this study[6]. The stack is fully engulfed by flame, with a total burning rate of 40 kg fuel per second, which corresponds to a burning rate per unit area of horizontal cross section, $\dot{m}''_{\text{fuel}}$, of 0.067 kg/m^{2}s and a total heat release, $\dot{Q}$, of 664 MW. The fire is affected by a cross-wind directed parallel to the longer side of the stack and modeled by its vertical velocity profile $u(z) = V_{\text{wind}}\left(1 - \exp\left(-z/z_{bl}\right)\right)$, where the wind velocities are $V_{\text{wind}} = 0, 2, 4,$ and 8 m/s, and the height of near-surface boundary layer is $z_{bl} = 2$ m.

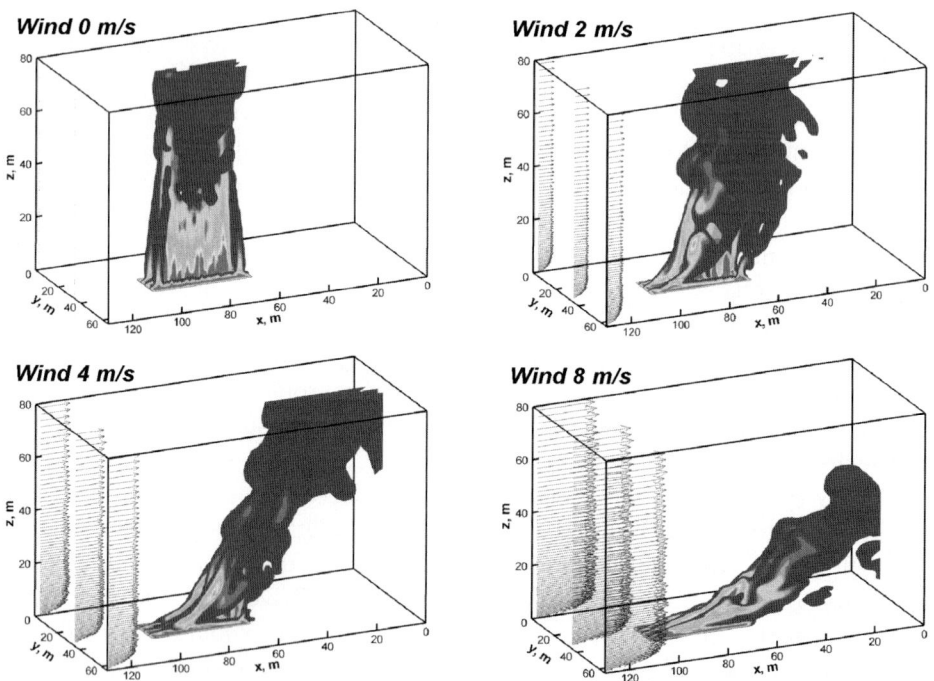

Figure 1. The wind-blown fire plume and computational domain ($V_{\text{wind}} = 0, 2, 4,$ and 8 m/s)

The computational grid used in the simulations of wind-blown flames contains $96 \times 72 \times 72$ control volumes (497,664 in total) covering the domain of 130 (length) $\times$ 65 (width) $\times$ 80 (height) m^3. The grid is uniform in horizontal directions above the fuel bed and the control volumes are weakly stretched towards the open boundaries of computational domain (the size ratio of the adjacent control volumes does not exceed 1.09, minimum grid size is $0.833 \times 0.625 \times 0.722$ m^3).

The numerical solution obtained in LES is a fluctuating realization of random filtered fields. The grid resolution and the accuracy of the numerical procedure must ensure that at least some (low frequency) part of the inertial

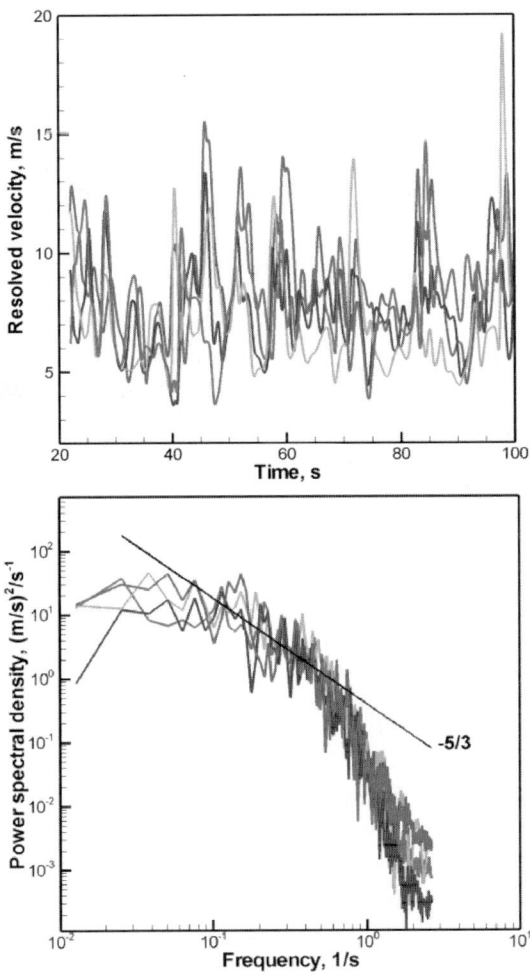

Figure 2. Velocity variation and power spectra in four different points of the flow

subrange of the fluctuation spectrum is resolved. This can be tested by comparing the power spectrum of the resolved velocity fluctuations with the inertial subrange of the Kolmogorov spectrum, which is known to decay according to "−5/3 law"[3]. Such a comparison is shown in Figure 2, where velocity time histories (Figure 2a) and its spectra (Figure 2b) are demonstrated for four points inside the plume. It can be seen that within the certain frequency range, the power spectrum does indeed obey the "−5/3 law", which indicates that the largest scales of the inertial subrange are resolved in this turbulent flow.

At this point, additional comments should be made. First, the number of required grid points in each direction was analytically estimated by Hartel[20] (using the model spectrum) to be of order of 80. The grid used in this work can therefore be regarded as the minimum acceptable limit for a quality LES. Second, the above considerations do not take into account the resolution requirements for the near-surface boundary layer. The near-field flow (several characteristic sizes of the burning surface) and its stability are likely to be sensitive to the vorticity produced near the ground surface. This could be more pronounced when the surface is of an irregular shape as is generally the case in realistic flows above obstacles.

The fire studied can be characterized by a dimensionless heat release rate $Q* = \dot{Q}/\left(\rho_{air}C_{Pair}T_{air}\sqrt{gD_{eff}}D_{eff}\right) = 0.15$, where $D_{eff} = 27.6$ m is the characteristic size of the fuel bed. The expected flame length L_f in stagnant atmosphere, calculated using the Heskestadt empirical relationship[2], is 22 m, see Eq. (1).

$$L_f = 0.235\dot{Q}^{2/5} - 1.02D_{eff}, \qquad (\dot{Q} \text{ is in kW}) \qquad (1)$$

Nondimensional wind velocities[2], $V_{wind}^* = V_{wind}/\left(g\dot{m}_{fuel}D_{eff}/\rho_{fuel}\right)^{1/3} = 0.625$, 1.25 and 2.5, respectively for the above values of V_{wind}. Application of the empirical formula[2] for the flame tilt angle θ (angle between the flame and the vertical) yields $\theta = 27°$ and $51°$ for wind velocities of 4 and 8 m/s, see Eq. (2).

$$\cos\theta = \min\left(1, 1/\sqrt{V_{wind}^*}\right) \qquad (2)$$

For statistically steady flames and plumes affected by the cross-winds of different velocity, numerically predicted tilt angles were compared to those calculated from the empirical correlation by the American Gas Association[2]. As shown in Figure 3, reasonably good agreement has been obtained. This substantiates applicability of the model for the case studied, in which flow vortical structure is analyzed below.

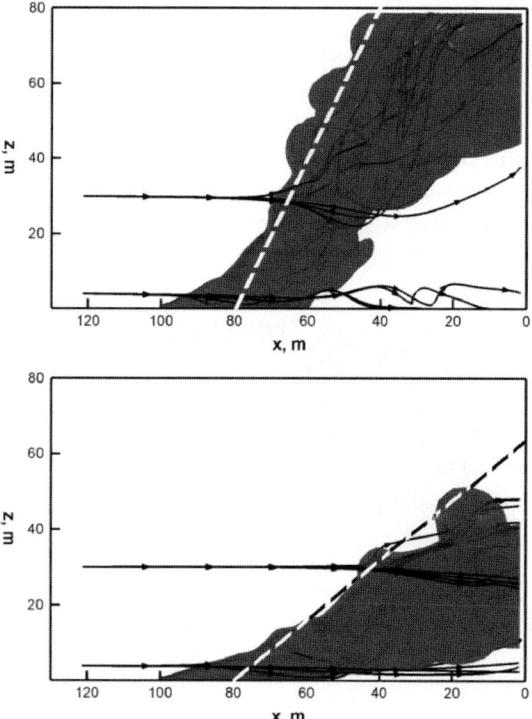

Figure 3. Temperature iso-surface 310 K and flow streamlines. Dashed lines correspond to the tilt angle calculated from the empirical correlation by the American Gas Association (see[2]). (a) V_{wind} = 4 m/s, (b) V_{wind} = 8 m/s

Jet flows and flames in a cross-wind present a highly important configuration carefully studied in a number of works. At least four different types of coherent vortical structures have been identified in previous experimental and numerical studies of jets in cross-flow[15,16]. Similarly, four distinct types of coherent vortical structures have been observed in our simulations:

1. Rolling up shear-layer ("hanging"[15]) vortices, developing as a result of the Kelvin-Helmholtz instability

2. Edge vortices, incipient at the windward corners of the fuel surface and located near the ground surface

3. Counterrotating vortex pair

4. Wake vortices

These vortical structures are designated accordingly by numbers 1–4 in Figures 4 and , where the case of cross-wind velocity of 4 m/s is demonstrated.

The first and third types of vortexes are very similar to those observed in experimental studies of jets in a cross-flow, e.g., Fric and Roshko[16]. Rolling up shear-layer vortices appear to be of the type that also develop in "puffing" buoyant flames in a stagnant environment. Furthermore, the time intervals between shedding of subsequent rolls predicted numerically in this work are in reasonable agreement with estimates for the characteristic period of puffing phenomenon given by the well-known empirical correlation for the pulsation frequency, f, presented by Eq. (3)[17].

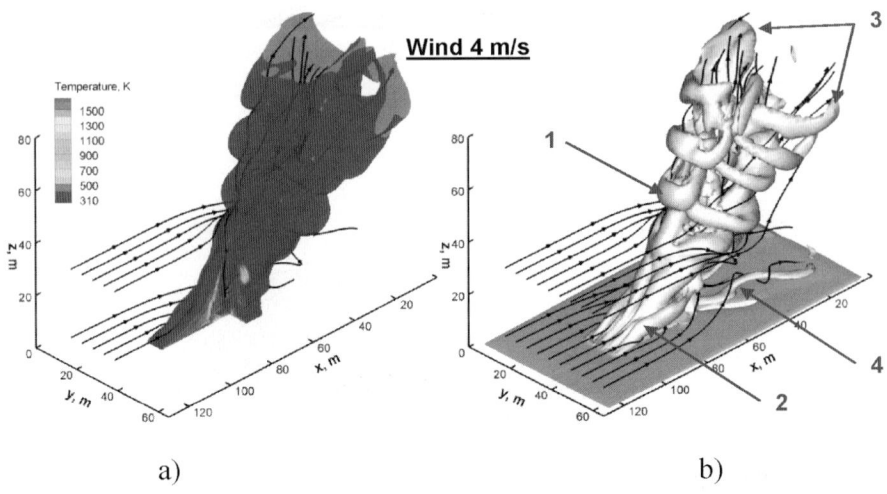

Figure 4. Instant resolved temperature (a) and vorticity (b, iso-surface 2 s^{-1}) fields

$$f^2 = \frac{const}{D_{eff}}, \qquad const = 2.3 \text{ m/s}^2. \tag{3}$$

Edge vortices have the same origin as so-called *horseshoe vortices* developing in the near-field of jets in cross-flow. *Wake vortices* are known to be either: (1) streamwise vortices which lie close to the ground surface[15] or (2) vertical vortices extending from the surface to the jet body[15,16]. In our simulations the streamwise wake vortices are clearly visible (designated by number 4 in Figures 4 and), whilst vertical wake vortices are not observed[16]. This is possibly because of the variations in flow geometry and jet-to-wind velocity ratios between this study and those of Yuan et al.[15] and Fric et al.[16], or insufficient grid resolution in the wake region. However, the wake vortices do not dominate the flow; this intrinsically transient flow is most strongly affected (in near-field region) by the *shear-layer vortices*, emerging and propagating downstream in a quasiperiodic manner and contributing to the formation of the *counterrotating vortex pair* (CVP). The vertical structure of CVP can be better

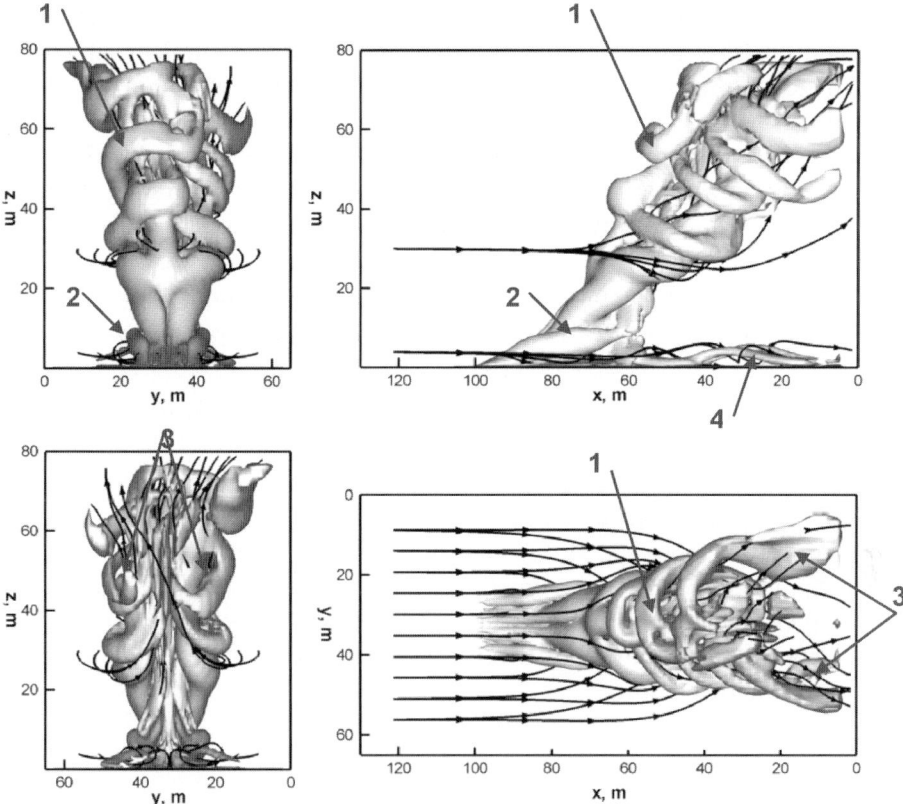

Figure 5. Vorticity iso-surface 2 s^{-1} and flow streamlines: windward, side, leeward, and top view
Wind 4 m/s

seen in Figure 5 (indicated by 3), where 2 s^{-1} vorticity iso-surface is shown
from leeward direction.

In Figure 4b, the first predicted shear layer vortices appear at a considerable
distance from the ground surface, although visual observations of this type of
fire indicate that they occur just above it. Such a delay can be attributed to the
underperturbed near surface flow, caused by both the underresolved boundary
layer and the excessive dissipative properties of the numerical scheme.

Another observation made from the numerical simulations is the sensitivity
of plume structure to the availability and geometry of obstacles located upstream at ground level. For example, the edge vortices did not develop in the
simulations (not show here) with an obstacle of just 2 m height (about 10% of
the flame length) extending through the entire computational domain at $x =$
100 m (windward edge of the burning area). This implies that the shape of the
ground surface, which can be very complicated in a realistic environment, is

highly significant. Furthermore, the formation of pulsating recirculating flow behind the obstacle increased the entrainment rate of the ambient air into the fuel rich zone, thereby increasing the temperature and emission of thermal radiation.

A quantitative prediction of flame radiative impact is another objective of this work. This data is crucial in the determination of safe separating distances, where the effect of cross-wind on the radiative heat fluxes is difficult to be accounted for in simplified engineering calculations. An empirical correlation (based on experimental data from large-scale pool fire experiments) is available[2] for the radiative heat flux q at ground level as a function of the radial distance from the center of the fuel bed:

$$q = q * \left(D_{\text{eff}}/r\right)^n, \qquad (4)$$

where $q* = 15.4$ kW/m^2 and $n = 1.59$. Among the engineering approaches available[1-3], the point source model has been applied to the case studied. According to this approach, all the energy is radiated from a single point of the flame axis, and the point is assumed to be at a distance half the flame length from the fuel surface. For the heat flux at the ground level ($z = 0$ m), along the line $x = \text{const}$, $y = r$ crossing the center of the fuel surface, the following relationship can be obtained:

$$q(r) = \frac{1}{\pi} \frac{f_r \dot{Q}}{\left(L_f/2\right)^2 + r^2}, \qquad (5)$$

where $f_r = 0.21 - 0.0034 D_{\text{eff}}$ is the fraction of radiated heat as recommended by the SFPE guide[2]. In Eq. (5), heat flux is assumed to hit a surface element positioned normally to the heat flux vector $\vec{q}_r$.

According to Eqs. (4) and (5), the distance that corresponds to the heat flux of 2.5 kW/m^2 (maximum tolerable limit for prolonged exposure of human skin) is 86 and 98 m, and the distance corresponding to 20 kW/m^2 (caused by the majority of the combustibles being ignited) is 23 and 33 m, respectively. These estimates are compared below with those obtained from CFD modeling.

Representative snapshots of the distribution of the radiative heat flux over the boundaries of the computational domain are presented in Figure 6 (propagation of radiation in dry air is assumed). In Figure 6b, the heat flux range just below 20 kW/m^2 is coloured in black, and transparent areas corres-pond to heat fluxes below 2.5 kW/m^2. Arrows in Figure 6a indicate the magnitude and direction of the radiative flux vector, $\vec{q}_r$, at a given location. Despite plume tilting by the wind, greater values of the radiative flux may be observed at the windward side of the flame. This is demonstrated by the distribution of heat fluxes across the fuel bed at ground level (Figure 6b). This is due to the higher flame temperature in this region, achieved by the more intensive air entrainment into the fuel rich zone of the flame. For the stronger cross-wind (e.g., 8 m/s), the leeward shift of high-temperature zone results in downwind redistribution of the heat flux.

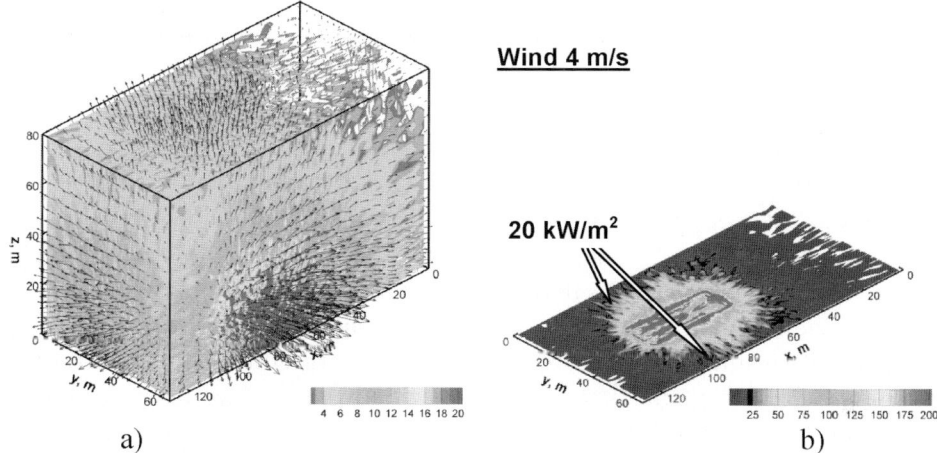

a)   b)

Figure 6. Radiative heat fluxes, kW: boundary surfaces of the computational domain (*left*), ground surface (*right*)

It is instructive to compare the numerical predictions obtained with the estimates obtained from the empirical relationship, Eq. (4), and from the point source model, Eq. (5). For example, it can be seen that the transversal extent of the region with heat fluxes greater than 20 kW/m^2 (boundary is indicated by thick arrows in Figure 6b) is in the order of 25–30 m in both directions, which is in agreement with the above made estimates using the empirical correlation,

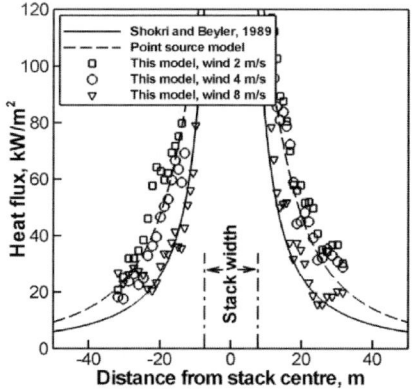

Figure 7. Radiative heat flux at the ground surface (along the line normal to the longest side of the stack, crossing the center of fuel surface)

Eq. (4) (r = 23 m), and the point source model, Eq. (5) (r = 33 m). The predicted region with heat fluxes above 2.5 kW/m^2 is also similar to that obtained from Eq. (4) (r = 86 m) and Eq. (5) (r = 98 m). Further comparison is demonstrated by Figure 7, where the profiles of the radiative heat flux obtained using different approaches can be seen to be in good agreement.

These comparisons suggest validity of the CFD model in prediction of the radiative heat fluxes targeting remote objects.

3. Conclusions

In the series of computational studies presented in this work, the behavior and formation of large-scale coherent structures in buoyant turbulent diffusion flames exposed to external crosswinds has been demonstrated. The computational studies were undertaken using LES, the application of which was complicated by the need to carefully consider both combustion and thermal radiation in a turbulent flame. The model suggested has been applied to simulate buoyant turbulent diffusion flames developing in the open atmosphere in the case of a massive wind-blown fire.

For flames in cross-winds, the grid resolution utilized has been shown to be sufficient to resolve the low-frequency region of the inertial subrange of the power spectrum of the velocity fluctuations. Four distinctive types of coherent flow vortical structures have been observed in the numerical predictions and compared to those reported in the experiments with jets in cross-flow. To further validate the flow predictions, plume tilt angles have been compared with empirical data and reasonable agreement has been found. The radiative heat fluxes produced by the flames have been predicted, analyzed and shown to compare favorably with the empirical correlation and simplified engineering approach available.

This work is a part of EPSRC project GR/S69122/01. The simulations have been conducted using computer facilities obtained due to financial support from the Royal Society (research grant RSRG 24350). Collaborative support from Greater Manchester Fire and Rescue Service, and essential contribution by Jim Marsden are gratefully acknowledged.

References

1. D. Drysdale, An Introduction to Fire Dynamics, Wiley, Chichester, New York 1999.
2. C.L. Beyler, In: SFPE Handbook of Fire Protection Engineering, 3rd edn., NFPA, Quincy, MA, 2002, pp. 3/268–3/314.

3. K.S. Mudan, P.A. Croce, Fire Hazard Calculations for Large Open Hydrocarbon Fires. In: *SFPE Handbook of Fire Protection Engineering*. 2nd edn., NFPA, Quincy, MA, 1995, pp. 3/197–3/240.

4. S.P. Burns, Turbulence Radiation Interaction Modeling in Hydrocarbon Pool Fire Simulations, Sandia Report SAND99-3190, 1999.

5. A. Yu. Snegirev, Statistical modeling of thermal radiation transfer in buoyant turbulent diffusion flames, *Combustion and Flame*, **136** (2004) 51–71.

6. A. Yu. Snegirev, J.A. Marsden, and G.M. Makhviladze, Radiative Impact of Wind-Blown Fires, *Proc. Interflam-2004*, **1** (2004) 289–300.

7. A. Yu Snegirev., J.A. Marsden, J. Francis, and G.M. Makhviladze, Numerical studies and experimental observations of whirling flames, *International Journal of Heat and Mass Transfer*, **47** (2004) 2523–2539.

8.   A. Yu. Snegirev, Large-eddy simulations of buoyant turbulent diffusion flames exposed to crosswinds and circulating flows. *European Combustion Meeting*, ECM2005, 3–6 April 2005, Louvain-la-Neuve, Belgium.

9. A. Yu. Snegirev, Wind-blown and whirling flames: numerical studies and experimental observations. Proceedings of the Second International Conference Fire Bridge 2005, Belfast, 9–11 May 2005, pp. 1–24.

10. F. Di Mare, W.P. Jones, and K.R. Menzies, Large eddy simulation of a model gas turbine combustor, *Combustion and Flame*, **137** (2004) 278–294.

11. M.G. Cooper, Contract Research Report 396/2001, HSE Books, 2001.

12. R.P. Cleaver, Cumber, P.S., and Fairweather, M., Predictions of free jet fires from high pressure, sonic releases, *Combustion and Flame*, **132** (2003) 463–474.

13. F. Muldoon, Numerical Methods for the Unsteady Incompressible Navier-Stokes Equations and Their Application to the Direct Numerical Simulation of Turbulent Flows, Ph.D. thesis, Louisiana State University, 2004.

14. H.R. Baum, K.B. McGrattan, and R.G. Rehm, Simulation of smoke plumes from large pool files, *Proceeding of the Combustion Institute*, **25** (1994) 1463–1469.

15. L.L. Yuan, R.L. Street, and J.H. Ferziger, Large-eddy simulations of a round jet in crossflow, *Journal of Fluid Mechanics*, **379** (1999) 71–104.

16. T.F. Fric, and A. Roshko, Vortical structure in the wake of a transverse jet, *Journal of Fluid Mechanics*, **279** (1994) 1–47.

17. A. Bejan, Predicting the pool fire vortex shedding frequency, *Journal of Heat Transfer – Transactions of the ASME*, **113** (1991) 261–263.

18. P. Poudenx, L. Howell, D.J. Wilson, and L.W. Kostiuk, Downstream similarity of thermal structure in plumes from jet diffusion flames in a crossflow, *Combustion Science and Technology*, **176** (2004) 409–435.

19. X. Zhou, K.H. Luo, J.J.R. Williams, Large-eddy simulation of a turbulent forced plume, *European Journal of Mechanics. B – Fluids*, **20** (2001) 233–254.

20. C. Hartel, In: Handbook of Computational Fluid Mechanics. R. Peyret (ed.), Academic Press, New York, 1996, pp. 283–338.

MATHEMATICAL MODEL OF FILTERED ISOLATION
IN A HIGH-TEMPERATURE FURNACE WITH ORIENTED
INJECTION OF THE COOLER

E. D. SERGIEVSKIJ
The Moscow Power Engineering Institute (TU)

Abstract. Cooling of high-temperature industrial energy installations by injection of air through permeable features of the design (filtered isolation), is an effective way of influencing the characteristics of flow and heat exchange. Heating of an air-coolant through a permeable wall of the housing of a working chamber leads to improved fuel economy since the heat removed from the system simultaneously heats one of the components of combustion, i.e., air. In this paper, the considered facility consists of housing with the current of the heat-carrier isolated in the flat channel with complex boundary and entrance conditions. The turbulent model which best represents the experimental data for the above mentioned conditions has been determined. Testing of the model has been carried out under basic injection conditions, and also for injections through a porous surface at different angles to the wall.

Keywords: filtered isolation, character of flow, heat exchange

1. Choice of Turbulent Model

Motulevich et al.[1] present experimental data for injection directed into the channel at various angles to the surface, which consider the joint influence of the gradient of pressure and the degree of turbulence of the main flow.

The dimensions of the working site of the aerodynamic installation (Figure 1) on which experiments were performed were $40 \times 78 \times 430$ mm^3 (rectangular section of the channel). The dimensions of the section of injection were 100×40 mm^2 (after injection air is carried out in the bottom part of the channel at a

N. Syred and A. Khalatov (eds.), Advanced Combustion and Aerothermal Technologies, 211–218.
© 2007 *Springer.*

distance of 0.25 m from the beginning of the channel). The mathematical model of the channel used the dimensions corresponding to those of the working site of the aerodynamic installation: on an axis $x = 0.43$ m (length); $y = 0.04$ m (width); $z = 0.078$ m (height).

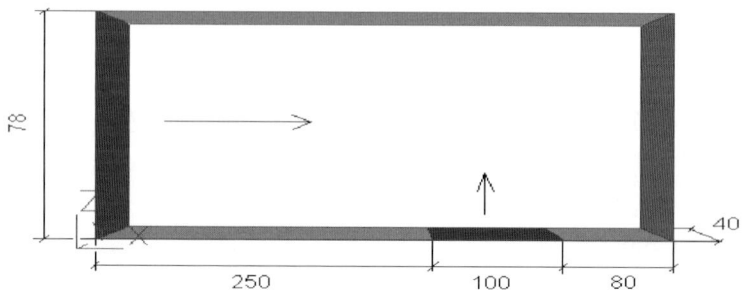

Figure 1. Mathematical model with the sizes of experimental aerodynamic installation

The temperature of the main flow was 20°C. The temperature of the main flow was 80°C. The velocity of air $U = 10$ m/s. The parameter of gradient of pressure $K = 0$. The degree of turbulence of the main flow was 5%. The angle of air-injection into the channel was varied between 0°, 15°, 90°, and 165°. The parameters of injection and suction varied in the range $F = (0.03)–(-0.004)$.

As certain errors were encountered, in producing the experimental data, we shall consider the behavior of the velocity profile for various turbulent models at different values of the parameter of injection.

In Table 1 three researched parameters of injection are presented with the resulting injection velocities. Values of injection on both the x- and z-axis are correlated depending on the angle of injection into the channel.

TABLE 1. Dependence of air on the angle and velocity of injection

$F = V/U$	V (m/s)	$\alpha = 15°$		$\alpha = 165°$	
		Vx (m/s)	Vz (m/s)	Vx (m/s)	Vz (m/s)
1.2×10^{-1}	1.2	1.159	0.311	−1.159	0.311
1.2×10^{-2}	0.12	0.116	0.031	−0.116	0.031
1.2×10^{-3}	0.012	0.012	0.003	−0.012	0.003

Six models of turbulence were compared: standard $k–\varepsilon$ model – KEMODL, $k–\varepsilon$ model at small Reynolds numbers – KEMODL-LOWRE, model Че KECHEN, model at small Reynolds numbers – KECHEN-LOWRE, KERNG and a model for modeling mixtures – MIXLEN.

It was found that most models correctly repeated the velocity profiles found as a result of experiments and two turbulent models exhibited the minimal error: the k–ε model – KEMODL and the k–ε model changed Chen-Kim – KECHEN.

Comparing the chosen turbulent models at various parameters of injection it was found that velocity profiles behave equally at small injection, and when parameter $F = 0.12$ the k–ε turbulent model produced the most similar result to the experimental profile.

2. Comparison of Calculations of Mathematical Model of Filtered Isolation and Experimental Data

In the study by Alimgazin et al[2], the experimental installation for the process of filtered isolation (Figure 2) was used to determine velocity profiles and bulk air temperatures.

Furthermore, a mathematical model of filtered isolation, using the geometrical dimensions corresponding to those of the experimental installation utilized by Alimgazin et al.[2] has been created. Calculations were carried out using the results of the numerical simulation as well as equilibrium calculations of the regime parameters of filtered isolation of the high-temperature furnace.

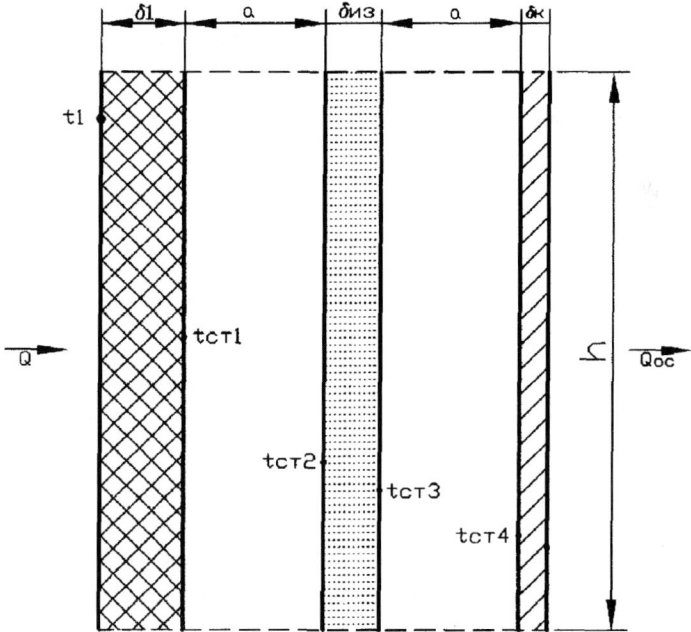

Figure 2. The basic circuit of isolation in experimental installation

Comparison of the temperatures determined numerically (tout = 57,423°C) and experimentally (tout = 49,272°C) has been carried out. The error was found to be less than 15%. As a result, we can say that it is possible to use the mathematical model created in PHOENICS to calculate regime characteristics of filtered isolation.

3. Comparison of Various Parameters of the Process of Filtered Isolation

A high-temperature glass-making furnace is considered where the temperature of the internal surfaces of the working chamber exceeds 1,500°C, hence, significant heat losses take place through the housing of the furnace.

Recycling of heat contributes to a radical decrease of heat losses in environment Qenv and to a decrease in temperature loadings on elements of the design by submission of air through the surface of a permeable wall.

In order to determine the most effective way of recycling a stream of heat flux through the surface of a wall by ventilation, it is necessary to consider qualitatively the three available techniques of the given process.

1. The charge of air moves in the channel (a collector of cold air) between the brickwork of a lateral wall of the glass melting furnace and the external casing (Figure 3), i.e., without application of filtered isolation.

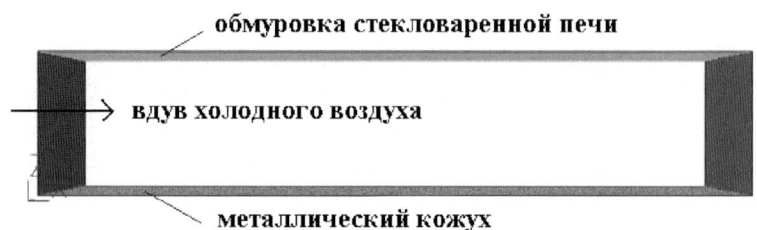

Figure 3. The first technique for cooling the external casing of a glass melting furnace

2. Part of the air (~20%) moves in a cold air channel between the permeable wall and the metal casing, and the rest of the air (~80%) moves in the hot air between the brickwork of the glass melting furnace and a permeable wall. The air is carried out through the hot air channel. Thus, part of the air passes through filtered isolation (Figure 4).

3. Cold air moves through the cold air channel which filters through the permeable surface into the hot air channel and removes heat from the brickwork furnace to be used for further technological processes (Figure 5).

Figure 4. The second technique for cooling the external casing of a glass melting furnace

Figure 5. The third technique for cooling the external casing of a glass-melting furnace

On the basis of the losses determined by equilibrium calculations of the regime parameters of filtered isolation of a high-temperature furnace, the bulk air temperature at the output from filtered isolation and the temperature of the metal casing of the furnace have been found. Heat from fuel combustion was used for preheating the cold and hot air channels. Thus, the bulk air temperature at the output from filtered isolation is the primary parameter which enables the estimation of the economy of the fuel going on heating air before being vented into the furnace. Results of calculations of the three limiting ways of cooling the furnace are submitted in Table 2.

TABLE 2. The summary table of results

Kind of isolation	t_{shell} (°C)	tout (°C)	Qoc (w)
Without a permeable wall	199.68	374.35	3.594×10^3
With partial filtration of air through a permeable wall	20.02	451.91	0.326
With full filtration of air through a permeable wall	20	453.77	9.526×10^{-8}

Analysis of the results produced by calculations of the three limiting ways of cooling furnace walls by the flow of air, shows that the application of filtered isolation considerably reduces losses of heat in an environment and the filtration of the total volume of air reduces them to zero. The temperature of air at the output from isolation also increases significantly to the point where the air is useful in other technological processes. Temperature stresses on the furnace housings are reduced. From this it is possible to draw the conclusion that the most effective of the three considered ways of cooling the furnace housing is the application of isolation with a full filtration of air through a permeable wall.

4. Implications of Filtered Isolation on the Economy of Natural Gas

In this section, comparison of the various methods of alternating the permeable and impenetrable parts of filtered isolation is carried out. In Figure 6 the distribution of temperatures in a hot air channel is shown for the process of filtered isolation with an impenetrable part at the beginning of the channel.

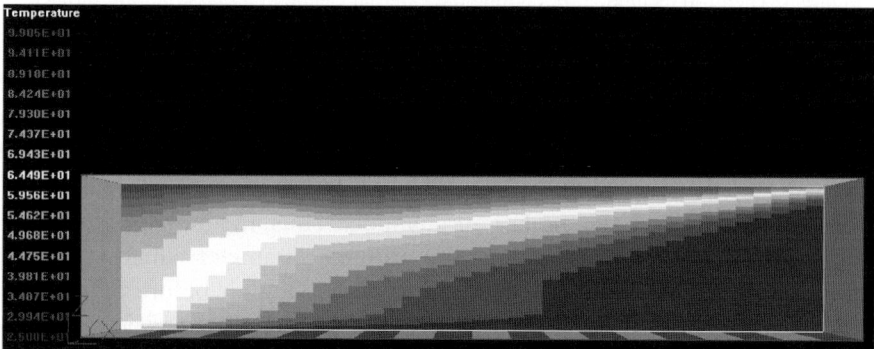

Figure 6. Distribution of temperatures on length of a collector of hot air

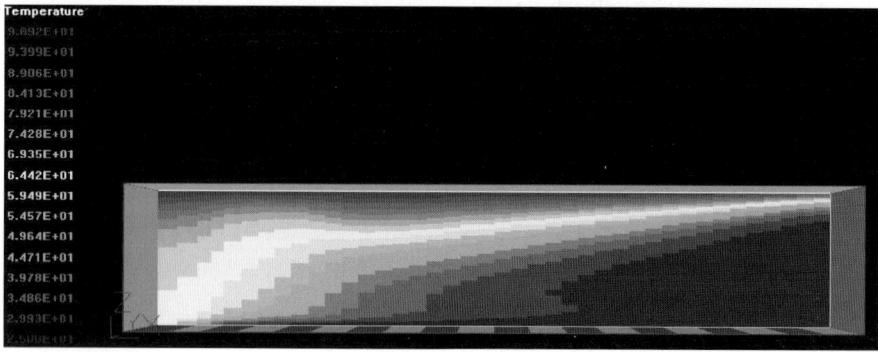

Figure 7. Distribution of temperatures on length of a collector of hot air

It is visible, that near to a wall face there is a zone of increasing temperature. Therefore, it was suggested that air be added directly along a face surface (Figure 7). Temperature values have decreased by 15%.

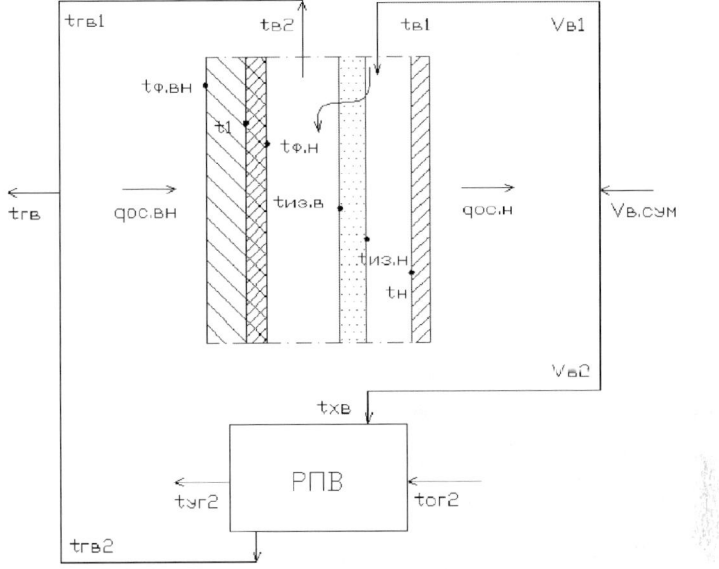

Figure 8. The block diagram of the filtered isolation used in glass melting furnace

5. Principle of Work of Filtered Isolation

Cold air from the fan acts in two streams: one in the recuperative heater, another in the cold air channel between porous isolation and a metal covering. Passing through the porous layer, air is heated up due to a stream of thermal losses from the furnace and enters the hot air channel between the layer of filtered isolation and the brickwork furnace. Further heated air mixes with air heated in the recuperative heater and moves to the burners. Through calculations of the model of filtered isolation, it is revealed that as the share of air injected by filtered isolation increases, the density of a thermal stream on an external surface of a wall decreases proportionally; approaching zero when the share of ventilated air $\Phi_{air} = 0.5$. As the value of Φ_{air} increases from 0 to 0.225, the density of heat flux on an internal surface decreases, but increases for values of Φ_{air} of 0.225 and above. By the example of a high-temperature furnace it is shown, that the process of filtered isolation improves the efficiency of installation, thus losses in an environment are reduced and the charge of fuel is reduced by 1.3 times.

6. Conclusions

1. The mathematical model, taking into account the features of heat transfer in the vicinity of the permeable wall is offered, enabling numerical research of the processes occurring in concrete designs of high-temperature installations.

2. Calculations of air heating in the channel of filtered isolation have been carried out and ways of eliminating overheating have been determined. Of the several techniques of filtered isolation, the best has been determined in terms of of the greatest bulk temperature and the minimum temperature of the metal casing of the furnace.

3. The conclusion about economy of fuel (the cost of fuel was reduced by a factor of 1.3) and decrease in thermal losses in an environment is made due to addition of filtered isolation to a high-temperature furnace.

References

1. V.P. Motulevich, E.D. Sergievsky, S.V. Terentev, and V.V. Shitov, Heat Transfer in Regions of Separated and Nonseparated Flows with Oriented Injection, *Heat Transfer* 90, IX Proc. Int. Heat Trans. Conf. Jerusalem, Israel, **3**, 351–356 (1990)
2. A.S. Alimgazin, V.P. Motulevich, E.D. Sergievsky, and V.I. Baranov, Numerical Investigation of Gas Dynamics and Heat Transfer in Thermotechnological Processes, International Symposium on Manufacturing and Materials Processing, **1**, 543–552 (1997)

BIOMASS AND ALTERNATIVE FUELS

DEVELOPMENT OF BIOMASS AND GAS COFIRING TECHNOLOGY TO REDUCE GREENHOUSE GASEOUS EMISSIONS

I. BARMINA, A. DESCNICKIS, A. MEIJERE, M. ZAKE*
Institute of Physics, University of Latvia, Salaspils-1, Miera 32, LV-2169, Latvia

Abstract. The prime objective of the present study is to explore the basic mechanisms that control and stabilize the biomass combustion, and to study the processes involved in the formation of polluting emissions from the cofiring of wet wood biomass (wood pellets) with a premixed swirling propane/air flame flow. The influence of the cofiring on the chemical/physical changes in the wood pellets, resulting in an ignition and burnout of volatiles, has been investigated for different moisture content in the pellets and different amounts of additional heat supply from the swirling flame flow. The results of the experimental investigations are discussed, describing the effect of the cofiring on the rate of unsteady drying, heating and thermal decomposition of the wet pellets, as well as the ignition and burnout of volatiles that will facilitate the control and optimization of polluting emission composition.

Keywords: cofiring, renewable, fossil fuel, greenhouse emissions

1. Introduction

Fossil fuel combustion is regarded nowadays as the main source of greenhouse gaseous emissions, causing global warming of the Earth's atmosphere and climate change. Unlike fossil fuels, which pollute the atmosphere, the renewable fuels (wood, waste biomass) used for the heat and energy production have less impact on the environment, since the biomass is considered CO_2-neutral: it consumes the same amount of carbon dioxide during growth as it releases during combustion. In this context, the development of a cofiring

---

* To whom correspondence should be addressed. Maija Zake, Institute of Physics, University of Latvia, Miera Street 32, Salaspils-1, LV-2169, Latvia; e-mail: mzfi@sal.lv

N. Syred and A. Khalatov (eds.), Advanced Combustion and Aerothermal Technologies, 221–230.
© 2007 *Springer.*

technology by using a renewable fuel (wood biomass) to replace in-part some of the fossil fuel, allows a reduction in the greenhouse CO_2 emissions from the heat and energy production systems, so saving the fossil fuel resources and producing a cleaner combustion with less harmful environmental impact. Additionally, the cofiring of wood biomass with a small amount of gas improves and stabilizes the boiler operation, preventing the problems that accompany the combustion of wood biomass having dissimilar structure and variable moisture content.

There are many ways to provide the cofiring of a renewable with a fossil fuel, e.g., the direct cofiring of wood biomass with coal[1] by feeding and combusting the biofuel together with coal; an indirect cofiring in which the wood gasification/pyrolysis products are cofired with coal[2]; and a cofiring of the wood biomass gasification/pyrolysis products with natural gas[3]. The cofiring of biomass with natural gas can thus be used to avoid the problems that accompany the burnout of wet wood biomass with high moisture content. The testing of natural gas cofiring systems demonstrates[3] that a small supply of natural gas (about 15–20% from the total heat produced) into biomass-firing boilers enhances the biomass power production when compared to acceptable NO_x and CO boilers.

Previous experimental studies of cofiring the wood biomass (wood pellets) with propane have shown[4] that with the low and constant moisture content in wood pellets the cofiring process provides a faster and more stable wood pyrolysis, resulting in a faster thermal breakdown of volatiles, a faster ignition and burnout (depending on the rate of additional heat supply) and produces carbon-neutral CO_2 emissions (up to 86%). At the same time, the enhanced combustion of wood biomass by cofiring with a fossil fuel (propane) increases the heat release from the fuel combustion, so increasing the peak temperature value inside the flame of volatiles, increasing the rate of NO_x production. Previous experimental study has also shown that the increased rate of wood gasification causes an intensive release of CO, wet flue gas, biomass ash tars and other condensable organic species that form deposit layers on heat surfaces and may cause fouling and slagging problems.

The main aim of the present study is to develop and optimize the cofiring of the fossil fuel (propane), to provide an effective and stable burnout of the renewable, at the same time minimizing the impact of cofiring on the formation of harmful polluting emissions and the impact of fouling and slagging on the heat production. The investigations are carried out using wet wood biomass of varying moisture content with a varying amount of additional heat supply into the wet wood biomass.

2. Experimental

Previously, the cofiring of wood pellets with the fossil fuel was investigated under the conditions in which the additional heat and air supply (to enhance and complete the burnout of wood pellets) had been injected into the bottom part of the wood biomass[4]. The results of these investigations show that the additional heat supply into the bottom part of the wood pellets enhances their volatilization, ignition and burnout, but also causes an intensive release of tars and CO (up to 10,000 ppm). In addition, the formation of harmful deposits on heat surfaces has been observed. To minimize these effects, the cocombustion process was modified, providing the tangential air supply (c) and additional heat supply from an external heat source – a swirling propane/air flame flow (b) into the upper part of the combustor (a), charged with a fixed amount of wood pellets (100 g) (Figure 1). At the initial stage of the gas cofiring, the injection of a high-temperature ($T \approx$ 1,200–1,500 K) propane flame flow into the top of the wood layer advances enhanced drying, pyrolysis, gasification and burnout of the wood pellets in the top part of the wood biomass. Due to the enhanced drying, gasification and burnout of the wood pellets, the unburned wood biomass gradually flows down to the bottom part of the combustor opposite to the primary air flow introduced into the combustor bottom, so completing the self-sustaining burnout of the wood pellets. For such conditions the cofiring of propane enhances the burnout of up flowing hot volatiles[4] downstream of the water-cooled channel sections (d), of $L = 360$ mm total length. Residual ash is removed from the bottom of the combustor.

The rate of stoichiometric propane supply into the swirl burner can be varied from 0.5 to 0.85 l/min. Hence, the heat released from the propane combustion in these experiments can be varied in a range from 760 to 960 J/s, whereas the additional energy supply from the propane combustion into the combustor is limited by 30% of the net amount of the total heat released during the burnout of wood pellets and volatiles. The flow rates of primary and secondary air flows could be varied in a range of 20–90 l/min.

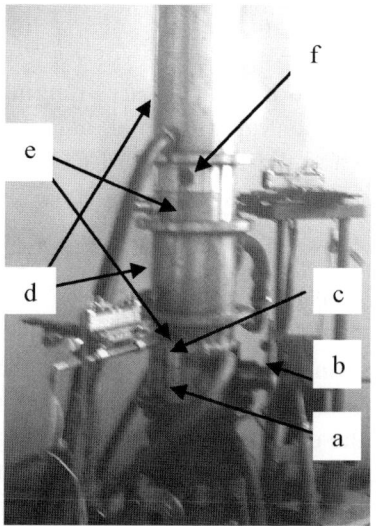

Figure 1.   Digital image of the experimental setup: a – combustor; b – premixed propane-air burner; c – secondary air supply above the wood layer, d – water-cooled channel sections; e – sections with peepholes for thermocouples; f – peephole for a gas sampling probe

Between the cocombustor and water-cooled channel section, and between the channel sections there are diagnostic sections with peepholes (e, f) for the local input of diagnostic tools (Pt/Pt-Rh thermocouples and gas sampling probes) into the flame of volatiles. The ignition of volatiles above the gasifier is registered by a photodiode. The average heat losses from the flame flow at different stages of the wood cofiring with propane are estimated from calorimetric measurements of the cooling water flow. Measurements taken during the parametric tests of flue gas included continuous monitoring of O_2, CO_2, CO, NO, NO_x, and NO_2, the temperature of the products and the efficiency of fuel burnout, using a gas analyzer TESTO-350XL. During these parametric tests the gas cofire rate, the primary and secondary air supply and the moisture content in the wood pellets (up to 30%) were varied.

3. Results and discussion

The cofiring of the wood fuel (wood pellets) with the swirling propane/air flame flow starts with an intensive unsteady heating, drying and thermal degradation of the wood biomass. Drying starts after the additional heat supply from the external heat source – the propane flame flow into a layer of the wet wood pellets and dominates within the temperature range of 370–420 K. Thermal degradation of the wood pellets is a two stage process. During the first stage of the thermal degradation (at the temperatures ranging from 700 to 1,200 K) the additional heat supply from the propane flame flow vaporizes the volatile components of the wood pellets and the vapor consists mainly of CO and hydrogen (1) (Figure 2).

$$C_6H_{10}O_5 + 6O_2 + mH_2O + \text{Heat} \rightarrow 6CO + 5H_2 + nH_2O. \tag{1}$$

The next stage of thermal degradation takes place at temperatures ranging from 1200 to 1,600–1,700 K, when the pyrolytic vapors ignite and are immediately burned with an intense heat release and the formation of CO_2 (2) while the mass fraction of free oxygen in the products rapidly decreases (Figure 2).

$$C_6H_{10}O_5 + 6O_2 + (m{-}n)H_2O \rightarrow 6CO_2 + (5 + m{-}n)H_2O + \text{Heat}. \tag{2}$$

By cofiring the dry wood pellets (moisture content 6%) with propane the dominant effect of cofiring on the composition of the products is observed at the initial stage of unsteady temperature rise (at the end stage of the nonflaming heterogeneous burnout of wood charcoal), resulting in an enhanced release of CO emissions up to 1,600–1,800 ppm and free hydrogen up to 500–600 ppm

(Figure 2). The peak value of CO and H_2 mass fraction in the products at the unsteady stage of the flame temperature rise correlates with a slight decrease in the rate of the temperature rise inside the flame of volatiles, indicating that the endothermic processes of thermal decomposition of volatiles (1) result in heat consumption from the external heat source. Hence, at this stage of the gas cofiring the rate of the thermal decomposition of the wood biomass can be controlled by varying the rate of additional heat supply from the propane flame flow into the wood biomass.

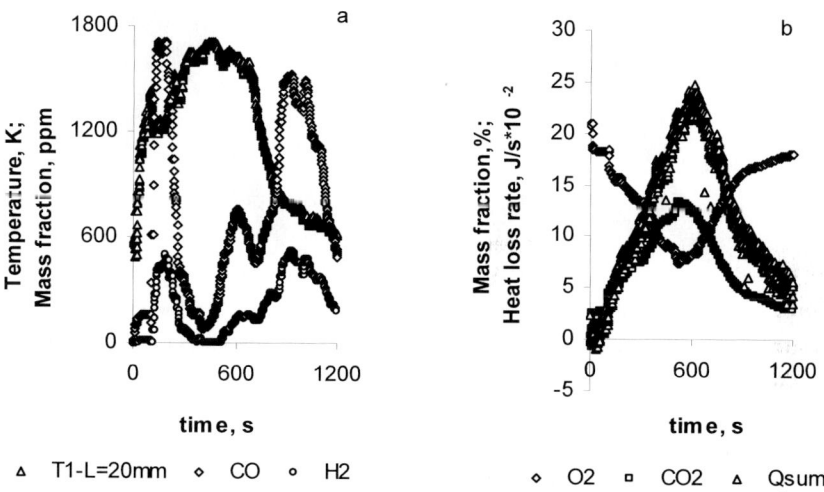

Figure 2. Time-dependent variations of the composition of the products, flame temperature, and heat loss at the constant additional heat supply from the propane flame flow (900 J/s) and at the constant moisture content in wood pellets (6%)

As follows from (2), the next stage of the thermal degradation ($T > 1,200$ K) of the wood pellets and burnout of the volatiles leads to an intensive release of CO_2, while the mass fraction of CO in the products rapidly decreases to the minimum value approaching to 80–200 ppm during the burnout of volatiles (Figure 2a). The peak value of CO_2 mass fraction (12–14%) in the products correlates with the peak value of the temperature inside the flame of volatiles and the peak value of heat losses from the flame (Figure 2b), while the mass fraction of free oxygen in the products at this stage of burnout decreases below 6–7%. In general, about 12–16% of the total CO_2 mass fraction released during the gas cofire is related to the greenhouse carbon emissions produced during the burnout of propane, while about 84–88% refers to the carbon-neutral emissions. In the case of a 20% gas cofire, the efficiency of the wood burnout is relatively high and achieves 86–87%. The NO_x (NO + NO_2) emissions at such a gas cofiring rate (20%), with the peak value of the flame temperature being 1,690–1,750 K and 12% of CO_2, approach 80–100 ppm and are in compliance with proposed NO_x limits (Figure 3a). Figure 3a illustrates that the dominant release of NO_x emission can be related to the formation of temperature-sensitive NO within the flame of the volatiles. The most intensive formation of NO_2 emissions (up to 20 ppm) is detected during the initial stage of the flame temperature rise, when the air excess within the flame ($\alpha \approx 1.3$–1.5) is fixed.

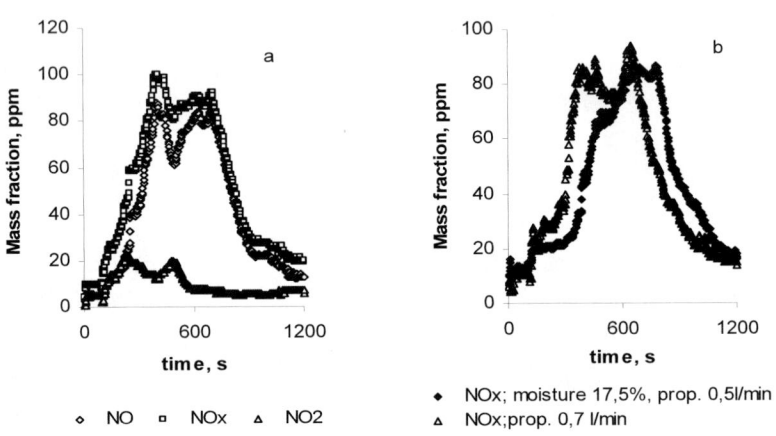

Figure 3. Time-dependent variations of the mass fraction of polluting NO_x, NO, and NO_2 emissions in the products by cofiring of dry wood pellets (6%) with propane (20%) (a) and variation of NO_x mass fraction by cofiring wet pellets (b) at different rates of propane supply

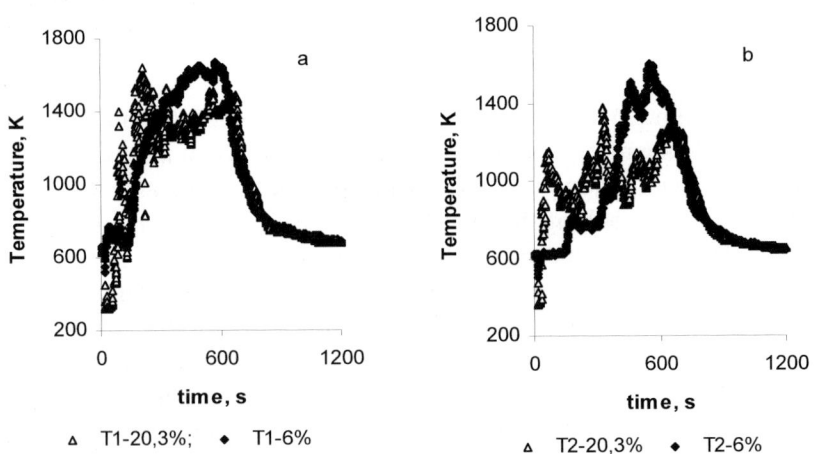

Figure 4. The effect of moisture content in wood pellets on the rate of the flame temperature rise at different stages of the flame formation at the constant gas cofire rate (20%); (a) $L = 40$ mm above the combustor, (b) $L = 170$ mm

The cofiring of wood pellets of a higher moisture content (up to 20%) with propane substantially disturbs the combustion conditions for the wood pellets, giving rise to temperature pulsations at the initial stage of the temperature rise and decreasing the peak value of the temperature within the flame of the volatiles at different stages of the flame formation (Figure 4a and b). Increasing the moisture content in wet wood pellets significantly affects the rate of the heat release and the rate of the heat production downstream of the water-cooled channel; decreasing the peak value of heat losses which are highly dependent on the moisture content in wood pellets (Figure 5a and b). In addition, it should be noted that the burnout of wet wood pellets of higher moisture content at fixed gas cofiring (20%) decreases the efficiency of the wood burnout to below 82%.

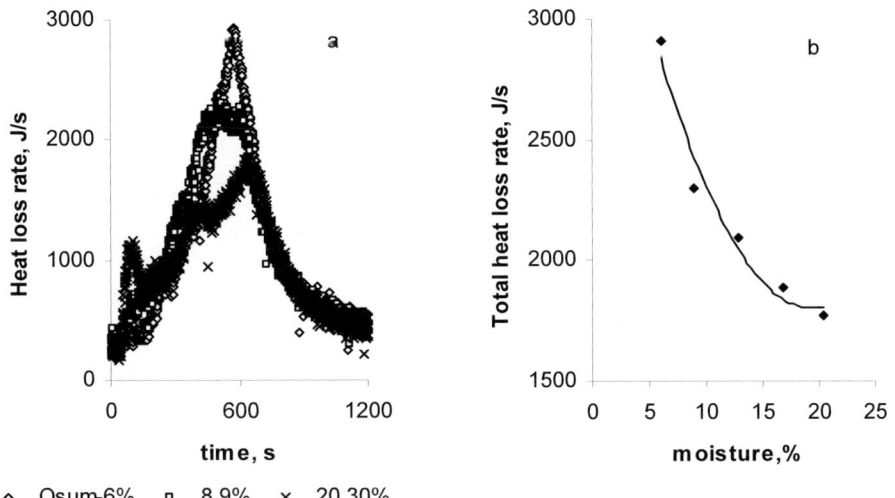

Figure 5. Time-dependent variations of the heat loss rate at different moisture contents in wet wood pellets: (a) dependence of the total heat loss rate on moisture content in wet wood pellets; (b) at the constant gas cofire rate (20%)

The continuous monitoring of the composition of emissions during the initial stages of the temperature and heat loss rise, clearly shows that the change of the combustion conditions for wet wood pellets directly affects the composition of the products. An increased moisture content in the wet wood pellets slows down the rate of wood burnout, decreasing the peak temperature value within the flame reaction zone, reducing the rate of CO_2 and NO_x formation downstream of the flame channel flow and decreasing the mass fraction of CO_2 and NO_x emissions in the products. The mass fraction of CO_2 in the products decreases from 14% for the burnout of dry pellets to 11% for the burnout of wet pellets at the 20% moisture content (Figure 6a), while the mass fraction of NO_x emission in the products under such conditions decreases from 120 to 90 ppm (Figure 6b). In fact, during the burnout of wet pellets, the release of unburned CO slightly increases, and the mass fraction of CO in the products for such conditions increases from 80 up to 210 ppm (Figure 6b). A very intensive release of CO by cofiring wet wood pellets is observed during the initial stages of the temperature rise, when the mass fraction of CO in the products rapidly increases and approaches 4,000 ppm (Figure 8b).

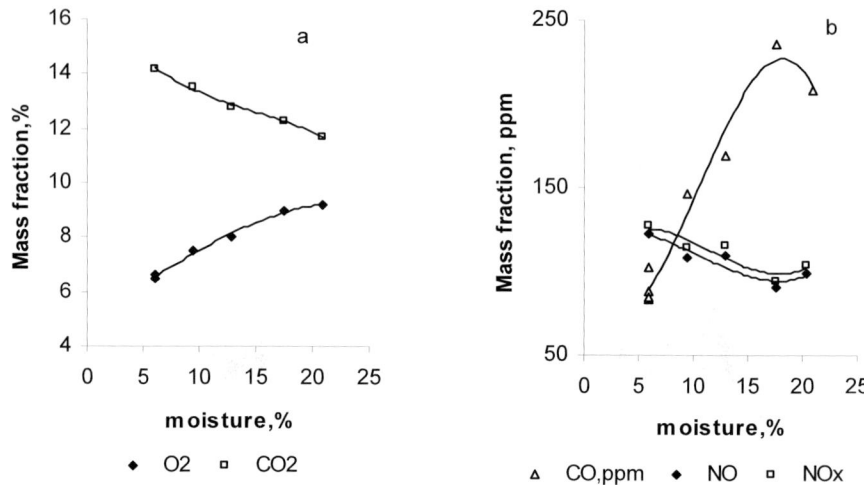

Figure 6. The effect of moisture content in wet wood pellets on the composition of emissions

With the aim of minimizing the release of unburned CO, the wet pellets were burnt out at different rates of gas cofire. The experimental study of the gas cofiring of wet wood pellets at a constant moisture content (17.5%) with different rates of propane supply into the swirl burner, and different rates of gas cofiring, have shown that a propane/air supply increase of 30% into the swirl burner increases the efficiency of the burnout of the wet wood pellets up to 87–88% by reducing to the minimum the biomass ash tars and other species that form deposit layers on heat surfaces. Moreover, the 30% gas cofiring rate increase provides a faster temperature and heat loss rise during the initial stages of the flame formation. The burnout of wood pellets develops at a higher flame peak temperature value, with a higher heat production in the flame of the volatiles, so increasing the heat losses (Figure 7).

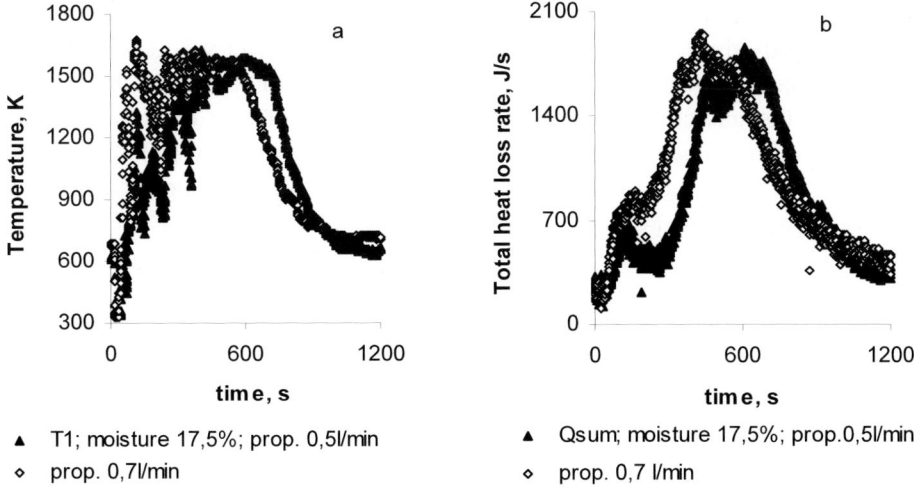

▲ T1; moisture 17,5%; prop. 0,5l/min   ▲ Qsum; moisture 17,5%; prop.0,5l/min
◇ prop. 0,7l/min   ◇ prop. 0,7 l/min

Figure 7. The effect of propane supply on the rate of the temperature (a) and heat loss rise (b) during the burnout of the wet wood pellets (17.5%)

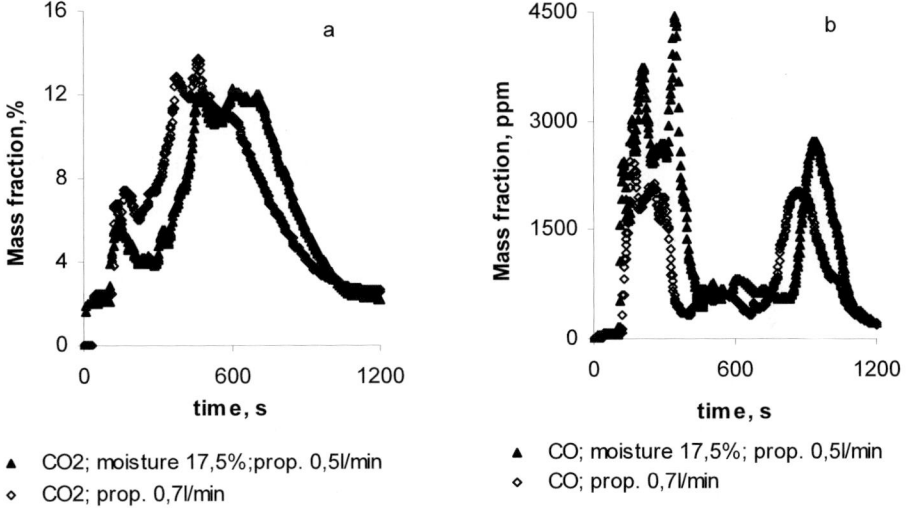

▲ CO2; moisture 17,5%;prop. 0,5l/min   ▲ CO; moisture 17,5%; prop. 0,5l/min
◇ CO2; prop. 0,7l/min   ◇ CO; prop. 0,7l/min

Figure 8. Time-dependent variations of the composition of the products at different rates of propane supply and different rates of gas cofiring of wet wood pellets (17.5% moisture); variations of the mass fraction of CO_2 in the products (a); CO (b)

The increased rate of gas cofiring of wet pellets (17.5% moisture content) causes a higher rate of CO_2 production, increasing the peak value of CO_2 in the products to 13.6% (at 6–7% of free oxygen in the products). The emission of CO is reduced from 4,000 ppm at 20% cofire to below 2,000 ppm at 30%

cofiring during the initial stage of the thermal decomposition of the wood pellets, so providing a cleaner method of combusting wet wood with a lower mass fraction of CO in the products (Figure 8a and b). The results show that an increase in the gas cofiring rate leads to a faster formation of NO_x emissions during the initial stage of the flame temperature rise, slightly reducing the net NO_x release during the burnout of wet pellets. The cofiring has only a relatively slight influence on the peak mass fraction of NO_x released during the burnout of volatiles (Figure 3b).

4. Summary

Investigations into the cofiring of wood biomass with propane has shown that a modest (15–30%) propane content increases the burnout rate of wet wood biomass, negating the losses that occur when using a wet wood fuel of 15–20% moisture content.

The increased rate of the wet wood burnout advances the heat production and increases the peak heat production rates by 12–14%. For the cofiring of renewable wood biomass with 25% fossil fuel (propane), the net release of the greenhouse CO_2 emissions is below 16% of the total amount of CO_2 released during the heat production, due to the carbon-neutral nature of biomass fuels. The faster heat production causes a reduction in the release of CO during the initial stage of the thermal decomposition of the wood pellets, with acceptable levels of NO_x (80–100 ppm) and CO emissions (80–200 ppm) during the wet wood combustion stage for gas cofire rates of 15–30%.

References

1. M. Sami, K. Annamalai, M. Woolridge, Co-firing of coal and biomass fuel blends, Prog. Energy Combust. Sci., **27**, 171–214, 2001.
2. D.I. Granatstein, Case Study on Biococomb Biomass Gasification Project Zeltweg Power Station, Austria, IEA Bioenergy Agreement, Natural Resources Canada/CANMET Energy Technology Centre (CETC), 2002, pp. 1–18.
3. H.B. Mason, L.R. Waterland, Optimization and Testing of Natural Gas Cofiring in Two Biomass Power Boilers, Final Report of Western Regional Biomass Energy Program FR-02-111, 8 July 2002, pp. 1–45, http://www.westbioenergy.org/reports/55036/55036.htm
4. M. Zake, I. Barmina, A. Meijere, Formation of polluting emissions at wood biomass co-firing with propane, Latv. J. Phys. Tech. Sci., **1**, 33–43, 2005.

INVESTIGATIONS OF INSTITUTE FOR ENGINEERING THERMOPHYSICS (THERMOGASDYNAMICS DEPARTMENT) IN THE AREA OF COMBUSTION AND GASIFICATION OF ALTERNATIVE FUELS

A. A. KHALATOV, I. I. BORISOV, S. G. KOBZAR,
G. V. KOVALENKO, O. E. KHLEBNIKOV
Institute of Engineering Thermophysics (IET), National Academy of Sciences, Ukraine

Abstract. The Ukraine is dependent on other countries to provide around 60% of its annual energy demand. The fuel balance of Ukraine is as follows: (1) natural gas is 41%, (2) coal is 24.3%, (3) oil is 18.4%, (4) nuclear is 12.5%, (5) others is 4.8%. The total annual energy balance is between 110 and 140 million tons of the "conditional" fuel, the annual import of natural gas is 53 billion cubicmeter, of oil is 13 billion tons, and of coal is 10 billion tons. The renewable and alternative energy sources must be developed to have an increased involvement in the countries energy balance, as the Ukraine has a significant supply of brown coal, peat and biomass. Currently, the use of renewable and alternative energy sources is only 0.5% (2003) of the countries total energy balance, with a projection to be 2.5% by 2020 and 3.5% by 2030. Over the last 10 years the Institute of Engineering Thermophysics has been involved in the national research program to increase the use of alternative energy sources in various industrial applications. This report provides the brief review of some accomplishments in the field, including the application of: (1) a wood gas generator of 50 kWt, (2) thermal energy generators using peat, brown coal and their blends with biomass, (3) a thermal energy generator using waste motor oil. The basic design of all test facilities used is given in this report along with the results of experimental studies.

Keywords: combustion, gasification, alternative fuels, brown coal, peat, biomass

N. Syred and A. Khalatov (eds.), Advanced Combustion and Aerothermal Technologies, 231–245.
© 2007 *Springer.*

1. Introduction

The continuing growth in the price of oil over recent years has stipulated the worldwide interest in alternative fuels. The Ukraine is very rich in biomass, peat and brown coal, the overall potential of alternative fuels in the Ukraine is estimated to be 3.5 million tons of the "equivalent organic fuel". Currently, the alternative energy sources are not in use in the country, with the share of peat and brown coal in the energy balance of Ukraine being 0.3% and 1.2%, respectively [1]. As of now, a great deal of the Ukrainian villages are not gasi-fied (no pipeline service). The involvement of biomass, peat, brown coal, as well as other alternative fuels into the countries energy balance, primarily for domestic and rural applications, is a very important goal of the Ukrainian energy policy.

The *brown coal* reserves in the Ukraine are estimated to be 3.5 billion tons, from which around 560 million tons were found out in the Alexandria region (Dnieper rRiver basin). The average ash content in the Ukrainian brown coal is roughly 20%, the humidity rate is from 43% to 62%, and the lowest caloric value is from 7.1 to 9.6 MJ/kg. The cost of excavating the brown coal in the Ukraine by the open-cast technique is around $30 for 1 t of the equivalent fuel [2].

The prospective *peat* reserves in the Ukraine are estimated to be around 2.2 million tons, the energy potential of the peatis 0.84 billion tons of the equivalent fuel. Currently, the annual peat production is 1.7 million tons (0.5–0.6 million tons of the equivalent fuel). The peat reserves will last for up to 200 years if the production levels of 1995 are to be kept. The peat production in the Ukraine was as its maximum in 1991 (7.5 million tons), however from 1995 due to various reasons the peat production has dramatically reduced [1]. The excavated peat is used as a local fuel (peat briquettes), primarily in the housing and rural areas. Using the currently available technique it is estimated that the production of 700,000 t of peat briquettes per year would be possible in the Ukraine. The cost of the thermal energy obtained from the briquetted peat is 10–30% lower than that for the traditional sorts of fuel.

The overall potential resources of the *wood waste*, including the rind in the forestry and woodworking industries are around 3.7 million cubicmeter, the equivalent of 984,000 t of the equivalent fuel per year.

The data presented in [1] shows that the forestry, woodworking, wood pulp and paper industries are actually unable to consume the overall amount of wood waste, so they could effectively be utilized on site for the energy production. The annual unused wood waste potential is around 2.9 millions cubicmeter, corresponding to 0.75 million tons of the equivalent fuel. The foreign experi-ence demonstrates that existing boilers running on coal could also combust the wood waste in mixtures with a quantity equivalent to between 4% and 10% of the total fuel without modification [4]. Wood comes to approximately 0.4% of the energy balance of the country [1].

The peat, brown coal and biomass are low-calorific fuels characterized by their high volatile content. The quality of their burning is at the highest at the beginning; however it decreases dramatically at the after-burning stage of the coke residue formation. As a result, the burning of alternative fuels in chambers of traditional design, especially in low-size units, exceeds the emissions levels of carbon oxide (CO) and nitric oxides (NO_x) established by the European pollution standards.

An additional problem is that brown coal extracted from the Dnieper River basin contains the halogen compounds ("salted coals"), leading to the corrosion of boiler equipment and some ecological problems due to the emission of dangerous chlorine organic compounds.

2. Aim

The application of biomass, peat, brown coal, as well as waste motor oil for power generation requires development of modern thermal equipment combining high performance combustion and low pollution emissions. It is well known, that the application of a bistage combustion process where the first stage is a fuel pyrolysis/gasification, and the second stage is the after-burning of the produced gas (hydrogen, carbon oxide, hydrocarbons), enables an increase in the thermal efficiency of the equipment. The ability to control the combustion process also allows a reduction in the generation of harmful pollutants.

This report demonstrates some primary details of the Institute of Engineering Thermophysics (IET, Kiev, Ukraine) research program towards the development and testing of small-scale thermal equipment running on biomass, peat, brown coal and their blends, as well as utilizing waste motor oil.

3. Results and Discussions

3.1 WOOD DOWNDRAFT GASIFIER (50 KWT)

The basic criteria defining the gasifier design are the following: (1) simple production technology and low manufacturing cost, (2) operation reliability and easy process control, (3) easy integration into an electric generator unit.

The design of the gasifier developed by IET is based on the prototype available at Cardiff University (UK), with some further investigations performed in [5, 6]. The gasifier unit is of the downdraught scheme, i.e., it uses the parallel movement of dry wood blocks and air flow. As such, this scheme produces the least quantity of tars in the pyrolysis zone, decomposing afterwards into the simple chemical compounds in the combustion zone.

The gasifier (Figure 1) consists of the housing 1 with an inside diameter of 600 mm, the cone 2 with a fire grate, the tube with nozzles (4) for the air flow supply. The housing was made of sheet steel of 12 mm thickness, enabling it to resist the high-temperature gradient with no material deformations. The entire gasifier height weighs 2.5 t. The hermetically closed doors 3 and 5, for the fuel loading and ash removal, are arranged at the top and bottom of the housing. The inner surface of the operating zone is covered with a refractory layer (15 mm) made on the magnesium oxide basis. The fuel is supplied both through a loading hatch and through a door in the upper portion of the housing.

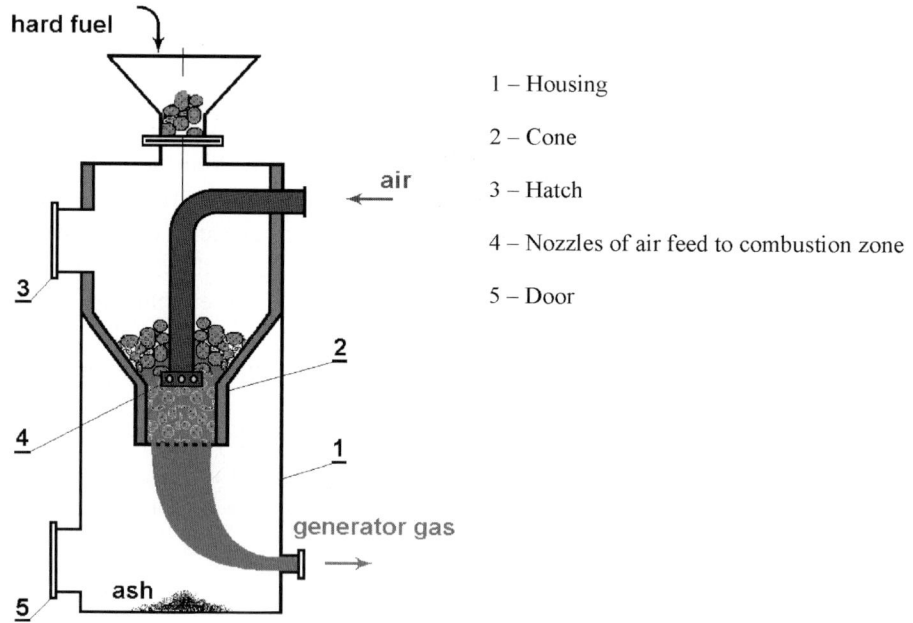

1 – Housing

2 – Cone

3 – Hatch

4 – Nozzles of air feed to combustion zone

5 – Door

Figure 1. The gas producer of an inversed type with a heating rating 50 kw

The results of experimental studies for gas production from wood are presented in [7]. The basic gasifier characteristics are as following: (1) fuel type: wood blocks from 15 to 35 mm in size; (2) fuel consumption: 15 kg/h; (3) air rate: 450 l/min; (4) gasifier thermal power: 50 kWt; (5) air fan power: 1 kWe.

The wood-gas produced by the gasifier is of the following composition (by volume fraction): **CO**: 21%, **H_2**: 17%, **CH_4**: 2%, **N_2**: 48%, **CO_2**: 12%. The temperature in the combustion zone (the cone neck) is between 950°C and 1,000°C, the tar-in-the wood gas concentration is from 1 to 2 g/m^3.

As a whole, the gasifier was manufactured to investigate the vortex-based clean-up system, however, currently it is designated to study the burning of other types of fuel, such as the peat, brown coal and briquetted fuels.

3.2   VORTEX-BASED CLEAN-UP SYSTEM

The vortex-based clean-up system uses the rotating gas–liquid flow, which generates the highly developed gas–liquid structure with approximately uniform bubble diameter (2–3 mm). Specific surface of the gas–liquid contact runs up to 1,000 m^2/m^3. Clean-up efficiency was in the region of 67–78%.

3.3   BIOMASS FURNACE WITH VORTEX EJECTOR

The biomass furnace with vortex ejector (Figure 2) was developed at the Institute of Engineering Thermophysics (patent of Ukraine [13]). It consists of a primary combustor (60 l in volume) with the ash pit and secondary cylindrically shaped combustor made of ceramic. The secondary air to the ejector vortex

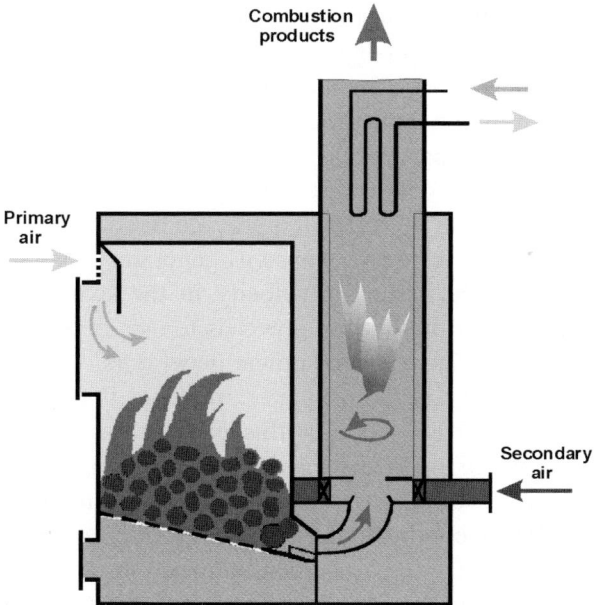

Figure 2. The heat generator with the vortex ejection of the producer gas

chamber is supplied by a fan. The wood gas from the primary zone enters through the thermally insulated channel into the ejector axial zone, where it is mixed with the secondary air to create a combustible gas–air blend. The resulting burning process occurs in the swirling flow. The high-temperature gas from the secondary combustion chamber is employed in the external heat exchanger.

The preliminary testing has shown that the vortex ejector provides a volumetric ejection coefficient (k_e) of around 2.0. Such a value of k_e enables the vortex ejector to "suck out" out the gasification products from the primary zone, which is operating under a small vacuum. From the operating point of view, such a scheme is more advantageous as there is no smoke leakage into the housing during the combustion chamber operation. Besides, the application of the vortex ejector enables the controlling of the furnace power by means of the flow rate coming through the swirl generator.

During the experimental program the following measurements were made: (1) temperature in the primary and secondary combustors; (2) static pressure in the connecting channel inlet; and (3) static pressure in the vortex ejector nozzle. The carbon monoxide concentration in the combustion products was measured by means of a gas-analyzer "TESTO-300 M". The loading time and time of burn-out of the wood fuel were kept constant.

The entire combustion process involves three basic stages: (1) biomass ignition (~20 min); (2) the burn-out of volatiles (~60 min); and (3) coke residue after-burning (~60 min). The thermal power during the burn-out of volatiles is around 21 kWt. During the after-burning a considerable temperature drop in the secondary combustor was accompanied by an irregular thermal power output. This fact is explained by the low amount of gaseous components in the chamber and the fact that the burning process close to completion. The thermal power during after-burning of the coke stage is roughly 13 kWt, as the secondary air only diluted the combustion products. To provide the best conditions for the air delivery one possibility is to reduce the swirl generator cross section, but at the same time to maintain the same air velocity in the swirl generator. Another method to produce a constant thermal power is to use a continuous fuel supply, like that applied in all modern thermal power generators.

The air excess rate ($\acute{a}_1$) in the primary combustor is 0.75, while the overall air excess parameter ($\acute{a}_\Sigma$) is 1.55. For this gasifier design the air rate in the primary chamber depends upon the mass flow rate coming through the vortex generator. An excessive amount of secondary air leads to considerably greater value of $\acute{a}_\Sigma$ and a lower combustion efficiency, while to small a rate does not create sufficient draft. Thus, the agreement between the burning process' in the primary and secondary chambers is a very important factor for a gasifier supplied with a vortex ejector.

The measurements made demonstrated a low concentration of carbon monoxide (CO) in the outgoing gases (below 4 ppm), confirming the high completeness of the burning. As the flame in the secondary chamber is a very transparent one (absence of fine particles), this fact confirms also the high burning completeness. As reported in [8], there is a solid correlation between concentration of gaseous harmful products (particulates, carbon monoxide, etc.) and solid particles in the flame.

The numerical simulation of the three-dimensional turbulent swirling flow in the vortex ejector was performed using the commercial software "PHOENICS v.3.5" (UK). A good agreement (below 20%) with the experimental data regarding the ejection coefficient has been obtained. The high flow mixing rate is explained by the intense flow turbulization at the border between the ejected gas and the hot gas. In this region a sharp "peak" in the turbulent kinetic energy was found due to the high flow mixing rate.

3.4   COMBUSTION CHAMBERS (BIOMASS; BIOMASS + PEAT; BIOMASS + BROWN COAL)

A limited research program has been performed to study the combustion of alternative fuels in combustion chambers of an improved design (two and three chambers). Enabling the achievement of the maximum thermal efficiency, as well as a reduction in the formation of harmful organic products (nitric oxides, carbon monoxide, dioxins, furans, and aerosols) and fine solid particulates of the submicron size that are not "caught" by the cyclones [8].

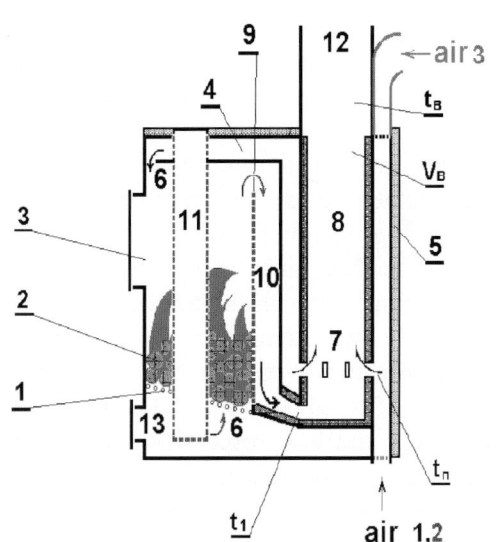

1 – Grate;   2 – Fuel

3 – Primary chamber;   4 – Air duct

5 – Thermal insulation;   6 – Primary air

7 – Secondary air

8 – Afterburning chamber

9 – Dashboard

10 - Additional (third) chamber

11 – Air box;   12 –Gas flue

13 – Ash pit

(Elements transmitting the dual-chamber furnace to three- chamber furnace are represented in dash lines).

Figure 3. Schemes of the investigated dual- and three-chamber furnaces

The alternate fuels (peat, brown coal, and freshly chopped wood) are of high humidity. This is why, the counterflow scheme is usually used in the bistage combustion chamber design. Due to the high heat transfer rate between the solid fuel and the gas, there is a large heat release due to oxidation which can be returned into the combustion zone together with heated reactants.

For the burning of low-grade fuels (wet, high-in ash, cracking during burning) the bistage design does not provide sufficient burning quality. It was found that this problem can be solved by the implementation of an intermediate duct installed between the primary chamber and afterburning chamber [12].

This paper presents results for the burning of alternative fuels (peat; brown coal) in two and three chamber combustor designs (Figure 3). For both cases the fuel (2) was periodically fed to the primary combustion chamber (3) onto the fire grate (1). The air entered to the duct (4) between the case and the thermal insulation (5). The incoming air was divided into the primary air (6), which was fed to the primary chamber (3), and the secondary air (7) which entered into the after-burning chamber (8). The air sharing was organized in such a manner as to burn the wood in the primary chamber with a lack of oxygen (pyrolysis). The secondary air (7) was directed to the chamber (8) through tangential slots, to provide intermixing and afterburning of the pyrolysis products. After combustion in the chamber (8), the burning products were thrown out to the gas duct (12). The unburned part of the fuel fell through the fire grate to the ash pit (13).

The difference between the two and three-chamber designs are as follow: In the three-chamber design the additional chamber (10) with a downwards flow of gases was introduced using a metal sheet (9). In the two-chamber furnace the air enters the primary chamber (3) through the upper slot, while in the three-chamber design the air is supplied under the fire grate by means of ducts (11). The longitudinal fins were welded onto the ducts (4) on the case wall to enhance heat transfer. (Dotted line illustrates details of the three-chamber design).

The following basic parameters were measured in the experiments: (1) the static pressure; (2) the temperature and air velocity at the combustion chamber inlet; (3) the temperature (t_1) at the afterburning chamber inlet; (4) the temperature (t_B), velocity and gas composition in gas flue; and (5) the weight, composition of the loaded fuel and the burning time. The briquetted peat (village Login, Zhytomir region) and wood waste (pine) were employed in the experimental program. The dimensions of the peat briquettes were $155 \times 67 \times 26$ mm^2, while the dimensions of the wood pieces were $50 \times 25 \times 10$ mm^3. The humidity content in the peat and wood pieces was 14% and 8%, respectively. The fuel weight loaded into the chamber was identical for all experiments (6 kg).

Figure 4 shows the temperature variation with time at the inlet of the afterburning chamber (peat briquettes, two- and three-chamber design). In the three-chamber design all experiments were conducted for two different shutter positions, one of which reduces the inlet air cross-sectional area. The gas composition at the exit from the afterburner is also given at several operating times. The carbon monoxide (CO) and nitric oxides (NO$_x$) concentrations were calculated according to the air excess coefficient $\acute{a}$ of 1.0.

At the beginning of the stage with the release of volatile products, a rapid temperature rise occurs. While at the coke residue burning out stage the temperature drops down dramatically. In all experiments during the initial stage the air excess coefficient ranged from 1.4 to 2.7, while during the afterburning stage it was between 6.4 and 9.0. The three-chamber design heats the primary air to temperatures of 113°C above that reached with the two-chamber design.

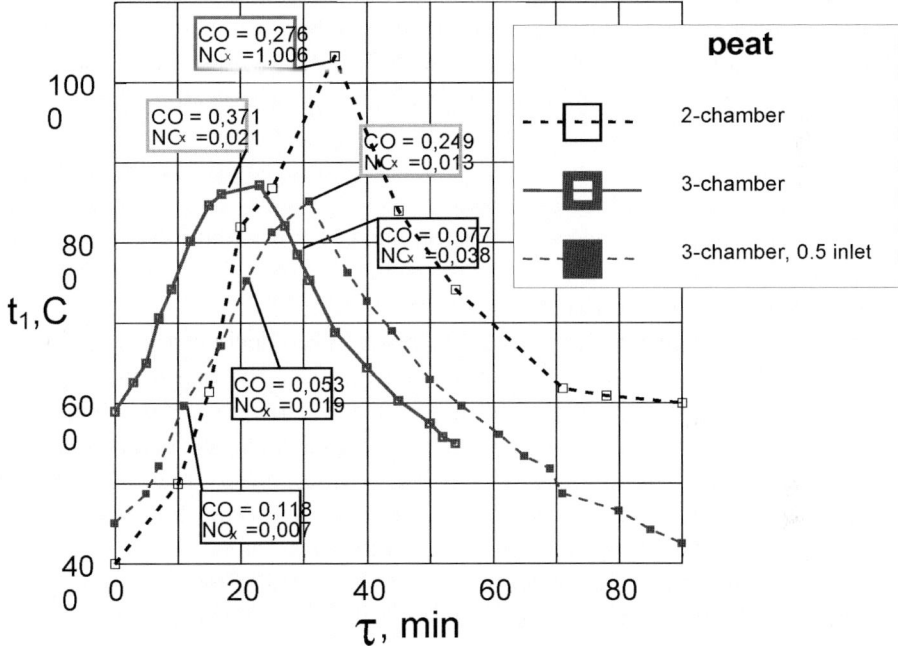

Figure 4. The temperature at the inlet to the afterburning chamber during peat combustion

The comparison of the curves for the two- and three-chamber designs in Figure 4 show that the arrangement of the additional chamber, with the primary air feed under the fuel bed and air preheating decreases the burning time by up to 1.5 times. It was found that the maximum temperature reached in the two-chamber design was around 160°C higher than that in three-chamber design. However, the concentration of nitric oxides in the three-chamber design was up to 20 times less than in the two-chamber design. The halving in size of the inlet air cross-section in the three-chamber design improved the parameters of the burning process, this is apparently not the limit as $\acute{a}$ only decreased to the value of 1.4 at the end of the process of the burning of volatiles.

Comparisons show that the concentration of the carbon monoxide is higher if only peat is burnt in the chamber, rather than a peat and wood blend. At the same time the amount of nitric oxides is a similar level for both cases. As seen, the CO concentrations at identical temperatures can be between 36% and 500% of the average. The growth in CO concentration is always observed for the cases with lower temperatures of preheating air entering the primary chamber. The maximum air preheating rate of 430°C provides the best results [14].

To conclude, the implementation of the intermediate chamber in the high-temperature zone (between primary chamber and afterburning chamber) leads to significant improvements in the solid fuel burning and to reductions in the emission of harmful products. The counterflow scheme of the fuel and the air enables the effective burning of low-grade fuels. The preheating of the air supplied into the primary chamber substantially improves the quality of the combustion process. Even in thermal generators of low power, the small chimney (of height up to 3 m) is able to overcome the hydraulic resistance of the additional chamber and the air heater.

The novelty of the engineering decision has been protected through a patent filed in Ukraine [15].

The high moisture content and low calorific value are the main disadvantages of brown coal. In small size power units the wet fuels are usually burned in the bed layer. According to the burning rate the bed furnace process has some advantages compared with the torch one. However, the enhancement of the burning process reached by means of increasing the air velocity through the fuel layer leads to the removal of particles, which do not have enough time to burn. Therefore, the presence of small-size fractions in the fuel bed does not allow an increase in the thermal load in the combustion zone, and to employ the advantages of the bed furnace process. The low strength of brown coal (Dnieper basin) does not make them competitive, even with peat briquettes. The strong dependence of the bed combustion quality on the layer porosity and temperature conditions is well known [7, 14, 16]. An attempt was made to provide acceptable layer porosity by "fuel briquetting", and the use of an artificial framework of waste materials, and also to improve the quality of combustion by the preheating of the secondary air.

The optimum (from a maximum power point of view) fuel size was determined in [16]. The cylindrical briquettes that were made of the brown coal by pressing (Novo-Dmitrovsk deposit of the Dnieper basin) had the following dimensions: the diameter was 39 mm and the height was 100 mm. The additive of 1% wood tar was used as a binder. The briquette layer porosity was measured to be 0.52. The framework to contain the coal pieces and prevent them breaking up during burning was made of steel chips – the waste product of turning. The layer of briquettes was alternated with a layer of the chip material having a thickness of 1 cm. The combined porosity of the layer of briquettes and the layer of the chip was 0.59. The LHV of the fuel and its moisture were $Q_1^w = 7.96$ MJ/kg, and $W^w = 38\%$.

The number of publications dealing with the burning of blended fuels grows each year [16, 17]. The joint burning of fuels having a different temperature of volatile combustibles enables not only the decrease in emissions of carbon monoxide (CO), but also the reduced generation of nitric oxides NO_x [18]. This is why in some countries the industrial application of combustors running with biomass and fossil fuels is encouraged legislatively [19].

An experimental program has been established in IET to study the joint burning of brown coal and wood in a small size combustion chamber. Figure 5 demonstrates the variation with time of the inlet temperatures for the afterburner for the burning of brown coal, wood and wood-brown coal blends. Thus, the velocity of the gases in the outlet gas flue for the combustion of 6 kg of the fuel stayed in a comparatively narrow range 0.90–1.5 m/s, in spite of the fact that the peat briquettes in the burning process were destroyed.

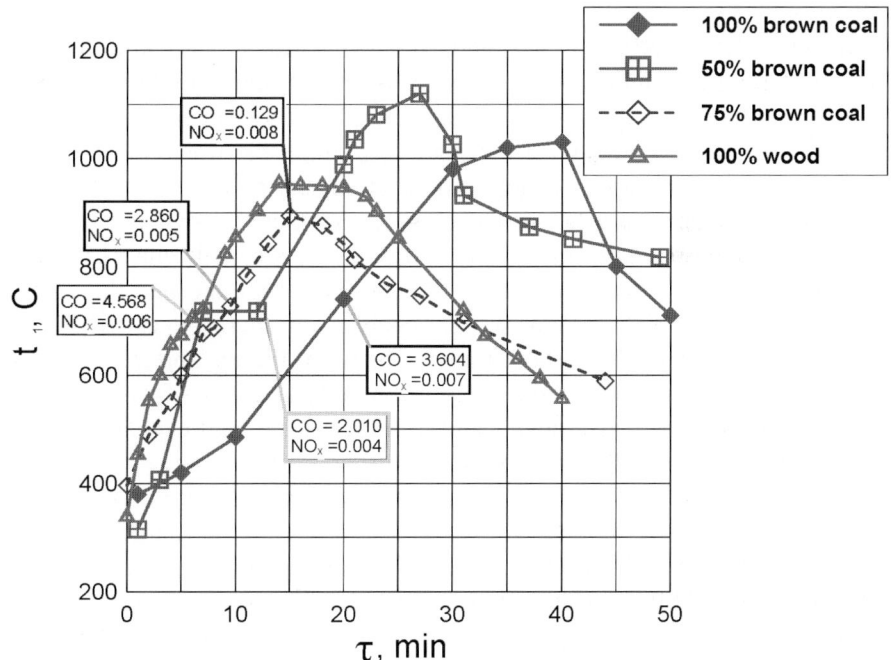

Figure 5. The temperature of gases at the inlet to the afterburning chamber during the combustion of the brown coal, wood, and its mixtures with wood

The chemical gas analysis data has shown that the concentration of NO_x and CO in the exhaust gases has a minimum value for the 50% wood mixture. At these conditions, the emission of CO is about 40% lower than that for all other wood concentrations. Thus, the briquetting of brown coal, the use of an artificial framework and the blending with wood would improve the burning process.

3.5 COMBUSTOR RUNNING THE WASTE MOTOR OIL

Specific vortex flow features has led to the use of vortex technology for the burning of waste oil. Garages, gas and service centers, transport companies and building enterprises often throw out the waste oil or spend money for its regeneration or utilization. Vortex combustion chambers running with waste motor oil enable it to be burned on site, producing thermal energy or hot air.

The complete and effective burning of the "heavy" fuel can be achieved: (1) by a high rate of mixing of the fuel and air; or (2) increasing the fuel-in-combustor residence time. Another way is the evaporation of the "light" components from the waste motor oil and their burning in the rotating flow. As well as the waste motor oil of petrol/diesel engines, the oil from the gear box, hydraulic and transmission system can also be utilized in such combustors.

The general view of the experimental set is shown in Figure 6. The basic component is the cylindrically shaped combustor made of steel, on the bottom of which is a plate containing the waste motor oil, supplied by the gear pump. The air enters tangentially into the combustor, a blade swirl generator is used to create the large-scale vortex over the plate containing the waste motor oil. The hot gas leaves the combustion chamber through the chimney.

A screen-baffle is installed in the chimney so that the swirling hot gas is lifted up initially and then dropped down through the slot-hole channel between the screen-baffle and combustor surface, to be discharged through the chimney blown by the cold air. Such a design enables an increase in the hot gas residence time, enhancing the heat transfer between the oil and the combustor wall. On the external wall of the combustor vertical tubes (d = 70 mm) are welded to improve the heat transfer and to increase the heat exchange surface. Four axial fans are used to blow the combustor with the cold air.

The basic parameters of the waste motor oil combustor are as follows: (1) the fuel tank volume is 35 l; (2) the cross section of the discharge flue is 200 cm^2, length is 65 cm and height is 160 cm; (3) the weight is 120 kg; and (4) fan electrical power is 800 W. The basic operating conditions are: thermal power is 9.2–13.3 kW, fuel rate is 1.23–1.80 l/h, heated air rate of 200 m^3/h and the air rate required for burning is 15–60 m^3/h.

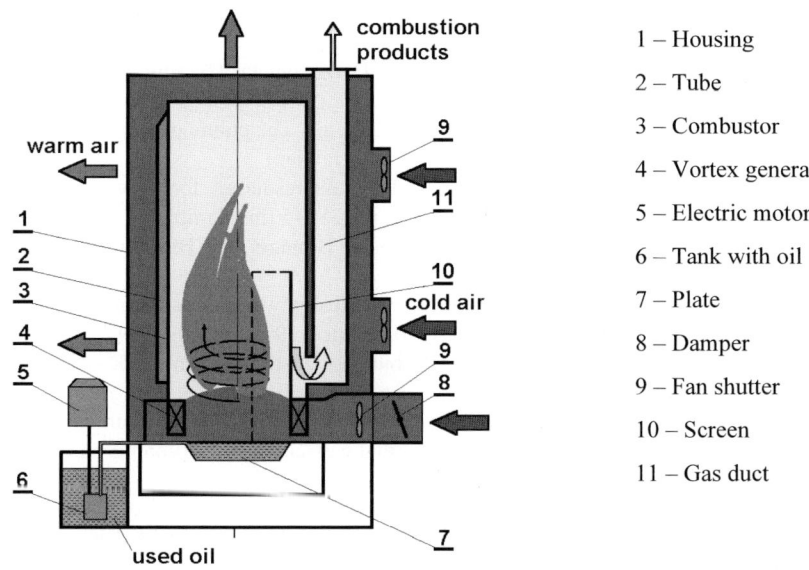

1 – Housing
2 – Tube
3 – Combustor
4 – Vortex generator
5 – Electric motor
6 – Tank with oil
7 – Plate
8 – Damper
9 – Fan shutter
10 – Screen
11 – Gas duct

Figure 6. The heat generator working on waste oil

The vortex combustor design has been protected by a patent in the Ukraine [21]. The design features and results of tests on the vortex combustor (temperature distribution inside combustor; wall temperatures; thermal efficiency, etc.) are presented in [20]. These results have confirmed the successful application of the vortex technology for the waste motor oil and other "heavy" fuels burning.

4. Conclusions

- A few pilot combustion chambers running on peat, brown coal, wood-blends, as well as waste motor oil have been developed and successfully tested.

- Two- and three-chamber combustor designs provide the effective burning of the "pure" alternative fuels and their blends with a wood.

- Due to the air preheating, the three-chamber design provides lower emissions of carbon monoxide and nitrogen oxides into the atmosphere, while running on wet low-grade alternative fuels.

- A large-scale vortex evaporates the "light" components from the rotating waste motor oil; such a technique prevents the generation of harmful pollutants.

References

1. Fuel and energy resources of the Ukraine. Stat. zb. Derzhkomstat Ukrainy. Kyiv, Ukraine, 1998, 384 pp. (in Ukrainian).
2. The state and future of the power engineering of the Ukraine. The public opinion. (Zbirka No. 2).-K.: Energetica ta electryfikatsia, 2005, 328 pp. (in Ukrainian).
3. Zhovmir M.M., Nedovesov V.I., Smirnov O.P. Biomass resources for power use in the Ukraine//Energetica ta elektryphikatsia, 2002, No. 6, pp. 38–45 (in Ukrainian).
4. Tillman D.A. Co-firing benefits for coal and biomass. Biomass and Bioenergy. 2000, 19, 363–364.
5. Hol, Why Kong. Gasification of rubber-wood for electrical power generation in a downdraft gasifier. The 3rd Asian Science and Technology Conference, 1992, pp. 100–104.
6. Zainal Z.A. Performance and characteristics of a biomass gasifier system. PhD dissertation. University of Wales, UK, 1996.
7. Borisov I.I., Khalatov A.A., Geletukha G.G., Kobzar S.G., Shevtsov S.V. Characteristics of the gas producer of the inversed type with the thermal capacity 50 kW running on the wood waste. Prom. Teplotechnica, 1998, **20**(1), 50–53 (in Russian).
8. Gaegauf C.K., Wieser U. Biomass burner with low emission of particulates. Biomass for Energy and Industry. Proc. of the 10th European Conf. C.A.R.M.E.N. Publisher, 1998, pp. 1509–1512.
9. Zuberbuhler U., Baumbach G. Low NO_x furnace engineering for residual and used wood combustion for the improvement of particle burn-out and efficiency in industrial. Biomass for Energy and Industry. Proc. of the 10th European Conf. C.A.R.M.E.N. Publisher, 1998, pp. 1389–1392.
10. Cowburn D.A., Holtman R.D., Berge N., Berg M. The reduction of emission from the combustion of biomass for domestic heating applications. Biomass for Energy and Industry. Proc. of the 10th European Conf. C.A.R.M.E.N. Publisher, 1998, pp. 1377–1379.
11. Launhardt T., Hartmann H. Organic pollutants from domestic heating systems using wood and herbaceous croups. Proc. of 1st World Conference on Biomass for Energy and Industry, June 2000, Sevilla, Spain, Vol. II, James & James, Ltd. (ed.), London, pp. 915–918.
12. Nussbaumer T. NO_x reduction in biomass combustion: primary and secondary measures. Biomass for Energy and Industry. Proc. of the 10th European Conf. C.A.R.M.E.N. Publisher, 1998, pp. 1318–1321.
13. Borisov I.I., Varganov I.S., Dolinsky A.A., Kalatov A.A., Khlebnikov O.E., Kobzar S.G. Furnace for wood waste combustion. Patent of Ukraine, No. 37875A, Bul. No. 4, 15.05.2001 (in Ukrainian).
14. Kovalenko G.V., Khlebnikov O.E., Khalatov A.A. Investigations of combustion of peat and its mixtures with wood in two-chamber and three-chamber design furnaces. Prom. Teplotechnika, 2005, **27**(3) 50–55 (in Russian).
15. Kovalenko G.V., Khlebnikov O.E., Khalatov A.A., Varganov I.S. The furnace for wood waste combustion. Patent of Ukraine, No. 74689, Bul. 16.01.2006 (in Ukrainian).
16. Khlebnikov O.E., Kovalenko G.B., Khalatov A.A. Investigations of the two-stage combustion of peat, brown coal and their blends with wood. Prom. Teplotechnika, 2005, **27**(2), 67–72 (in Russian).
17. Kovalenko G.V., Khlebnikov O.E., Khalatov A.A. The bed combustion of the briquetted brown coal and their blends with wood. Prom. Teplotechnika, 2005, **27**(6), 56–59 (in Russian).

18. Ryohei Miura and others. Research & development for coal and woody biomass co-firing technology in Japan. The 2nd World Conference on Biomass for Energy, Industry and Climate Protection, 10–14 May 2004. Rome, Italy, pp. 1223–1226.
19. Hotchkiss R., Matts D., Riley G. Co-combustion of Biomass with Coal – The Advantages and Disadvantages Compared to Purpose-built Biomass to Energy Plants. VGB Power Tech, 2003, **12**, 80–85.
20. Khlebnikov O.E., Khalatov A.A., Khrienko A.N. Thermal generator running on the waste oil. Prom. Teplotechnika, 2003, **25**(5), 54–56 (in Russian).
21. Khlebnikov O.E., Borisov I.I., Varganov I.S., Khalatov A.A. Device for the waste oil combustion. Patent of Ukraine, No. 38269A, Bul. No. 4, 15.05.2001 (in Ukrainian).

COMBINED ENERGETIC AND ENVIRONMENTAL FUNDAMENTALS OF SELECTION OF THE FUEL TYPE TO ENSURE THE PRODUCTION PROCESSES: THEORY AND APPLICATIONS

P. SANDOR[*]
Research and Development Company for Combustion Technology, 3515 Miskolc-Egyetemvaros P.O. Box 3, Hungary

B. SOROKA
Gas Institute, Nat. Ac. of Sciences, Degtyarivska 39, Kiev 03113, Ukraine

Abstract. Procedures have been proposed for comparing the relative influence of the properties and composition of fuel and oxidant on both power efficiency and emission of harmful (toxic + "greenhouse") substances (NO_x, CO_2) from fuel combustion. Thermodynamic principles were found to optimize the distribution of a set of various fuels by a number of processes under conditions of metallurgical production from the standpoint of minimizing hydrocarbon fuels utilization. The recommended approaches have been developed and accomplished under industrial conditions, to determine main power and main environmental performance of the furnace unit, as well as to evaluate the energetic and environmental consequences of substituting one fuel for another (natural gas, coke-oven gas and a mixture of the two with low calorific blast furnace gas).

Keywords: boiler, coke-oven gas, environmental, fuel, Monte Carlo procedure, natural gas, NO_x formation, power efficiency, reheating furnace

1. Introduction

Power efficiency and environmental pollution are interdependent problems of fuel combustion. Regarding the specific fuel utilization process or unit (boiler,

---

[*] To whom correspondence should be addressed. Research and Development Company for Combustion Technology, 3515 Miskolc-Egyetemvaros P.O. Box 3, Hungary. e-mail: sandordr@vnet.hu

N. Syred and A. Khalatov (eds.), Advanced Combustion and Aerothermal Technologies, 247–258.
© 2007 *Springer.*

industrial furnace) both aspects have the same means of analysis in terms of the combined consideration of the transport (fluid dynamics, mass – and heat transfer) and chemical kinetics of combustion and the formation of harmful substances. A suitable modern approach to the engineering challenge is provided by computational fluid dynamics (CFD) procedures and universal computer codes, such as the various versions of FLUENT, PHOENIX, CFX, etc.

However, there are some global questions which do not need mathematical modeling and the appropriate numerical simulation. Consequently, the conception approach cannot be found in the form of CFD codes because this technique is performed by setting rigorous boundary conditions individual to each unit.

Nevertheless, the general characteristics of the complex problem of the power efficiency and environmental performance of fuel furnaces may be found by agreement with the laws of equilibrium thermodynamics.

2. Comparison of the Influence of Fuel and Oxidant in Combustion Processes

The heat transfer process determines the energetic status and efficiency of industrial furnaces and boiler furnaces. One can evaluate the influence of the properties and composition of the components of combustion using a thermodynamic approach. A similar approach accounts for the formation of NO_x, which is one of the main harmful substances produced by gas combustion. In spite of the significance of chemical kinetics, the equilibrium estimation gives the first reliable approximation for the computation of relative NO_x (NO) emission for various initial conditions.

Because of the mainly thermal mechanism of NO formation and the leading role of Arrhenius temperature dependence of the reacting medium, the value of effective temperature should be under consideration.

We have developed two techniques for thermodynamic calculations:

- Using the thermodynamic equilibrium concentrations of the components at the theoretical (adiabatic) combustion temperature T_T

- The same under more realistic (effective) temperatures T_{eff}, computed by simplified calculations, accounting for temperature correction $\Delta T = T_T - T_{eff}$.

2.1   EVIDENCE OF REPRESENTABILITY OF THERMODYNAMIC APPROACH

The general conclusion regarding the applicability of the thermodynamic approach as the first approximation for computation of the power and environmental efficiency of fuel combustion is positive. It must be stated that the thermodynamic calculation method makes a quantitative approximation of the energy status of the combustion process or unit, whereas, in the case of NO_x prediction, the possibilities of the thermodynamic approach are concerned with the quantitative estimation of the relative influence of changing the composition and properties of the fuel and oxidant, as well as the process temperature. Thus, the determining temperatures are: T_T and T_{ex}, in the calculation of efficiency for NO_x formation $- T_{react} \approx T_{eff} = T_T - \Delta T_{rad}$.

Previous studies by Soroka and Sandor[2] experimental proof of the adequacy of the simple thermodynamic approach was collected by comparing the influence of type of fuel, composition and initial temperature of the oxidant. Within the framework of present paper, the *detailed Monte Carlo procedure* (MC) for the account of radiative heat transfer[3,4] was used to calculate the temperature profile at any point within a multisection furnace chamber of a TVG-8 boiler equipped with two-side heat reception water-tube screens. Monte Carlo radiative heat transfer calculations were used to predict actual temperature profiles at the furnace height and to choose the maximum temperature values depending on the recirculation ratio r with which equilibrium NO_x concentrations were concerned.

The data (curve 2) to be validated with the experimental area 1 in Figure 1 was found by means of MC, for some types of boilers. It can be stated, based on the results of comparing the thermodynamic equilibrium NO_x values at T_T with those which accounted for heat losses, that divergence between experimental (curve 4) and predicted (curve 3) data is higher than in the case of using MC to accurately define the characteristic temperatures.

2.2   DEVELOPMENT OF THERMODYNAMIC PROCEDURES FOR ENERGETIC AND ENVIRONMENTAL ANALYSIS

The following two criteria are proposed for estimating the complex (combined) power and environmental efficiency of fuel use, in accordance with modern requirements: specific fuel consumption b_f and specific yield of nitrogen oxides C_{NOx} per unit of production capacity (furnace output G_{fur}). It has been supposed that C_{NOx} is a more representative value than the concentration of nitrogen

oxides [NO_x] in the combustion products. Taking into account the environmental consequences of the consumption of carbon contained in fuel, C_{CO2} must
be added to these characteristics – the specific issue being CO_2 in the combustion products.

Efficiency of fuel consumption and of the formation of harmful substances
(including toxic and greenhouse ones) has been determined by means of
computation of the composition and properties of the thermodynamic equilibrium
combustion products for various types of combustible (natural gas, coke-oven gas
and their mixtures) and oxidants by support of some given conditions (Figure
2–4). From the standpoint of saving fuel, it has been stated that combustion
preheating by conservation T_T = idem is more efficient (lower b_f) than oxidant
enrichment with oxygen. Comparing these conditions, the specific NO_x yield –
C_{NOx} – is practically the same for air preheating and for air enrichment with
oxygen.

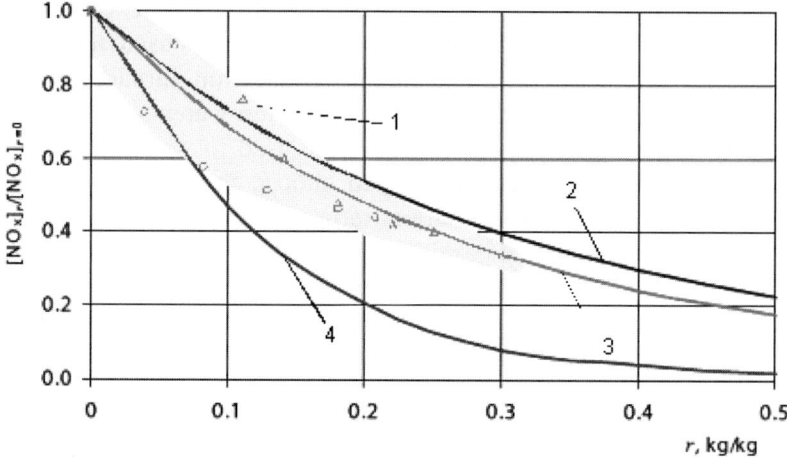

Figure 1. Dependence of relative reduction of NO_x concentration on mass recirculation ratio r of
the natural gas/air combustion products. 1 – Area of recalculated experimental data for the boilers
under consideration in accordance with Sigal[5]; ○ – boiler TGMP-314A arranged with the
modified burners. Temperatures, °C: fuel gas – 20, combustion air – 340, recirculation gases –
350. Excess air factor $\lambda = 1.02$–1.04. Δ – boiler TS-35. Recirculation gases (combustion products)
temperature 20°C; 2 – Our computation data for TVG-8 boiler characteristic temperatures: T_{eff} =
T_T–ΔT_{rad} Temperatures, °C: fuel gas – 25, combustion air – 25, recirculation gases – 150. Excess
air factor $\lambda = 1.2$. 3, 4 – our computation data and experimental data by M. Flamme[6,7] –
respectively – for the type 09 burner under natural gas flow rate – 20 m³/h. Temperatures, °C:
fuel gas – 20, combustion air – 915, recirculation gases – 100. Excess air factor $\lambda = 1.05$

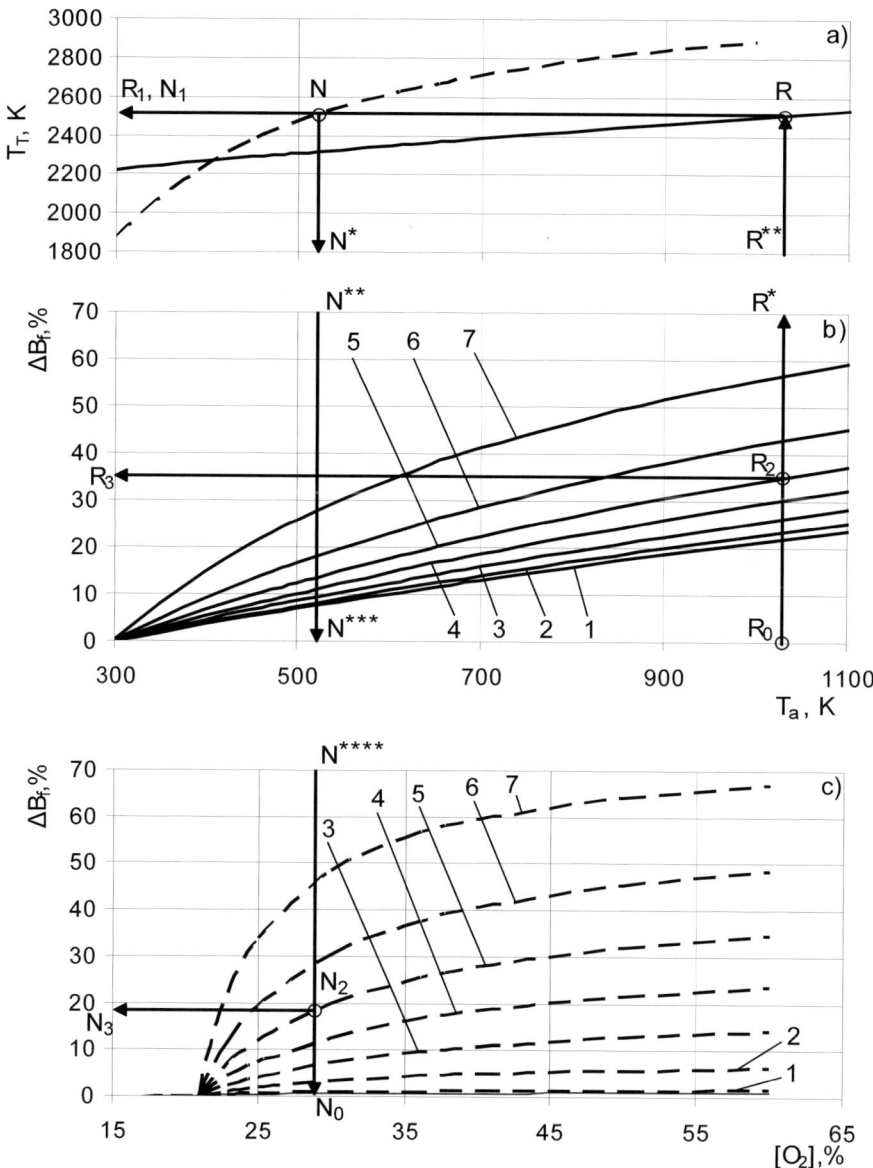

Figure 2. The adiabatic combustion temperture T_T for natural gas – air – O_2 mixtures (a) and fuel saving ΔB_f, % (b, c) in dependence on preheated air temperature T_a (a, b – continuous lines) and in dependence on cold air enrichment by oxygen $[O_2]$ (a, c – dotted lines). The flue gases temperature T_{fl}, K: 1 – 298, 2 – 570, 3 – 840, 4 – 1,110, 5 – 1,380, 6 – 1,650, 7 – 1923. Line R_0-R^*-R^{**}-R and indicators RR_1, R_2R_3 – for preheated air as an oxidant. Line N-N^*-N^{**}-N^{***}-N^{****} and indicators NN_1, N_2N_3 – for O_2 enriched oxidant

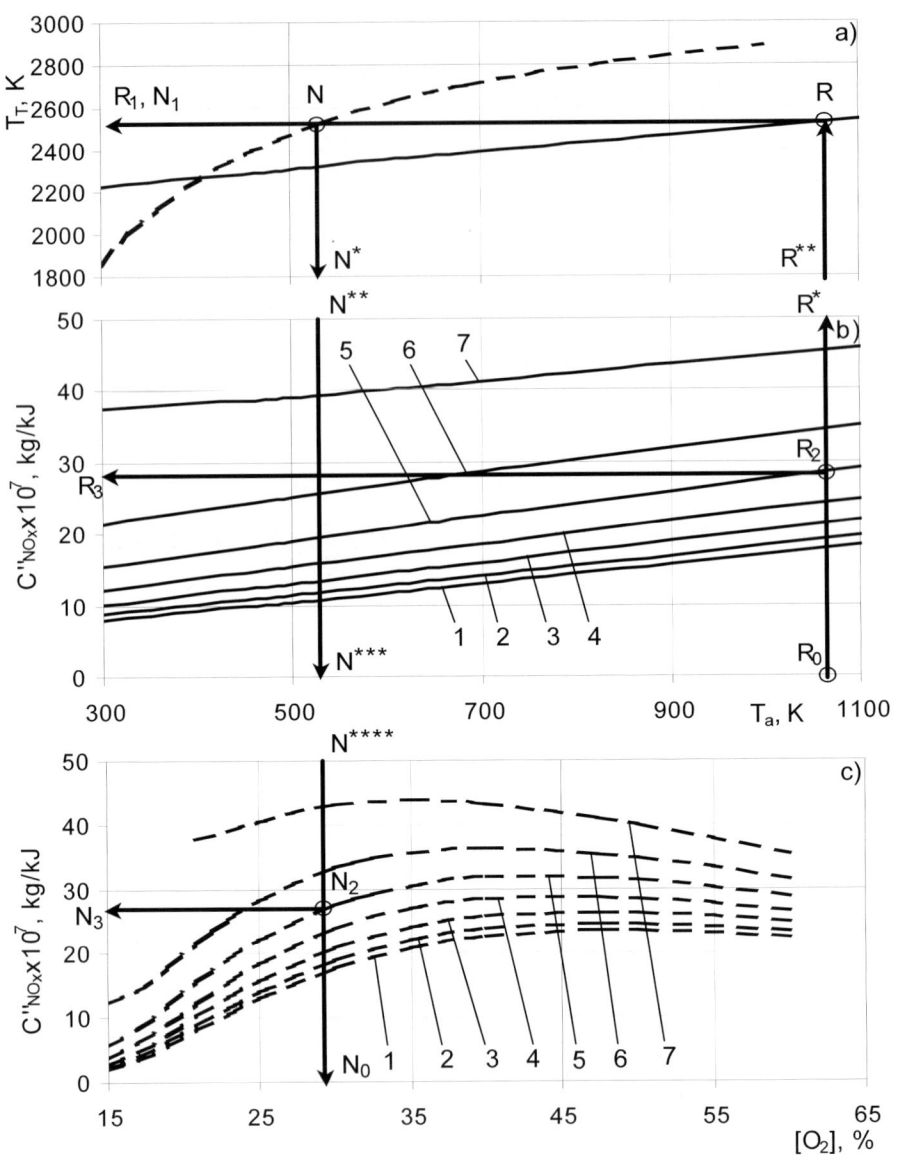

Figure 3. Dependence of the theoretical (adiabatic) combustion temperature T_T on preheated air temperature T_a (I) and on cold air enrichment by oxygen $[O_2]$ (II) – (a) dependence of specific NO$_x$ formation $C''_{NOx} \cdot 10^7$, kg/kJ on T_a; (b) and on $[O_2]$; (c) for various flue gases temperature T_{fl}, K: 1 – 298, 2 – 570, 3 – 840, 4 – 1110, 5 – 1380, 6 – 1650, 7 – 1923. Line R_0-R*-R**-R and indicators RR$_1$, R$_2$R$_3$ – for preheated air as an oxidant. Line N-N*-N**-N***-N**** and indicators NN$_1$, N$_2$N$_3$ – for O_2 enriched oxidant

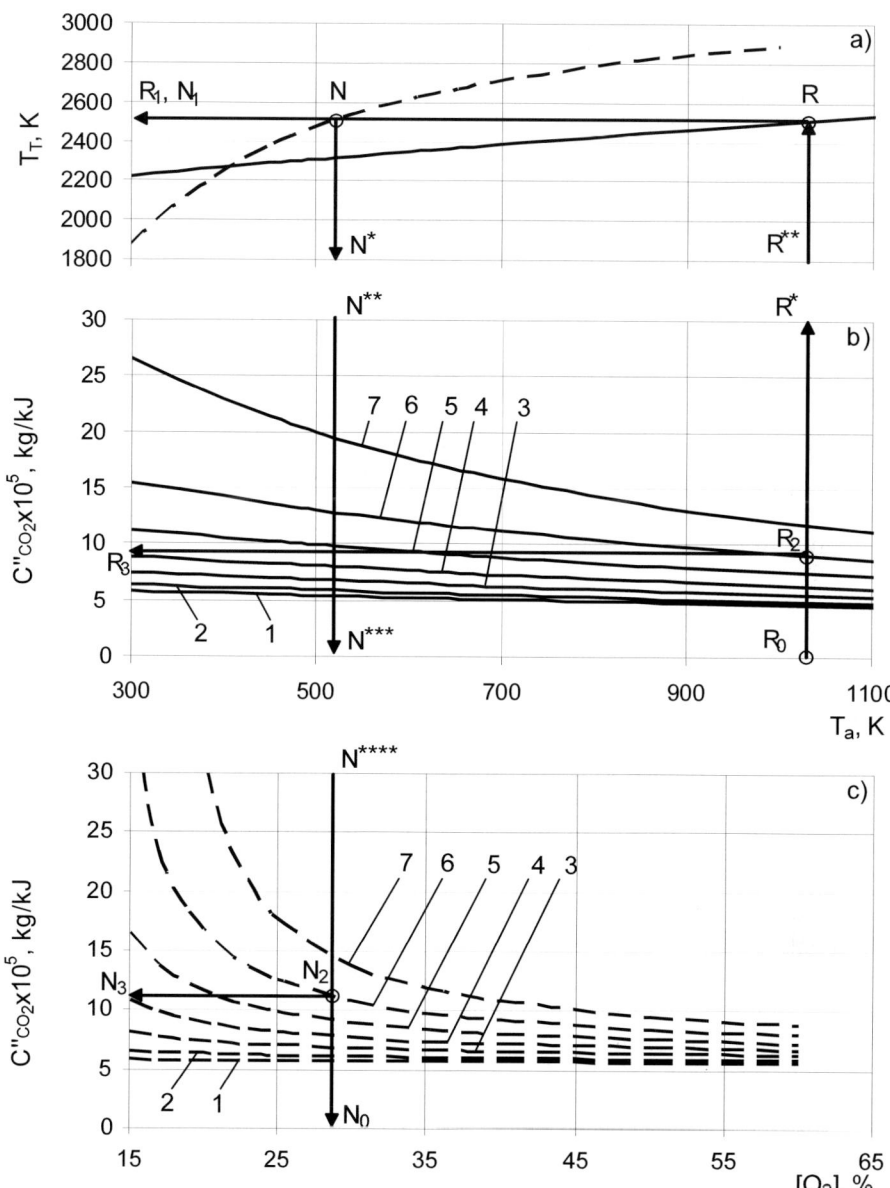

Figure 4. The adiabatic combustion temperture T_T for natural gas – air – O_2 mixtures (a) and specific CO_2 yield $C''_{CO2} \cdot 10^5$, kg/kJ, (b, c) in dependence on preheated air temperature T_a (a, b – continuous lines) and in dependence on cold air enrichment by oxygen $[O_2]$ (a, c – dotted lines). The flue gases temperature T_{fl}, K: 1 – 298, 2 – 570, 3 – 840, 4 – 1,110, 5 – 1,380, 6 – 1,650, 7 – 1,923. Line R_0-R^*-R^{**}-R and indicators RR_1, R_2R_3 – for preheated air as an oxidant. Line N-N^*- N^{**}-N^{***}-N^{****} and indicators NN_1, N_2N_3 – for O_2 enriched oxidant

The difference between C_{NOx} values produced by coke-oven gas and natural gas combustion at similar furnace operating conditions are too small, in spite of the fact that the concentration $[NO_x]$ becomes substantially higher by coke-oven gas combustion. C_{CO2} calculations demonstrate that in the case of O_2-enriched oxidant, C_{CO2} levels increase in comparison with preheated combustion air as an oxidant by conservation of T_T (Figure 4). Using coke-oven gas, the relative situation regarding C_{CO2} in the cases of preheated and O_2 enriched oxidants is similar to that for natural gas firing.

3. Option and Distribution of Fuel Types in a Multiusers System

Selection of a fuel type to perform thermal industrial processes must take into account the technological specifics of the process on one hand, and power need and efficiency as well as environmental consequences – on the other. Estimation of the power requirements is founded first of all, on the first and second laws of thermodynamics, i.e:

- Account of the energy balance (conservation condition)

- Comparison of the fuel (combustible and oxidizer) potential, estimated at the theoretical combustion temperature T_T, with the standard temperature (characteristic temperature) of the process T_p.

Let the following energy saving task need solving. The general production system to be supplied with the set of available fuels comprises some technological processes. We have at our disposal various types of fuel and oxidant ensuring different values of power potential for different combinations of combustible mixture. Bringing the fuel type into accord with each of the processes under consideration takes primary significance under limiting conditions of the fuels of the highest heat value, mainly of hydrocarbons.

So the effectiveness function is

$$\left\{ \frac{\sum\limits_{(HC,i)} B_i^{(HC)} Q_{l,i}^{(HC)}}{\sum\limits_{(i)} B_i Q_{l,i}} \right\} \rightarrow \min \qquad (1)$$

and all the more

$$\left\{ \frac{\sum\limits_{i} B_i^{(HC)}}{\sum\limits_{i} B_i} \right\} \rightarrow \min \qquad (2)$$

According to Prigogine[8], the thermodynamic force $F = T_a^{-1} - T_s^{-1}$ is considered as a measure of the correspondence of the selected fuel type to the thermal conditions of the real irreversible thermal process taking place within the examined system. Here, B, Q_l – fuel consumption, low heat value – correspond to, T_a, T_s – the temperatures of heat receiver and heat source (combustion products), the last two being assumed as $T_a \sim T_p$; $T_s \sim T_T$ by an order of the value. Subscripts: i – type of fuel, HC – for hydrocarbons.

Optimization of the appropriate fuel system with n fuel users (in the case under consideration – the manufacturing processes as the consumers) is proposed to be performed by maintaining minimum entropy output P in time as a criterion, given by Eqs. (3) and (4):

$$\begin{cases} P = \dfrac{dS}{dt} > 0 & (3) \\[2mm] P = \sum\limits_{i=1}^{n} F_i \dfrac{dX_i}{dt} \to \min, & (4) \end{cases}$$

where S is the entropy of the system distributed by the constituents participating in energy (heat) exchange, $X_i \equiv Q_i$ is the amount of heat necessary to perform the ith process, dQ_i/dt is the resulting heat flux received by the working medium within the framework of the ith manufacturing process.

The procedure of optimum fuel utilization including four combustibles – fuel (combustible) gases – has been carried out under the conditions of the metallurgical enterprises at "DUNAFERR RT", Hungary, where the complete production cycle encompasses five parallel metallurgical and power processes (technologies)[9].

The main purpose of optimization is to save the hydrocarbon fuels by substituting them for low calorific gases by implementation in low-temperature processes (e.g., heating the boiler).

As a result of the outlined procedure, the portion of natural gas as the most valuable fuel at the enterprise was successfully minimized. Natural gas now makes up only 23.21% of the total amount of fuel gases whereas the least calorific gases – the blast furnace and converter gases – constitute 41.93% in total. Coke-oven gas, being a high-temperature fuel by T_T value and providing 34.86% of total energy consumption, covers almost the whole fuel demand of the rolling and coke productions.

4. Experimental Validation of the Prediction Data: Main Characteristics of the Industrial Furnace

Fuel consumption and environmental pollution caused by the operation of thermal units may be minimized by a combination of preliminary thermo-dynamic computations and testing the equipment with the available fuel types to determine the associated variation in the furnace output G_{fur}.

The theoretical statements were tested and validated at "DUNAFERR RT"[2,9] rolling mill plant through the investigation of two pusher type multizone reheating furnaces equipped with radiant flat-flame burners with a nominal output of 170 t/h. The following gaseous fuels were compared by firing the furnaces: natural gas, coke-oven gases as well as mixtures of natural gas and coke-oven gas or natural gas and blast-furnace gas. In the past, combustion air was preheated to temperatures in the region of 300–350°C; however, during the 1990s new recuperators were installed, providing air preheating up to 500°C.

The main characteristics of the furnaces are: in relation to thermal efficiency – specific consumption of fuel heat b_f, MJ/t, and for provision against environmental pollution – emission of specific toxic substances C_t, particularly C_{NOx}, kg/t and C''_{NOx}, kg/kJ.

Information on fuel using unit operation can be summarized in terms of two main characteristics: power - the dependence of b_f on G_{fur}; and environmental – the dependence of C_{NOx} on G_{fur}.

As a result of complicated power and environmental tests on the reheating pushing furnaces at "DUNAFERR RT" (Hungary)[2], the main power and environmental characteristics of the furnaces have been found to be qualita-tively similar.

The following conclusions, which are in general agreement with theoretical findings, have been determined as a result of the commercial tests:

- The higher the potential of the combustible mixture, the lower the specific fuel heat consumption b_f. As a result, by furnace firing with coke-oven gas, b_f is approximately 20% less than by firing with natural gas under similar conditions. Use of the mixture of natural gas with blast-furnace gas is not efficient from the point of view of saving natural gas. Bringing the blast-furnace gas fraction up to 50% and increasing the amount of natural gas in the mixture creates overexpenditure of the absolute rate of natural gas instead of expected savings. The following values of specific fuel heat consumption b_f were determined for a furnace output of 160–200 t/h and air preheating of 300°C; 1.55 MJ/kg – by use of natural gas; 1.62–1.96 MJ/kg – by use of a mixture of natural gas with blast-furnace gas, including 1.45–1.77 MJ/kg – the share discharged at the expense of natural gas.

- The higher the potential of the combustible mixture, the higher the NO_x concentration ($[NO_x]$) at the furnace exit under conservation of other parameter values. For example, firing the furnace with natural gas can produce an average of 40% less $[NO_x]$ (ranging from 115 to 285 ppm) than firing the furnace with coke-oven gas: 270–520 ppm. However, consideration of NO_x formation in terms of C_{NOx} reduces this divergence. The difference between C_{NOx} emissions is two or more times less on average than the difference between $[NO_x]$ emissions for the mentioned gases: in terms of $[NO_x]$, the minimum difference is 33%, the maximum is 60%; in terms of C_{NOx}, the differences are 14% and 35%, respectively. Thus, it can be concluded that C_{NOx} represents a much more stable and conservative characteristic than $[NO_x]$ and may be recommended for environmental estimation of the firing units.

- It has been found that the b_f–G_{fur} dependence (main power performance) and the C_{NOx}–G_{fur} dependence (main environmental performance) are similar. The strong dependence of the main furnace indicators, b_f and C_{NOx}, on the furnace output G_{fur}, has been demonstrated, with the region of optimal output $G_{fur,opt}$, corresponding to the minimal values of the mentioned indicators.

5. Conclusion

The theme of this paper is affirmation regarding the opportunity of applying equilibrium thermodynamics approaches by generalized evaluation to the power efficiency and environmental performance of fuel used in processes and plants.

In terms of thermodynamic analysis of fuel using processes and plants, the main applied recommendations are:

- The most representative parameters are the specific fuel consumption b_f and specific harmful substances C_f formation (e.g., C_{NOx}, C_{CO2}).

- Combustion air preheating is more effective than air enrichment by O_2 – from the standpoint of saving fuel.

- Both of the above-mentioned approaches to air-oxidant preparation for combustion produce approximately the same consequences from the standpoint of environment protection (NO_x yield).

Data collected by firing tests of the reheating furnaces at "DUNAFERR RT" and the results produced by comparative analysis based on b_f and C_{NOx} computative values have been found to perfectly coincide for the use of natural and coke-oven gases as the furnace fuel.

References

1. Б. Сорока, Приближенный метод учета влияния состава и параметров топливной смеси на образование оксидов азота в процессах горения, *Экотехнологии и ресурсосбережение*, No. 6, 1993, с. 47–53
2. B. Soroka, P. Sandor. Combined power and environmental optimization of the fuel type by reheating and thermal treatment processes, *Proceedings of the 21st World Gas Conference*, Nice, France, 6–9 June 2000, 15 pp.
3. B. Soroka, K. Pyanykh, V. Zgursky, M. Khinkis, H. Abbasi, J. Rabovitser, Mathematical modeling of low-emission combustion processes basing upon Monte-Carlo procedures, *Preprints of 5th European conference on industrial furnaces and boilers*, 11–14 April 2000, Espinho-Porto, Portugal, Vol. II, 12 pp.
4. B. Soroka, V. Zgurskyy, K. Pyanykh, P. Sandor, Development Of Monte-Carlo Calculation Technique For Detailed Prediction of The Thermal State Of The Industrial Furnaces, *6th HiTACG Symposium*, GWI Essen/Germany, 17–19 October 2005, 9 pp.
5. И. Я. Сигал, Защита воздушного бассейна при сжигании топлива. Л.: Недра (1977), 294с.
6. M. Flamme, Neuester Stand der NO_x-Minderung am Beispiel von Industriebrenners, *GWF Gas Erdgas,* **134** (1993), 10, 533–541.
7. M. Flamme, H. Kremmer, NO_x Emission and Reduction Potential from Industrial Gas Burners with Air Preheat Temperatures of Up To 1,000°C, *Natural Gas Policies and Technologies European Conference.* Part II: Infrastructures and Technologies Conference Proceedings, 14–16 October 1992, Athens, Greece, pp. 136–144.
8. D. Kondepudi, I. Prigogine, *Modern Thermodynamics: From Heat Engines to Dissipative Structures.* Wiley, Chichester, New York/Weinheim/Brisbone/Toronto/Singapore (1999). И. Пригожин, Д. Кондепуди, *Современная термодинамика от тепловых двигателей до диссипативных структур*: Пер. с англ. Ю.А. Данилова и В.В. Белого. – М.: Мир. – 461с. (2002, in Russian).
9. B. Soroka, P. Sandor, Combined power and environmental optimization of the fuel furnace, Proceedings of the 1st Conference *Industrial Furnaces and Refractory Materials*, 7–8 November 2000, Dom techniky, Kosice, pp. 179–183.

REDUCTION AND TREATMENT OF WASTE IN POWER GENERATION

A. J. GRIFFITHS[*], K. P. WILLIAMS
*Centre for Research in Energy, Waste and the Environment,
Cardiff School of Engineering, The Parade, Cardiff,
CF24 3AA, UK*

Abstract. Reduction and treatment of waste within power generating activities offers substantial benefits. These are related to the amount of power that can be generated from waste, such as municipal solid waste (MSW), and the by-product ash, which can be used successfully in the construction sector. As high as 550 kWh of electrical power can be produced from 1 t of MSW with calorific values varying from 10 to 15 MJ/kg depending of the condition of the waste. The ash associated with the combustion process also has potential resource possibilities.

Keywords: waste, MSW, power; energy

1. Introduction

Consider the current world scenario; global warming is an accepted fact of life. The North Pole has lost one third of its ice cap and the USA has seen some exceptional adverse weather conditions. China and similar countries are growing at a phenomenal rate and are now sucking in a disproportionate fraction of the world's natural resources. The outcome is the recent sustained up-serge in oil prices. However, there appears to be no slowing down of consumption. In the UK, for example, both electricity and gas prices have also increased and the need to import natural gas is upon us. Security of supply is now also on the

---

[*]To whom correspondence should be addressed. Prof. A. J. Griffiths, Cardiff School of Engineering, Cardiff University, Cardiff, CF45 3AA. e-mail: anthonyaj2@cf.ac.uk

N. Syred and A. Khalatov (eds.), Advanced Combustion and Aerothermal Technologies, 259–271.
© 2007 *Springer.*

energy agenda and the need to use indigenous resources is making both political and economic sense[1].

TABLE 1. Installed power capacity within the European Union from 1991 to 2001[2-4]

Country	Installed capacity for 1991 (GW)	Installed capacity for 2001 (GW)
European Union	562,156	670,867
Thermal power capacity with the European Union	327,728	390,009
France	104,348	115,513
Germany	118,227	119,389
Hungary	7,193	8,392
Poland	28,288	30,672
UK	70,054	79,729
Japan	–	129,600
USA	–	1049,000

The unfortunate fact is that whilst technological progress over the past three centuries has facilitated much of the population growth, chiefly with measures such as improvements in sanitation, ready supply of energy, medicine and intensive farming, its by-product has been the considerable rise in all forms of pollution. To sustain this growth, material and energy consumption has rapidly grown to reflect both the impact of population and that of increased worldwide industrial globalization; with growth comes waste. However, the rate of consumption has effectively followed gross domestic product (GDP) and thus, developed countries have consumed a disproportionately large fraction of the scarce resources.

Table 1 shows the declared power generation of a number of countries within the European Union for 2001. Between 1991 and 2001, the installed capacity of the European Union increased from 562,156 to 670,867 GW i.e., a 19% rise. Furthermore, in 1991 France, Germany and the UK installed 52% of the declared capacity but by 2001 this had fallen to 47%. This indicates that the other member states were becoming more power hungry and their citizens more materialistic. In the same time frame France, Germany, and the UK increased their declared capacity by 14%, 1% and 11%, respectively. It is interesting to note the very small German rise. This was in part due to the bringing together of East and West Germany and the recession of the German economy. It can also be seen that the three new European members also increased their declared capacity by about 10%. During the same period the thermal installed capacity has remained high at 58% of the total. It is expected to remain at this level for the foreseeable future and hence combustion systems will be fundamental in supplying the increased power consumption. The capacity of Japan and the USA is indicated as a bench mark. The USA has 1,049,000 GW of capacity, which is 57% greater than the total European capacity. About 60% of the capacity is generated from fossil fuels. In the case of Japan, over 66% of the electrical capacity is from fossil fuels, which are mainly imported.

Table 2 highlights the equivalent municipal solid waste (MSW) generation. There is a strong link between consumerism and waste generation. The waste generated per capita for France, Germany and the UK is above the European average, while Hungary and Poland is below the average. The USA at 777 kg/capita is 51% greater than the European average, while Japan at 412 kg/capita is about 20% lower. This is seen as partly being cultural.

TABLE 2. Treatment and disposal of municipal solid waste in 2001[2]

Country	Municipal waste arising (million tons)	Waste generated per population (kg/capita)	Percentage landfilled (%)
European Union	238.91	516*	54
France	32.17	545	43
Germany	48.84	610	25
Hungary	4.81	475	88
Poland	11.08	287	96
UK	34.85	590	80
Japan	52.36	412	7
USA	229.2	777	56

*Estimate for 2001

In 2001 Japan generated 52.36 million tons of waste, which was similar to that of Germany. However, with substantially different topographical conditions, it would be expected that Japan would need to use extensive waste reduction measures to help resolve the problems of disposal. For comparative purposes, MSW production in the USA, the UK, France, Germany are also shown, the variations being due to population differences. The waste per capita for each of the countries identified show significant variations, which are assumed to be a function of cultural and life style differences. Furthermore, there is a substantial variation in the amount of waste that is landfilled, which will be a function of the availability of disposal sites as a function of land mass, the use of other reduction technologies, recycling and local control and legislation. The rise in municipal waste is shown in Table 3. This table highlights the changes in waste from 1995 to 2001. The data are shown as kg/capita and therefore normalizes the population size.

TABLE 3. Municipal solid waste collected per capita during the period 1995–2001[2]

Country	1995 (kg/capita)	2001 (kg/capita)	Change since 1995 (%)
European Union	459*	516*	12.4
France	501	545	8.9
Germany	533	610	14.4
Hungary	505	475	−5.9
Poland	285	287	<1
UK	499	590	18.2

*Estimated for the year based on previous consumption during the period

As can be seen, all the major countries show a substantial increase since 1995 normally increasing from about 500 to 600 kg/capita, i.e., an average 3% year on year rise in waste. The only exceptions are France, Hungary, and Poland. In France, the growth is about half that of the other countries identified, in Hungary there is a 5.9% decrease and for Poland a rise of less than 1%. The level of waste production in the latter two countries would have been subject to population reduction and poor economic performance. However, if in general the average growth is maintained until 2020, the amount of waste collected/ generated will double, thus causing further pressure on the disposal of this ever increasing problem. Reuse, recovery and recycling strategies are unlikely to meet this increased potential growth, hence other options have to be urgently considered and thermal treatment processes offer viable alternatives. In some cases, landfill costs have started to show dramatic increases. For example, in the UK costs have more than doubled since 2003. Thus many countries are at a crucial junction in tackling the ever-increasing waste problem.

Thus, a fresh look at its strategy indicates that processes such as "energy from waste (EFW)" could now play a more important role. This neglected energy source could also have a major impact in power generation at a local level and hence resolve some of the potential future generation problems, while at the same time meeting the EU landfill directive and other world wide legislation as well as conserving landfill void space. The calorific value of MSW is about 10 GJ/t (about one third of that for coal), which when burnt in a modern, well-controlled mass burn incinerator will give about 20% efficiency for conversion to electricity. Hence, in terms of power, about 550 kWh/t of MSW can be generated. Much higher conversion efficiencies are possible with improved recycling and refining through processes such as mechanical/ biological treatment (MBT) to create an energy-rich, consistent material, well suited as a fuel. Here, potential conversion to electricity efficiencies as high as 40% are possible. Table 4 highlights the critical components that affect steam raising capacity and hence power production. As expected, the amount of moisture and ash (noncombustible material) has a critical impact on calorific value. Lee et al.[5] suggest that the potential for energy generation from both the MSW and commercial and industrial residual sectors could rise from 19% of the total power consumption in 2005 to over 25% in 2020.

TABLE 4. The key components of MSW that affect calorific value

	MSW in as received state			
Water content (%)	15	18	25	39
Noncombustion (%)	14	16	20	28
Combustion (%)	71	66	55	33
Calorific value (mJ/t)	15	14	12	9
Steam generated (t/t of MSW)	4.3	3.9	3.2	2.3

2. Municipal Solid Waste

Generally, between 80% and 85% of MSW arises from the residential/domestic sector. Typical household waste classifications in an urban and a rural authority are shown in Table 5[6]. Paper/card represent large fractions within the waste stream. However, at 30% and 52% for urban and rural households, respectively, putrescibles represent a significant fraction that will impact on potential calorific value. It is also noticeable that the fraction of plastics at a typical 8–10% is high. This trend has been increasing since the early 1990s. The use of MBT plant, as described earlier, will have an impact on the potential residual streams which constitute refuse-derived fuel (RDF). Typical compositions arising from such plants are shown in Table 6[7]. Today, there are a range of processes that can use a variety of waste streams, thus offering a range of energy outputs and hence a flexible solution in terms of waste reduction and more importantly energy generation.

TABLE 5. Household waste classification for an urban and rural authority[6]

Category	Classification for an urban authority (%)	Classification for a rural authority (%)
Paper/card	24	25
Putrescibles	30	52
Glass	8	3
Textile	4	2
Ferrous	4	2
Nonferrous	1	0.5
Plastic	10	8
Fines	6	1.5
Miscellaneous	13	6

TABLE 6. Percentage composition of RDF and process residues from typical MBT plant

Constituent	Ecodeco* (%)	Grontmij[7] (%)	Herhof[7] (%)
Paper and card	35.1	25.0	55.0
Putrescibles	2.1	15.0	0.0
Glass	0.9	4.0	1.0
Fines	20.3	0.0	0.0
Ferrous metals	2.2	1.0	0.0
Dense plastics	23.0	15.0	5.0
Plastic film	0.0	5.0	4.0
Textiles	14.0	0.0	5.0
Miscellaneous combustibles	1.6	30.0	30.0
Miscellaneous noncombustibles	0.8	5.0	0.0
Dry lower heating value (mJ/kg)	16.5	17.2	17.4

*Measured at Cardiff University

3. Thermal Treatment Processes

3.1 GENERAL

The different types of technologies used within an integrated waste management route can be classified based on material input, processing temperature and related by-products. Generally, they are attached to some form of recovery system such as bulk material recovery (MRF) and/or MBT plants.

Figure 1. Manufactured fuels using MSW. (a) MSW fluff used to manufacture RDF from MBT plant; (b) pellets manufactured using extrusion

Product streams from these plants are generally a fluff or pellet as shown in Figure 1, to which this fuel is then fed directly to the thermal process. The process could be a mixture of fuel preparation and thermal conversion, which may be sited near to bulk recovery plants or as a centralized facility. A generalized integrated waste management flow incorporating thermal treatment is shown in Figure 2.

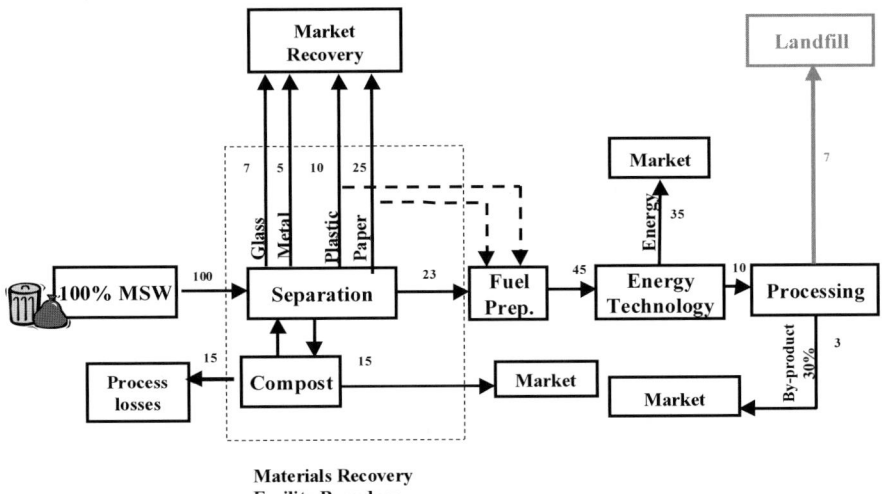

Figure 2. Schematic Illustration of an Integrated waste management scheme (with nominal percentage flows)

As can be seen the flow includes separation, composting, thermal treatment and processes for using the secondary residue. It is assumed that aggregate and cement production are the main beneficiaries of thermal residue. The figures quoted in the diagram represent an ideal material balance and indicate that landfill is still required, amounting to about 7% of the initial MSW. Figure 3 shows a typical temperature process path for waste treatment, which can be summarized as:

- Anaerobic digestion – organic fraction of the waste stream
- Sterilization – this is a low-temperature pretreatment process using "as received" MSW followed by a range of waste manipulation processes
- Gasification and pyrolysis – flexible process producing a range of outputs including syngas
- Direct combustion – most common form is incineration

Anaerobic digestion produces a reasonable calorific value gas which can be directly fired to produce power as well as heat energy. It is worth noting that the sterilization process is a pretreatment process and as such does not produce power directly. It does however, condition the residual waste for use in energy generation. The char derived from the process does not lend itself for further processing due to the high levels of carbon left in the material.

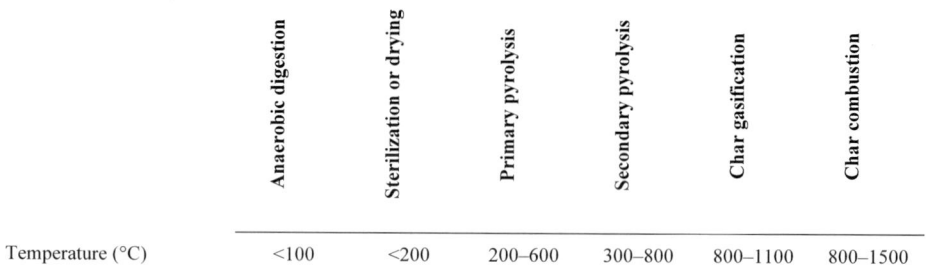

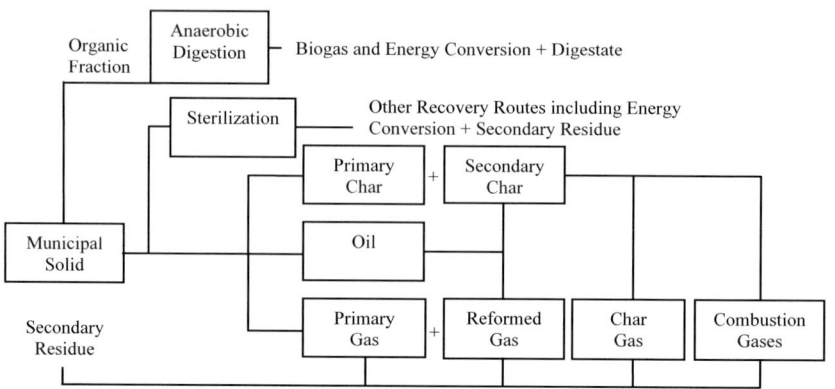

Figure 3. Thermal treatment technologies according to processing temperature and available energy form

3.2 ANAEROBIC DIGESTION

Anaerobic digestion has a number of environmental advantages and was initially designed to handle farm slurries. The use of this technology is gaining acceptance in processing the organic fraction of MSW after which the digestor can then be disposed of. However, there is further interest in using this as an amendment to the traditional composting process. Additional studies are required to define the processing envelope for such material. The syngas produced can either support the process or be converted into electricity by a suitable prime mover/generation plant. The most critical factors that affect the economic viability of digestors are feedstock, gas yield and efficiency of utilization. There are three main types of digestion process, depending on temperature. These are:

1. Psychrophilic digestion occurring at a temperature between 5 and 15°C. This low-temperature process requires less time for heating, but needs

more time to breakdown the waste into syngas. This is seldom used as it is difficult to maintain the reactor at a constant temperature.

2. Mesophylic digestion occurring at temperatures between 25 and 45°C. This requires some heat input.

3. Thermophilic digestion occurring at temperatures between 55 and 75°C, which requires a much larger heat input for the process to occur. The heat input can be as much as twice that of the mesophylic process. However, the material is broken down more quickly.

The size of the digestor is a function of the process temperature and the economic retention time, which is the time needed to release the maximum energy, taking into consideration the energy potential of the syngas and the energy required to heat the digestor. These systems are normally used in conjunction with a boiler or combined heat and power (CHP) plant and the by-product is a liquor and solid residue.

3.3 STERILIZATION

Sterilization is commonly regarded as a prerequisite to a "dirty MRF" where the MSW is accepted in the "as received" state. The process typically uses autoclaves to heat and sterilize the organic fraction of the waste at a temperature of up to 175°C using steam. The process is undertaken in batches and recycling is conducted once sterilization has been completed. The pathogens are killed off and the recyclate fractions such as cans and glass, etc. are cleaned during the process. Due to the processing temperature, the plastics are generally agglomerated for further processing say as feedstock for a CHP system. The organic fraction, including paper, forms a "mush" which can be composted. This type of process has good potential and has gained considerable interest. However, for those countries that have specific recycling and composting targets, there has been some concern over the recovery rates, especially that for composting, since a large fraction of the waste is used to generate energy.

3.4 GASIFICATION AND PYROLYSIS

The processes are extremely complex and are highly sensitive to moisture content, quality of the feedstock and processing temperature. The basic differences between the two processes are twofold. Firstly, pyrolysis is undertaken at an equivalence ratio (actual fuel/air required for complete combustion) less than 0.2, typically below 0.15, whereas gasification usually has an operating ratio in the ranges of 0.2–0.4. Secondly, the processing temperatures for pyrolysis are lower than gasification, as shown in Figure 3.

Pyrolysis is divided into fast, slow, ablative, and flaming processes depending on the rate of change of temperature with time. Gasification is divided into fluidized bed, suspension, up draft, down draft, and cross-bed processes. Carbon monoxide, hydrogen, and methane are the typical gases generated within the reaction chamber.

These processes are gaining rapid acceptance in the market place as a viable alternative to incineration, especially at throughputs less than 100,000 t/year. The economics are crucial for market acceptability since operating costs can be high, so gate fees, markets for the residues and CHP output are critical factors for commercial viability.

Of the two, gasification is a more robust process to operate, having a wide range of operating conditions and techniques already highlighted. The UK Energy from Waste (EFW) Association has highlighted that average capital costs range from 8 to 93 £M (up to 360,000 t per annum plant capacity) with equivalent operating costs being about £45/t. The bonuses are:

1.   Planning hor izons are 4–6 months

2.   Design to commissioning time scales are only 12–30 months

If the waste is used merely as a nonhomogeneous fuel source, there are major impacts on the calorific value, quality and variability of the fuel gas produced for further use. Best operational conditions are obtained when the material is sorted and preprepared so that uniformity can be maintained. When the syngas, which can be hydrogen rich, is produced to drive a gas turbine, cleanliness and composition are crucially important to maximize power output. The downside is that there is a considerable lack of commercial systems operating using MSW.

At Cardiff University research within this field has been undertaken for a number of years and operational prototypes have been developed. Furthermore, it has a Waste Research Station that in conjunction with large-scale composting research facilities is allowing sustainable waste management techniques to be developed. On the ground, data collection is feeding into research programs that clearly indicate a need for EFW processes to be continually developed to meet the changing face of waste generation.

3.5 DIRECT COMBUSTION

The most well-known process for this technology sector is incineration. This operates with equivalence ratios greater than 1 and incorporates the general stages of drying, partial pyrolysis/gasification followed by final char burning eliminating the majority of carbon in the ash. This ash can be used as feedstock for processes such as cement manufacture and is therefore a valuable source of income. The key variables controlling the process are residence time, temperature, moisture content and size distribution of the material. The general types of process that use this technology are grate firing, suspension firing, and the use of fluidized bed combustion systems, which can be either of a standard, circulating or revolving bed design. Currently, the favored technology for large-scale plants (greater than 100,000 t per annum) has been the moving grate design, although there has been some variation to include partial suspension above the grate. This technology has been "tried and tested" with a range of fuels including MSW. Generally, a large amount of environment legislation exists alongside this technology so that the back end processes are very sophisticated to ensure discharges are within limits set by a particular country.

The Energy from Waste Association has also confirmed that mass burn incineration is at present the main thermal process with typical capital costs ranging from 16 to 100 £M depending on plant capacity (up to 500,000 t per annum plant capacity), with associated operating costs typically £45/t. Typically, planning time horizons range from 24 to 60 months. While design to commissioning time scales varying from 54 to 96 months. This clearly shows reluctance to promote this process and more importantly, the loss in time will ensure that, for example, the UK recycling and waste strategy is not developed on time.

"Mass burn incineration" is seen as a dirty phrase, stemming historically from plants being poorly operated. Modern mass burn incinerators are highly complex chemical plants with much of the investment arising from the complex treatment of the flue gas, solid and liquid effluents and clearly, rebranding is required. The downside, apart from public perception, is thermal efficiency. Typically, mass burn incinerators operate at about 20% for power generation, much higher if CHP is possible (although difficult to economically justify). Furthermore, typical plant capacities are in the range of 250,000 t per annum which is significantly greater than that generated by the largest unitary Authority in Wales for example. Current designs meet all emission/pollution standards. However process economics are doubtful at small scale.

Table 3 highlights the breakdown of MSW processes for those countries indicated earlier. In terms of incineration there is a wide variety of usage, ranging from 77% in Japan to only 7% in the UK. Historically the UK used landfill as its solution to the waste problem and there has always been a reluctance to use incineration since it has faced poor acceptance and public hostility. In Luxembourg 44% of its waste is incinerated and like Japan it reflects the lack of landfill disposal sites. In France, Germany and the Netherlands a range of typically 20–30% incineration is used. Again, there is some public disquiet over the use of this technology, but realizsation of the need to reduce the mass going to landfill and the lack of land availability has overshadowed any public concern.

As shown in Table 7, Japan is driven by technological solution to dispose of its waste. Although, incineration is the dominant technology there is considerable interest in

- Gasification with ash melting/vitrification attached to the overall process
- Thermal decomposition technology for hazardous "off gases"
- Conversion of waste plastics into oil
- Stoker furnace technology for improved incineration

TABLE 7. The amount of MSW processing for a range of countries for 2001[2]

Country	Landfilled (%)	Incinerated (%)	Recycled, composted or other treatment (%)
European Union	54	16	20
France	43	32	25
Germany	25	22	53
Hungary	88	10	2
Poland	96	0	4
UK	80	7	12
Japan	7	77	15
USA	56	15	29

4. Ash By-product

Fly ash from coal-fired power stations has been traditionally used in cement kilns for many years. Thus, with the potential influx of bottom ash from incinerators which could be between 15% and 25% of the input material, there is a large resource that could be used to benefit the construction industry. Many companies such as South East London Combined Heat and Power Ltd. (SELCHP) based in east London already produce over 100,000 t of bottom ash and have been working together with Hanson to produce a range of aggregate materials and block materials for specific applications. The next step is road manufacture using a mixture of bottom ash with asphalt as well as cement replacement. The industry is at a critical juncture where acceptability in the market place and support legislation is the vital necessary stimulant.

5. Conclusion

With the projected increase in MSW generated over the next 20 years, many countries are facing real problems in providing robust workable strategies for processing this waste. Thermal treatment offers a significant opportunity to support governments' drives to reduce waste going to landfill in conjunction with recycling and composting. Many of these plants also have net energy outputs, which for many countries could have a significant impact on power generation, since the mix and match of energy as well as security of supply are fundamental issues. The only certainty will be the mix of solutions in which thermal treatment and hence "energy from waste" will play a crucial part. Thus, for the "thermal treatment" sector to grow and hence succeed, political stimulation and public confidence will need to be developed alongside further continuing research and development in the technology.

References

1. A.J. Griffiths and K.P. Williams, Municipal Solid Waste. The Unmentionable Alternative Energy Source, *Waste Management Yearbook*, McMillan Scott, 2006, p. 69.
2. Energy, Transport and Environment Indicators – Data 1991–2001, Office for Official Publications of the European Communities, p. 161 (2004).
3. Energy Information Administration (1 April 2006), www.eia.doe.gov.us
4. Energy production in Japan (April 3, 2006), www.en.wikipedia.org.jn
5. R. Lee, D. Fitzsimons, and D. Parker, Quantifications and Potential Energy from Residuals in the UK, *Institution of Civil Engineers* (March 2005), p. 46.
6. N.E. Owen, A.J. Griffiths, K.P. Williams, and T.C. Woollam, Classification of Household in Ceredigion, Cardiff University, Report No 3118, (January 2006).
7. E. Archjer, A. Baddeley, J. Schwager, and K. Whiting, *Mechanical Biological Treatment, A Guide for Decision Makers. Processes, Polices and Markets*, Juniper Consultancy Services, Ltd., 2005.

BOILERS, FURNACES AND RELATED SYSTEMS

CLEANING OF FLUE GASES OF THERMAL POWER STATIONS FROM ASH AND SULFUR DIOXIDE BY EMULSIFIER: WET CLEANING UNIT

S. BORYSENKO, L. MALYY, G. BYKOVCHENKO,
L. BADAKVA
Yuzhnoye State Design Office, Dnepropetrovsk, Ukraine,
e-mail: info@yuzhnoye.com

Y. KORCHEVY, A. MAYSTRENKO
Institute Coal Power Technologies, Ukraine, Kiev,
e-mail: cetc@i.kiev.ua

Abstract. This paper presents the results of the development of a unit for the wet cleaning of the flue gases of thermal power stations (TPSs) where coal is used, from ash and sulfur dioxide – the emulsifier.

The results of experimental investigations of ash catching, verified by industrial tests, are presented in the article. The ash catching degree for provision of the optimum flow rate parameters is more than 99.9%.

The usage of water solutions of reagents (NaOH, Na_2CO_3, NH_3, $Ca(OH)_2$) of low concentrations as a spraying liquid allows to exploit the emulsifier for sulfur dioxide SO_2 catching with 80–95% efficiency. Flow rate parameters for the achievement of the required efficiency have been determined experimentally.

The experiments have been carried out with prepared model gas, containing a mixture of air (heated up to 160°C) and ash or SO_2 in the concentrations simulating their content in the natural flue gas.

It has been determined that there is an increase in efficiency of the ash catching process when there is increase of the velocity and the flow rate of the spraying liquid.

As a result, we have determined a range of flow rate parameters when stable operation of the emulsifier is provided with a 99.9% efficiency, with no higher entrainment of splashes.

Sodium hydrate NaOH, soda ash Na_2CO_3, ammonia NH_3, slaked lime $Ca(OH)_2$, limestone $CaCO_3$, magnesium oxide MgO, and carbamide $CO(NH_2)_2$ have been selected as absorbents.

N. Syred and A. Khalatov (eds.), Advanced Combustion and Aerothermal Technologies, 275–284.

The application of four reagents (NaOH, Na_2CO_3, NH_3 and $Ca(OH)_2$) has produced positive results when those reagents had been used as absorbents in the emulsifier's conditions.

The effect of increasing the flow rate parameters (needed for increasing the efficiency of the gas-cleaning processes) on the growth of hydraulic resistance of the emulsifier has been also determined during the experiments. Results of those investigations prove the fact that achievement of the maximum required cleaning efficiency and the capacity of smoke exhausters should be taken into account for determination of the flow rate parameters.

The emulsifier is a highly efficient, comparatively cheap unit of small dimensions for the wet cleaning of flue gases from those TPSs where coal is used for operation.

Keywords: coal, flue gases, harmful effluents, effective wet cleaning

1. Introduction

The power production of countries where coal is utilized as a fuel for thermal power stations (TPSs) is connected with a problem of efficient flue gas cleaning from harmful effluents. The basic effluents are ash, sulfur oxides and nitrogen oxides.

TPS become the basic sources of pollution of the environment when an intensive growth of power production is connected with the growth of industrial manufacturing and power consumption, with the modern tendency being to reduce the number of operating atomic power stations. Effluents in some countries achieve millions of tons annually; exceeding the allowed norms by factors of 10 and 100 times.

The task of high-efficient cleaning of the exhausted flue gases is especially urgent for those countries where local low-grade coal is used for power production. Those types of coal are of higher ash and sulfur content. Coal washing is not reasonable from the economical standpoint, if we take into account the low level of efficiency of that technology.

There is an urgent necessity to modify the operating TPS by the installation of high-efficiency environmentally protective equipment together with a necessity to develop a new generation of boiler equipment which can provide realization of the ecological norms.

Yuzhnoye State Design Office together with the Institute Coal Power Technologies scientists and designers have built a unique unit of low dimensions for wet cleaning of TPS flue gases. That unit was called an emulsifier.

The throughput capacity of the emulsifier is 50,000 m^3/h for gas. The general view of the emulsifier is given in Figure 1. The emulsifier has been initially designed as an ash catching unit. It is a set of single filtering elements (pipes) connected by two pipe plates (an upper and a lower plate).

Figure 1. General view of emulsifier

The principle of operation of the filtering elements (FE) is based on the interaction of the twisted gas flow with water supplied on the wall of the pipes. The separation of the solid phase contained in the flue gas is achieved by the moistening, weighting and removal in the form of slime under the action of the gravitational forces that take place in the field of centrifugal forces. In this case, a gas–liquid column with intensive mass exchange is formed above the swirler.

Flue gas cleaning from SO_2 takes place when water solutions of the reagents are supplied on the wall of the pipe as a spraying liquid; those reagents can react with sulfur dioxide SO_2 and form stable products of reactions.

A method of installation of the emulsifiers in the gas duct with a support on the lower plate is shown in Figure 2. The required number of emulsifiers installed is determined by the flow rates of flue gases in each case.

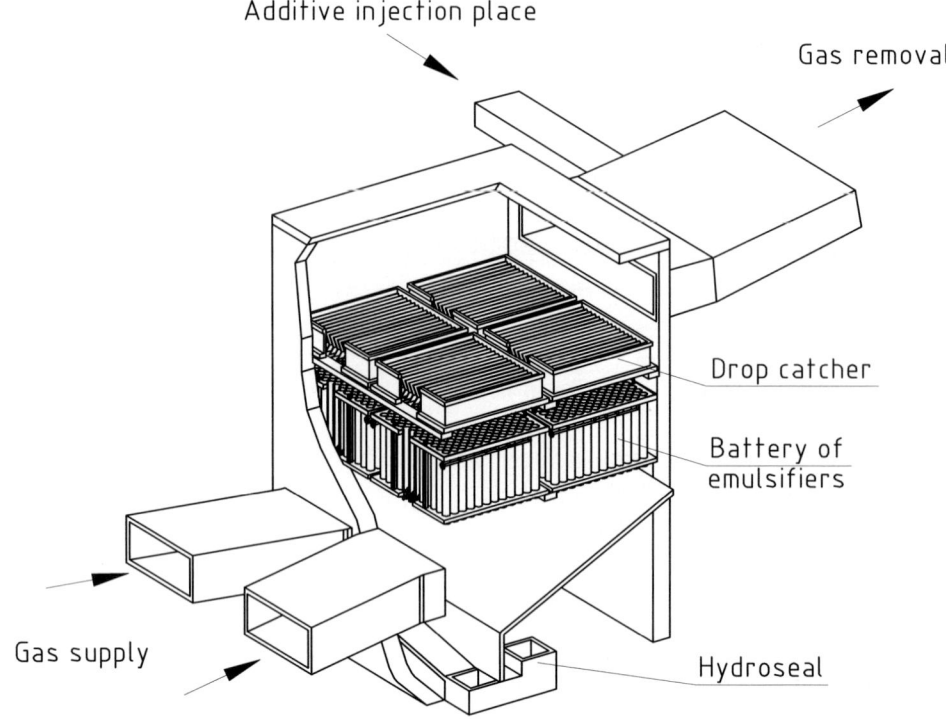

Figure 2. The scheme of installation of emulsifiers at thermal power stations

2. Results

Industrial tests of the pilot samples of the emulsifiers at numerous TPSs in Russia, Kazakhstan and Ukraine have verified high ash cleaning efficiency – 99.6–99.8%.

It was found necessary to study the flow rate parameters in detail, as it is these parameters that provide a stable cleaning process with a constant high efficiency and absence of the higher entrainment of splashes.

Two projects for studying the flow rate parameters of the emulsifier and the ash and SO_2 catching technologies have been executed with financial support from the Science and Technology Centre of Ukraine (supported by the USA and Canada).

The experiments were performed over a 6–12 m/s range of the gas flow velocities in the FE pipe, at 0.1–0.4 l/m^3 specific liquid flow rate. The variation in concentration of the reagents in the supplied water solutions was also investigated.

The experiments have been carried out with the model gas, which was a mixture of air (heated up to 160°C) and ash or SO_2 (depending on a task of an experiment) in the concentrations simulating their content in the natural flue gas.

It has been determined that the ash catching process takes place with increasing efficiency when there is an increase in the velocity and the flow rate of the spraying liquid. The relationship between the ash catching efficiency and changes of these parameters is given in Figure 3.

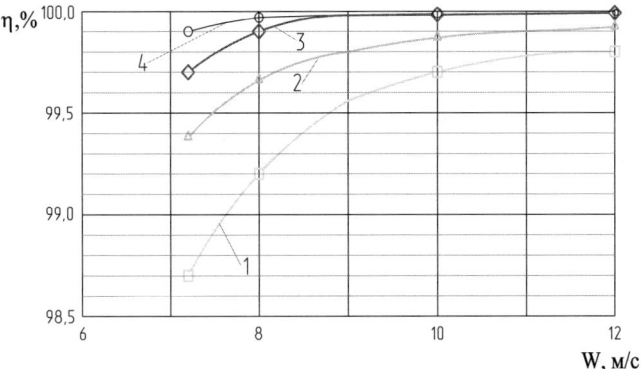

1 – At 0.1 l/m³ specific water flow rate;   2 – at 0.2 l/m³ specific water flow rate;
3 – at 0.3 l/m³ specific water flow rate;   4 – at 0.4 l/m³ specific water flow rate.

Figure 3. Dependence of ash cleaning degree on change of velocity and flow rate of spraying liquid

As a result, we have determined a range of flow-rate parameters when stable operation of the emulsifier is provided with 99.9% efficiency, with no higher entrainment of splashes:

- Gas flow velocity in the pipes – 9.5 ± 1 m/s

- Specific liquid flow rate – 0.25 ± 0.5 l/m³

A checkup of the SO_2 catching processes is performed in the range of the above flow rate parameters using the model gas (air + SO_2).

Some experiments have been carried out with ash blowing into the model gas flow. The influence of ash on the SO_2 catching process was not found to be considerable; it was helpful for simplification of the model gas.

Sodium hydrate NaOH, soda ash Na_2CO_3, ammonia NH_3, slaked lime $Ca(OH)_2$, limestone $CaCO_3$, magnesium oxide MgO, and carbamide $CO(NH_2)_2$ were selected as absorbents.

Limestone and magnesium oxide have been excluded from the above list after the first series of tests, because the SO_2 catching efficiency did not exceed 30–40% when limestone and magnesium oxide had been used. Another undesirable phenomenon was the formation of sediment in the supplying pipelines.

All published documents of the application of carbamide for gas cleaning from SO_x and NO_x prove the fact that the efficient cleaning processes are realized when the temperature in the reaction zone is more than 80°C.

It has been determined during the experiments that the temperature in the gas–liquid column above the swirler in FE does not exceed 42–45°C. All the measures for increasing the temperature which could be realized in TPS conditions (where the emulsifiers are in operation) have not provided any positive effect.

The efficiency of cleaning the SO_2 from the flue gas did not exceed 15–20% when carbamide was used. For comparison, SO_2 catching degree was 10–15% when water spraying was used.

The application of the four reagents (NaOH, Na_2CO_3, NH_3, and $Ca(OH)_2$) has produced positive results when those reagents had been used as absorbents in the emulsifier's conditions.

During the experiments with the model gas (with ~2% of SO_2 content), it has been determined that the maximum cleaning efficiency without its further growth is achieved by the use of the following solutions when the concentration of the reagents is increased:

- NaOH – 0.62%

- Na_2CO_3 – 1.64%

- NH_3 – 0.8%

- $Ca(OH)_2$ – 2%

If the SO_2 percentage in the flue gas is changed, then the required cleaning degree is achieved by the application of water solutions of the above reagents when the molar relation "reagent/SO_2" is 0.8–1.2. It has also been determined that the increase of SO_2 catching efficiency takes place when the gas velocity in the FE and the flow rate of the spraying liquid are increased. The diagrams given in Figures 4–6, reflect results of investigations into the application of water solutions of reagents NaOH, Na_2CO_3, and $Ca(OH)_2$. An 80–85% cleaning degree (and higher degree) is achieved when the velocities are 10.0–10.5 m/s and the flow rates of the spraying liquid are not more than 0.3 l/m^3.

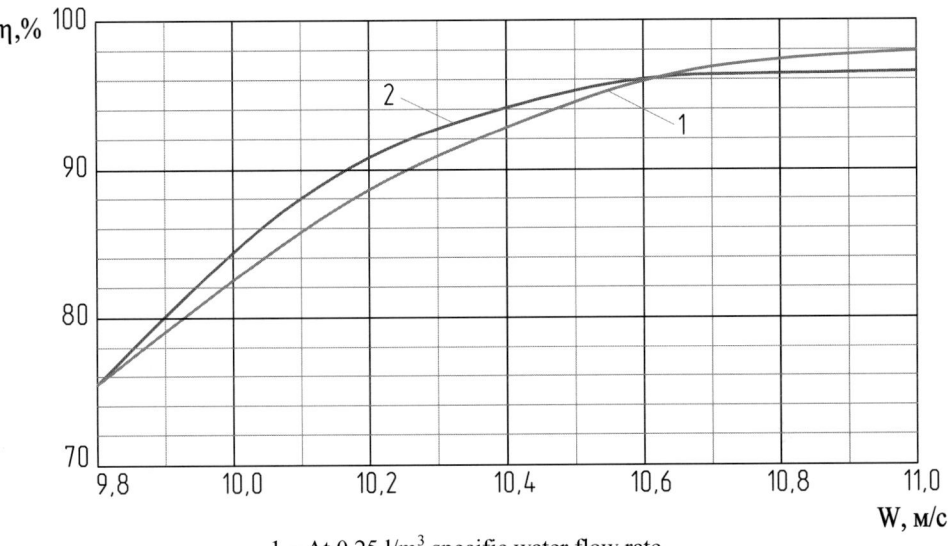

1 – At 0.25 l/m^3 specific water flow rate

2 – At 0.3 l/m^3 specific water flow rate

Figure 4. Dependence of flue gas-cleaning degree from SO_2 by solution NaOH (Concentration is 0.62%) on gas velocity

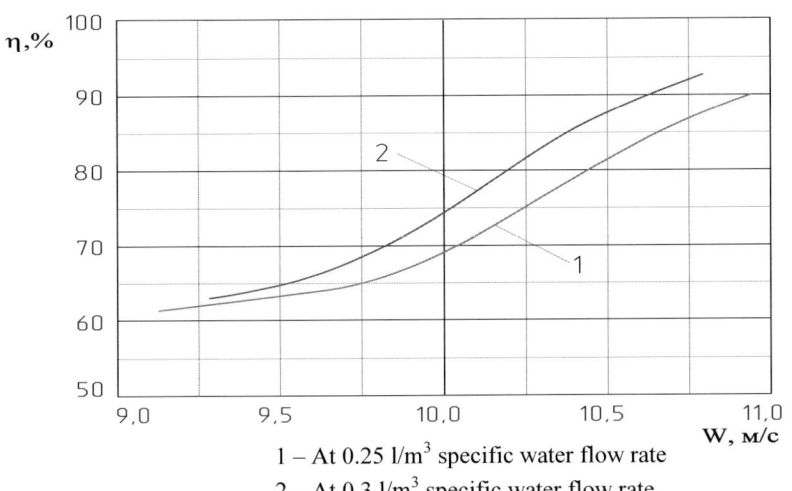

1 – At 0.25 l/m^3 specific water flow rate

2 – At 0.3 l/m^3 specific water flow rate

Figure 5. Dependence of flue gas-cleaning degree from SO_2 by solution Na_2CO_3 (concentration is 1.64%) on gas velocity

The experiments with ammonia solution have shown that the efficiency of more than 95% is achieved when the concentration of ammonia water is 0.8% at the minimum velocity (~9 m/s) and when the specific liquid flow rate is 0.2 l/m^3.

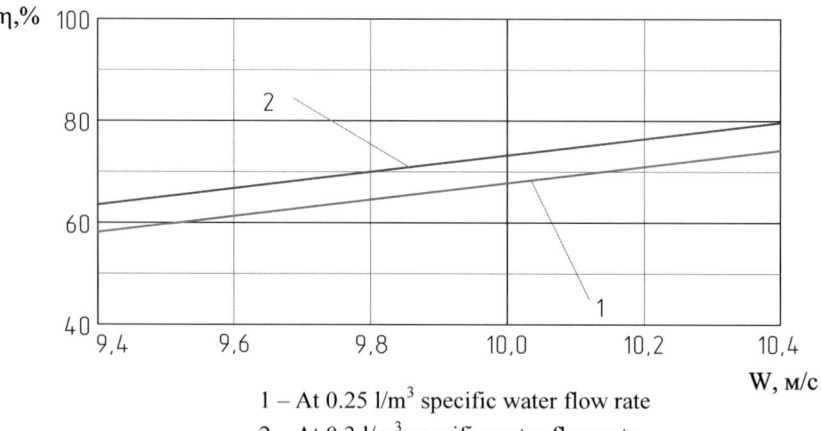

1 – At 0.25 l/m^3 specific water flow rate
2 – At 0.3 l/m^3 specific water flow rate

Figure 6. Dependence of flue gas-cleaning degree from SO_2 by solution $Ca(OH)_2$ (concentration is 2%) on gas velocity

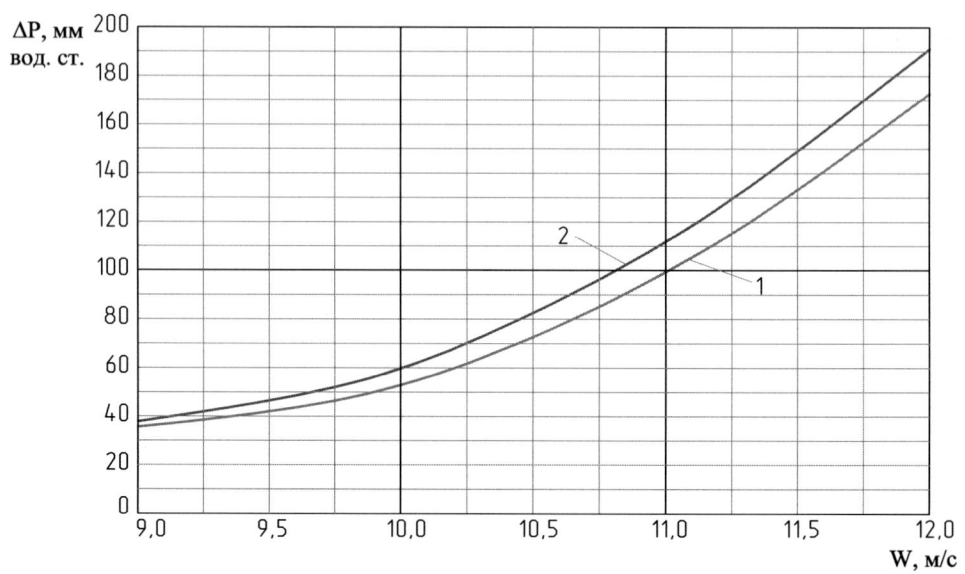

1 – At 0.25 l/m^3 specific water flow rate
2 – At 0.3 l/m^3 specific water flow rate

Figure 7. Dependence of emulsifier's resistance on flue gas velocity

A positive result has been obtained during the experiments when $Ca(OH)_2$ was blown into the model gas flow of 10.0–10.5 m/s velocity and spraying was performed by water with 0.3 l/m^3 specific flow rate. In that case, SO_2 catching efficiency was ~80%.

The effect of increasing the flow rate parameters (needed for increasing the efficiency of the gas-cleaning processes) on the growth of hydraulic resistance of the emulsifier has been also determined during the experiments in accordance with Figure 7. Results of those investigations prove the fact that the achievement of the maximum required cleaning efficiency and the capacity of smoke exhausters should be taken into account for determination of the flow rate parameters.

3. Conclusions

The emulsifier is a highly efficient, comparatively cheap unit of small dimensions for the wet cleaning of flue gases from those TPSs where coal is used for operation.

A high ash-catching degree (99.6–99.8%) has been verified by industrial tests. That catching degree can be more than 99.9% if the optimum flow rate parameters of the cleaning process are provided.

It has been determined experimentally that the emulsifier can be successfully used for SO_2 catching with 80–95% efficiency when water solutions of NaOH, Na_2CO_3, NH_3, and $Ca(OH)_2$ in low concentrations are used for spraying. Specific features of the emulsifier are:

- Low power consumption (the specific flow rate of the spraying liquid does not exceed 0.3 l/m^3 when hydraulic resistance is not more than 100 mm of water column)

- Multipurposeness (the emulsifier can be used for simultaneous cleaning from ash and SO_2 and for their separate catching)

References

1. Rusanov A. A. Dust and ash catching manual. Moscow: Energia, 1975.
2. Idelchik I. E. Manual of hydraulic resistances. Moscow: Machine-building, 1975.
3. Alekseyenko S. V., Okulow V. L. Twisted flows in technical applications (review). Thermal physics and aeromechanics, **3** (2), 1996.
4. Borysenko S. V., Malyy L. P., Bykovchenko G. I., Kulakov A. N. Is smoke of the native land sweet? Made in Ukraine. No. 3, Dnepropetrovsk, 2000.

5. Borysenko S. V., Malyy L. P., Bykovchenko G. I. Results of experimental investigations of the unit for emulsion cleaning of flue gases. Development of the improved design of the emulsifier. Energetics and electrification. No. 4, Kiev, 2000.
6. Zaitsev V. A. and others. Cleaning of flue gases of thermal power stations. Chemical industry. No. 3–4, Moscow, 1993.
7. Ostrovetsky R. M. and others. Flue gas cleaning from ash, sulphur oxides using emulsifiers. Proceedings of the first American–Ukrainian conference Atmospheric air protection from harmful TPS effluents. Kiev, 1996.
8. Updating of existing boiler equipment for transferring it to new technologies of burning Ukrainian low-grade coal with account of ecological requirements. Summary report of project #181 STCU, 1998.
9. Cleaning of flue gases from sulphur oxides by the emulsion cleaning unit – emulsifier. Final report of project #2501 STCU, 2003.

ADVANCED HIGH-EFFICIENCY AND LOW-EMISSION GAS-FIRED DRUM-DRYING TECHNOLOGY

Y. CHUDNOVSKY*, A. KOZLOV
*Gas Technology Institute, 1700 South Mount Prospect Road,
Des Plaines, IL 60018, USA*

Abstract. Drying is one of the most energy intensive industrial processes. Papermaking, drying of food, textile and pharmaceutical products require a significant amount of energy to evaporate the excess water from the product slurry. The most popular drying method is drum drying which traditionally uses condensing steam for drum surface heating. The use of steam requires the drums to meet the ASME codes for pressure vessels, which limits the steam pressure and consequently the shell temperature, reducing the drying capacity. In most cases drying is also the temperature-critical aspect of product processsing. Gas Technology Institute (GTI) together with its industrial partners has developed an advanced high efficiency gas-fired drum dryer concept based on a combination of a low-NO_x ribbon flame and an advanced heat transfer enhancement technique (US Patent 6,877,979). The new approach allows a significant increase in the surface temperature of the drum dryer, thereby increasing the drying rate. The pilot-scale unit was designed, fabricated, and successfully tested for the papermaking application.

Keywords: combustion, natural gas-fired, paper drying, drum dryer, efficiency improvement, dimples

1. Introduction

The drying of paper requires the evaporation of 1–2 lbs of water for each pound of paper or paperboard produced. The conventional drying method uses a series of metal cylinders, each approximately 5–6 ft in diameter and up to 30 ft long. These cylinders are heated from the inside by condensing steam. The use of

---

*To whom correspondence should be addressed. Yaroslav Chudnovsky, Gas Technology Institute, 1700 South Mount Prospect Road, Des Plaines, IL 60018, USA

N. Syred and A. Khalatov (eds.), Advanced Combustion and Aerothermal Technologies, 285–298.

steam requires the drums to meet the ASME codes for pressure vessels, which limits the steam pressure to about 160 psig. Consequently, the shell temperature and the drying capacity are also limited. In practice, most cylinders operate at an even lower pressure and temperature, further reducing their drying capacity.

Paper drying is the most energy-intensive and temperature-critical aspect of papermaking. It is estimated that about 67% of the total energy required in papermaking is used in the drying process. Paper machine speeds – and therefore production rates – are frequently limited by the dryer capacity. For this reason, a great deal of activity recently has been devoted to the development of new, high-efficiency, high-rate paper drying equipment and technologies.

Gas Technology Institute (GTI) together with industrial leaders (Boise Paper Solutions, Groupe Lapperierre & Verreault and the Flynn Burner Corporation) and with funding support from the US Department of Energy and the US natural gas industry has developed a high-efficiency gas-fired paper dryer (GFPD), based on a combination of a low emission ribbon burner and an advanced heat transfer enhancement technique. The GFPD is a high-efficiency alternative to traditional steam-heated drying drums and it is expected to exceed the performance of existing paper drying systems.

The new approach allows a significant increase in the surface temperature of the dryer (up to 600°F), thereby significantly increasing the drying rate for the state-of-the-art operations. Successful deployment of the GFPD will provide large energy savings to the industry, according to energy efficiency increases from 60–65% (steam operated) to 75–80% (gas operated). In addition, it will help the paper industry to increase production from dryer-limited paper machines by an estimated 10–20%, resulting in significant capital costs savings for both retrofits and new capacity.

2. Concept Development

Several gas-fired drum-drying concepts have been investigated during the last decade by British Gas, ABB Drying, Gastec NV, etc. GTI has generated two innovative gas-fired concepts and upon completing an extensive laboratory bench-scale evaluation the GFPD concept was finalized, as represented in Figure 1.

The low-NO_x Flynn ribbon burner model T-534 was selected as the combustion device, providing a high uniform sheet flow of the combustion products parallel to the drum surface. The attractive vortex heat transfer enhancement (VHTE) technology was employed in order to enhance heat transfer from the combustion products to the rotating drum and paper web. A

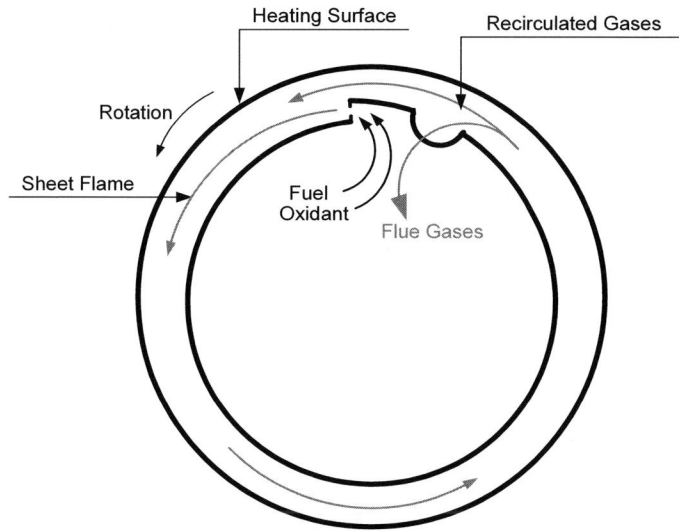

Figure 1. Final GFPD concept (US patent 6,877,979)

special deflector guide was used to provide a partial recirculation of the combustion products back to the flame root for combustion stability improvement as well as for further NO_x reduction.

Waste heat recuperation was considered during the concept development in order to significantly improve the overall thermal efficiency of GFPD. A recuperator is to be incorporated into the drum design in order to provide the efficient heat recovery of the exhaust gases.

3. Design and Engineering

The pilot-scale unit was designed and engineered with the strong support of industrial partners. The pilot-scale GFPD was comprised of a stationery combustion system (including the burner, ignition/flame safety, and recuperator) and a rotating shell as shown in Figure 1. All the above-mentioned features were reflected in the GFPD design: sheet flame, heat transfer enhancement, partial internal recirculation of the combustion products, and flue gas recuperation. The GFPD sides were insulated to minimize the heat losses and two viewing ports were placed on the side to observe the flame.

The ribbon burner was mounted inside the dryer on stationery a support that was insulated to separate the heating zone from the recuperation zone of the dryer. The rotating outer shell was profiled inside with a dimpled pattern to enhance the heat transfer from the combustion products to the rotating shell. Dimensions for the annulus, mixture supply line, and exhaust pipe as well as the geometry of the guide plate were calculated based on hydraulic design and the estimated losses.

4. Laboratory Simulation

The GFPD outer drum and most of the internals were manufactured and assembled by GL&V USA, Inc. with strong GTI engineering support. Figure 2 demonstrates the general view of the fully assembled GFPD and simulation rig for laboratory evaluation.

Figure 2. Pilot-scale GFPD and simulation rig equipped with the air/gas supply and heat recuperation

An in-house simulation of the paper drying was conducted at GTI Applied Research Laboratory to work out the combustion system prior to the field installation at the WMU Pilot Paper Plant, as well as to determine the operating envelope of GFPD. Figure 3 demonstrates a P&ID of the simulation rig.

While being in-house evaluated, the GFPD was fired in the range of 20–150 MBtu/h with controllable surface cooling by a wet fabric covering about one third of the drum surface. The combustion system demonstrated a reliable and stable operation throughout the entire test range.

The pilot-scale test procedure (including cold startup) was determined and the drum surface temperature of 500°F was achieved at the maximum firing rate.

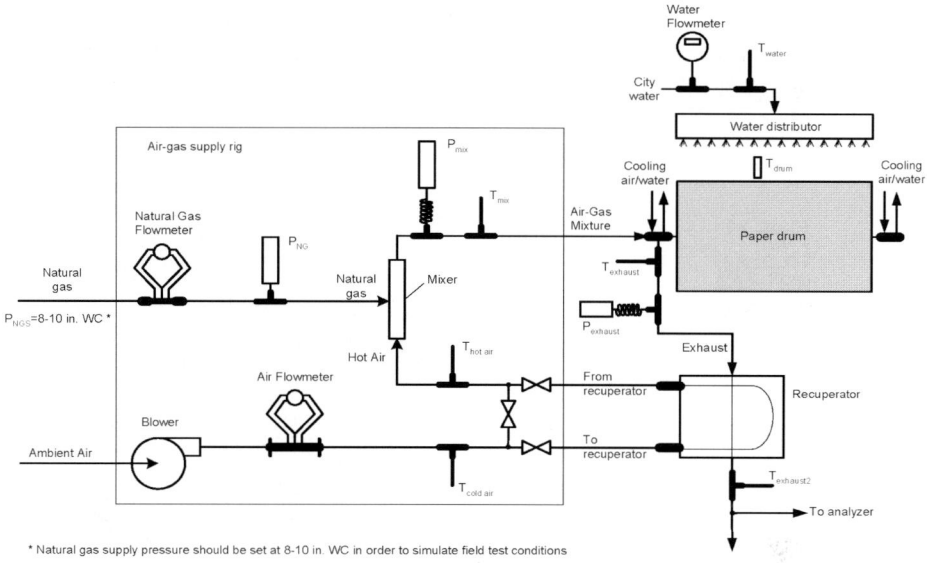

Figure 3. Piping and instrumentation diagram for laboratory simulation rig

5. WMU Pilot Paper Plant

The Pilot Paper Plant (PPP) at WMU's Department of Paper and Printing Science was selected as a test site for both the baseline (conventional steam heated) and the retrofit (natural gas-fired) testing. WMU PPP offers unique capabilities in various areas of papermaking and has technical staff with an extensive experience of maintenance and operation of the papermaking equipment.

The heart of WMU PPP is the Fourdrinier paper machine (Figure 4), which is capable of manufacturing a wide variety of papers and boards of 23 in. width and at speeds up to 150 ft/min. The paper grades manufactured range from 16 to 400 lbs of basic weight. Facilities for stock and additive preparation are sized to make extended continuous operation at production rates of up to 160 lbs/h. The paper machine has two independently controlled drying sections (with 13 drum dryers in total) that are completed with machine calendar, pope reel, and Accuray 1190 computer system with automatic basic weight and moisture control.

Figure 4. Overall view of WMU pilot-scale paper machine

6. Measuring System

The measuring system was incorporated into the existing drying section layout as shown in Figure 5. Mini-IR temperature sensors were installed above the selected test drum (#3 or #7) to measure the paper inlet and outlet temperatures as well as drum surface temperatures across the drum length. Two Aqua Moisture sensors were installed to measure the paper moisture content prior to and after the test drum. In order to fit the moisture sensors into the existing drying train, the inlet sensor was equipped with a low-profile IR attachment that allowed the installation of the sensor coaxial to the test drum. The steam supply lines were equipped with ABB steam flow meters and control valves to measure and analyze the heat balance in the drying sections during the baseline and pilot-scale test runs. The temperature of condensate leaving the test drum was measured by a K-type thermocouple in order to evaluate the heat losses of the steam-heated can. Steam header pressures were also measured at the inlet of the drums ##1–3, 6–9, and 13.

All the sensors were connected to the FieldPoint data acquisition system. The data was then monitored in real-time and recorded by using the National Instruments' LabView software on a laptop computer.

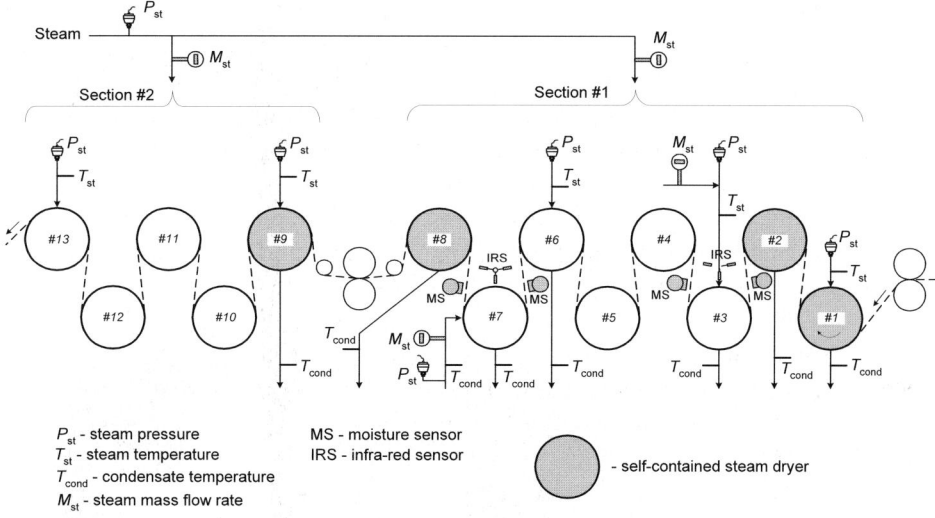

Figure 5. Measuring system layout

The following parameters were measured during baseline testing:

Pressure

Steam supply lines
Steam inlet of drums 1,3,6,7,9,13

Temperature

Steam inlet of drums 1,3,6,7,9,13
Steam condensate at drum 1,2,3,7,8,9
Paper between each drum (manual)

Surface of the test drum (IR)
Paper web before/after test drum (IR)
Surface of all drums (manual)

Flow rate

Steam supply for 1st section
Steam supply for 2nd section
Steam supply for test drum

Moisture

Paper web before/after test drum

Machine Parameters

Machine speed
Dry end moisture
Basis weight

The exhaust gas composition (O_2, CO, CO_2, NO_x) was measured by a portable Horiba PG-250 gas analyzer and recorded by a National Instruments' LabView data acquisition system. Paper samples were taken before and after each dryer to evaluate the physical paper properties such as basis weight, burst, caliper, Gurley stiffness, elongation, tear, and tensile strength.

7. Baseline Tests

Steam drums #3 and #7 were selected for the baseline evaluation (Figure 6). Drum #3 was selected per WMU recommendation as the most thermally loaded drum in the drying section. Drum #7 was selected for evaluation of GFPD performance at lower inlet paper moisture levels. The baseline test was assumed to measure the drum surface temperatures and paper moisture profiles across the paper web at different heat input levels. The obtained information was used as benchmark data for the comparative analysis with the gas-fired retrofit test results.

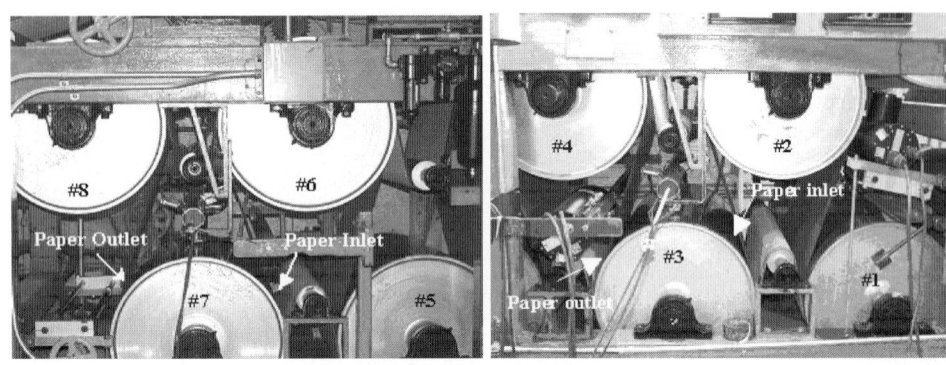

Figure 6. Steam drums #3 and #7 instrumented for baseline test

The paper moisture level at the inlet of the test drum was kept relatively constant in the target ranges of 10–60% for all baseline runs.

According to the baseline test plan/matrix all the data was collected, processed and analyzed. The drying rates and drum-efficiency values were calculated for the steam-heated dryers. Table 1 displays the dryer performance (results are presented for three different machine settings). The Table values represent an inlet paper moisture content ranging from 52% to 55% for the drum #3 and from 12% to 30% for the drum #7. It should be noted that the drying rate and efficiency depend on many parameters including heat input, drum surface temperature and inlet moisture content.

TABLE 1. Drying rate and efficiency for steam drums

Dryer location	Steam flow	Steam temp.	Drum temp.	Inlet moisture	Moisture removal	Drying rate	Drying eff.
	(lb/h)	(F)	(F)	(%)	(%)	lb/(h*ft^2)	(%)
Test #1 (baseline)							
Entire machine	642	235	229	57	54.6	1.3	27
Steam dryer #3	98	259	249	52	8.5	2.4	43
Steam dryer #7	49.7	260	252	12	5.3	0.8	17
Test #3 (baseline)							
Entire machine	653	231	222	57	54.2	1.4	32
Steam dryer #3	76	237	228	55	5.0	1.5	44
Steam dryer #7	42.3	237.8	230	30	3.9	0.8	27
Test #4 (baseline)							
Entire machine	706	239	232	57	54.4	1.3	28
Steam dryer #3	94	257	247	54	6.5	1.9	45
Steam dryer #7	51.5	258.1	251	21	10.8	1.9	45

8. Pilot-Scale Testing

During the pilot-scale test the GFPD drum was installed in the same two locations as for the baseline test. The drum was placed in the dryer #3 position to test at inlet paper moisture levels of around 55% (wet end) and in the dryer #7 position to evaluate the GFPD performance at inlet paper moisture of 10-40% (dry end). The instrumentation of the paper machine for the GFPD test was the same as during the baseline test with the exception of steam pressure and temperature measurements at drums #3 and #7. Figures 7 and 8 show the instrumentation diagram and GFPD view during GFPD firing. An advanced air/gas flow management and mixing system, equipped with high accuracy vortex flow meters and a portable gas analyzer, was used to control and monitor flame stoichiometry and exhaust emissions.

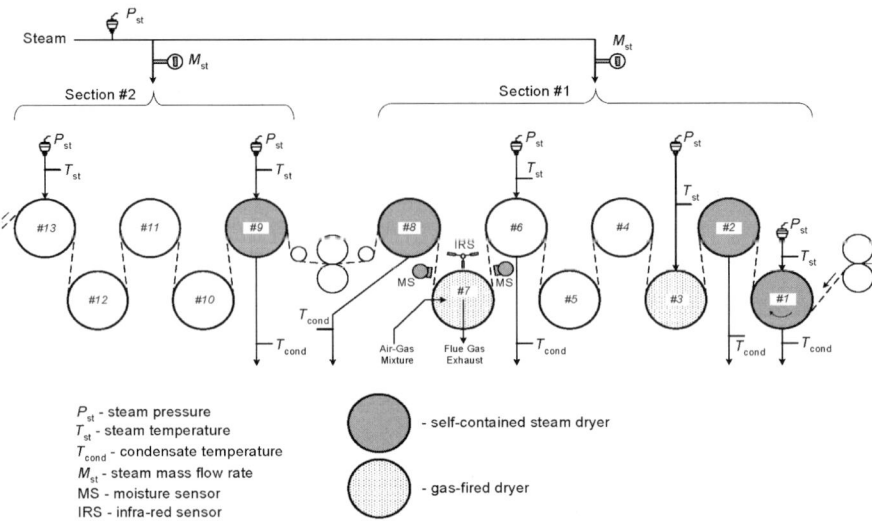

Figure 7. Piping and instrumentation diagram for GFPD testing

Figure 8. GFPD layout at high-temperature operating conditions

9. Drum Surface Temperature

In order to vary the test drum surface temperatures the burner was fired at rates ranging from 17 to 130 MBtu/h. The average drum surface temperatures ranged from 264 to 467°F, with a standard deviation ranging from ±3–±9%. The highest average drum surface temperature was 467°F with a standard deviation of ±17°F.

The GFPD surface temperature was limited to about 450°F due to the operating limit for the high-temperature felt (monofilament fabric – Kleenflex 69) that was donated by Albany International Company.

10. Drying Rate and Efficiency

The pilot-scale evaluation demonstrated an increased drying rate for the GFPD of about 4–5 times over the average value measured for the steam heated dryer (see Figure 9). However, the maximum operating temperature of the felt material limited the GFPD drying rate.

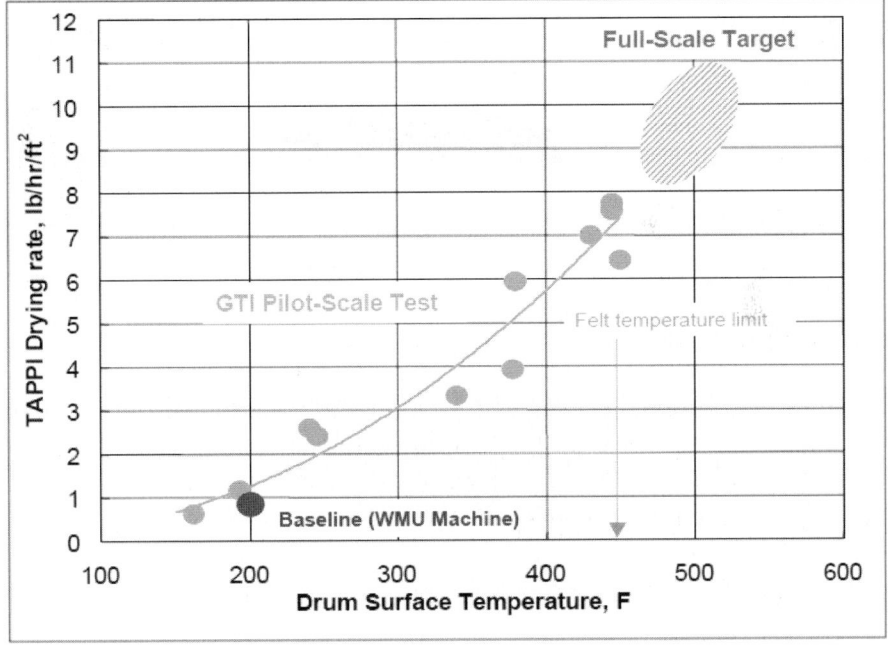

Figure 9. Drying rate vs drum surface temperature at inlet moisture level 55%

The drying efficiency was also calculated for the corresponding GFPD drying rates. At the highest firing rate of 130 MBtu/h, a maximum drying efficiency of about 60% was achieved at paper inlet moisture levels of 55%. On the dry end, a maximum efficiency of 71% was achieved with a corresponding drying rate of 4.9 lbs/h-ft^2, at a surface temperature of 430°F and an inlet paper moisture level of 43% (see Table 2). It should be noted that higher firing rates (>130 MBtu/h) would lead to a much higher efficiency.

TABLE 2. Obtained results for selected GFPD tests on the dry end[a]

Dryer location	Steam/gas flow rate (lb/h)	Steam/Gas temp. (F)	Drum temp. (F)	Inlet moisture (%)	Moisture (%)	Drying rate lb/(hr ft^2)	Drying eff.[b] (%)
Test #1 (gas-fired)							
Entire Machine	476.9	220	199	58	54.5	1.4	39[c]
GFPD (dryer #7)	2.7	154	279	41	5.7	1.3	51
Test #2 (gas-fired)							
Entire Machine	463.2	220	204	57	54.2	1.4	38[c]
GFPD (dryer #7)	4.4	164	352	38	11.0	2.4	51
Test #3 (gas-fired)							
Entire Machine	378.5	213	204	59	54.8	1.4	44[c]
GFPD (dryer #7)	6.4	165	430	43	19.7	4.9	71

[a]Based on average of four paper samples
[b]Calculated based on heat needed for water evaporation and heat input from steam and combustion
[c]Taking into account heat input from steam and natural gas combustion

11. GFPD Emissions

The natural gas emissions produced by the GFPD were constantly monitored during the testing and were all found to be within acceptable limits. The air/gas ratio was controlled to allow 2–3% O_2 in the exhaust products. Below are the ranges of exhaust emissions measured (corrected at 3% O_2):

- NO_x ~ 30–114 vppm

- CO ~ 3–30 vppm

- CO_2 ~ 10.2–11%

12. Paper Quality Results

Several paper strength properties were measured in order to determine the effect of a higher temperature drying on the paper quality. The paper quality analysis was performed by the WMU Paper Quality laboratory and is presented in Table 3. The data for the baseline and GFPD tests were obtained based on the averaging of 10 test samples and demonstrate that there is no tangible difference in the paper strength properties between the baseline and GFPD tests, which were carried out at different temperatures.

TABLE 3. Strength properties of the paper samples (linerboard)

	Drum temp. (F)	Basis weight (g/m^2)	Bursting strength (kPa)	Caliper 1/1000 (in.)	Gurley stiffness (GF units)	Elongation (%)	Tearing resistance (gf)	Tensile strength (kN/m)
Baseline test	238	213	323	14.8	3.9	1.7	240	7.0
GFPD test	430	210	351	15.0	4.5	1.8	233	7.9

13. Major Conclusions

The results of GFPD testing at the WMU Pilot Paper Plant successfully demonstrated the advantages of a natural gas-fired approach over the steam-heated one while producing an industrial linerboard (126 lbs/3,000 ft^2). It was proven that drum dryer surface temperature could be increased up to at least 500°F, resulting in a significant improvement in drying rate over the existing steam-heated drum without sacrificing product quality.

The GTI pilot-scale tests at WMU showed that the maximum *drying efficiency* for the GFPD (calculated as the ratio of heat required for evaporation/heat released by natural gas) is 75% compared to the 55% maximum *drying efficiency* (calculated as the ratio of heat required for evaporation/heat from steam condensation) for the comparable steam-heated dryer. The *overall efficiency* of the GFPD (calculated as the ratio of useful heat/total heat input) can be boosted up to 85% through the use of internal flue gas recuperation (e.g., for combustion air preheat). The maximum value of the steam dryer *overall efficiency* can be projected as 68% (assuming 15% losses in the boiler, 10% in the transmission system and 10% in the dryer can).

The TAPPI drying rate for gas operation was achieved at about 4–5 times higher over the conventional steam operation, based on the GFPD pilot-scale evaluation at WMU. The maximum increase in drying rate potential on the full-scale paper machine would depend on the actual operating conditions and paper grade, and is subject to determination during the GFPD field trial.

The conventional steam-heated drum at WMU consumed about 70 MBtu/h of saturated steam heat to reach 250°F on the drum surface, compared to less than 50 MBtu/h of heat that is necessary to reach the same surface temperature for the GFPD. Moreover, to reach 500°F on the gas drum surface (which is not possible in the case of a steam heating operation) the heat input would need to be increased by only 40–50% over the steam heated drum (up to 120 MBtu/h). It is believed that the gas-fired operation could allow the reaching of a higher TAPPI drying rate, even at the same drum surface temperature due to the absence of a condensate level on the bottom of the drying can that increases thermal resistance to the paper web.

Preliminary estimates indicate that the GFPD drying system could lead to a production increase of up to 20% without a dry end extension. The production rate increase will depend on drum the operating temperature and the number of drums installed/replaced. If the paper machine is not drying limited, the system may be operated for the purpose of reducing energy consumption.

14. Full-Scale GFPD Development

The obtained results provide the basis for designing a full-scale unit to be further evaluated for the merit of technical and economical benefits, and showcasing this challenging approach for the drying-, steam- and space- limited paper producers.

Acknowledgment

The project team is grateful to US Department of Energy (award DE-FC36-01GO10621), NASEO/MNSEO (award DE-FC36-03GO13026) and Members of Gas Industry Sustaining Membership Program for their financial support to this work as well as industrial partners for significant technical contributions to this development.

THE ORGANIZATION OF INTERNAL RECIRCULATION OF SMOKE GASES IN REVERSIVE WATER-COOLED CHAMBERS OF COMBUSTION OF BOILERS FOR THEIR MODERNIZATION

V. DEMCHENKO, A. DOLINSKIY, A. SIGAL
Institute of Engineering Thermophysics NAS of Ukraine.
01001, Kiev, Ukraine, P.B. B-22, e-mail: demchenko@at-eat.com

Abstract. The given work develops the theory and studies the practical application of multichamber combustion in flue water-heating boilers. The prototype design of the secondary screen-radiator, to alter the geometry of the combustion chamber is offered. Directions for the modernization and the intensification of the heat exchange process in combustion chambers are considered and determined. It is theoretically proven, that an offered method of modernization of the combustion chamber, raises the efficiency due to an increase in the convection and radiating heat exchange and reduces the harmful emissions to atmosphere due to a change in the aerodynamics, the thermodynamics and the kinetic processes taking place in the boiler. A mathematical model is developed to determine the volume of recycled gases and the speed of aerodynamic streams in the combustion chamber and fire tube of the boiler. Calculations of the material balance of aerodynamic streams and thermal calculations for the combustion chamber with a secondary screen-radiator installed are carried out. Experimental studies under laboratory conditions are executed, from which results confirming the correctness of the chosen method for an intensification of the heat exchange process are received. The optimum geometrical sizes and construction materials for secondary screen-radiators are found. The coordinates of installation of such screen-radiators in the combustion chamber to receive the maximal positive results are determined. Changes of structure of the flame are investigated; with a sharp decrease in the emissions of NO_x and a reduction in time of heating the boiler waters is recorded. The use of CFD modeling allowed a more detailed study of the aerodynamics, thermodynamics and chemical kinetics processes that occur in combustion chambers with a secondary screen-radiator installed. The initial

N. Syred and A. Khalatov (eds.), Advanced Combustion and Aerothermal Technologies, 299–315.

results of the analytical and CFD models agree to within 5%, which testifies to their adequacy.

Keywords: the chamber of combustion of the boiler, a secondary screen-radiator, the efficiency, harmful emissions, flue boilers

1. Introduction

The economical usage of fuel and energy resources is becoming more and more important every year. The reducing stocks of fossil fuels in the bowels of the Earth, have resulted in shortages and as a consequence a significant rise in price. This in turn results in a reduction in the caloric content delivered to the consumer of alternate organic fuels. It is therefore necessary to search for opportunities for mixtures of natural gas and biogas, and the replacement of diesel fuel by the fulfilled oils and so forth. How such changes are negatively reflected in ecology terms and result in the decrease in efficiency, capacity and a reduction of service life of the boiler equipment should be investigated.

The cold season of 2005–2006 in Ukraine, has evidently shown the complexity and importance of the problems connected with the stability of heating supplies to settlements and affects overall energy savings. Especially, as it is actually the municipal services of Ukraine that consume up to 70% of the fuel and energy resources of the country as a whole (Figure 1).

According to regional administrations carried out in the Ukraine, the operational life of 57% of boiler-houses exceeds 20 years, 40% maintain boilers with an efficiency less than 82%. Currently, there are around 10,800 boilers with productivities from 100 kW up to 1 MWt that have been in operation for more than 20 years. Therefore about 14,000 boilers with capacities of up to 1 MWt require replacement. There are questions of how to modernize and adapt the existing equipment to work on low-calorie types of fuel, and at their greatest possible efficiency.

It is necessary to note, that less than 30% of the total number of boilers have reversive water-cooled combustion chambers and work with ventilator burners. The majority of boilers are equipped with combustion chambers of great volume and work with atmospheric burners. In last years scheduled replacement of the code for the modernization of boiler equipment, a rule on smoke-pipes boilers being equipped with ventilator type burners has opened additional prospects for the use of an alternate method.

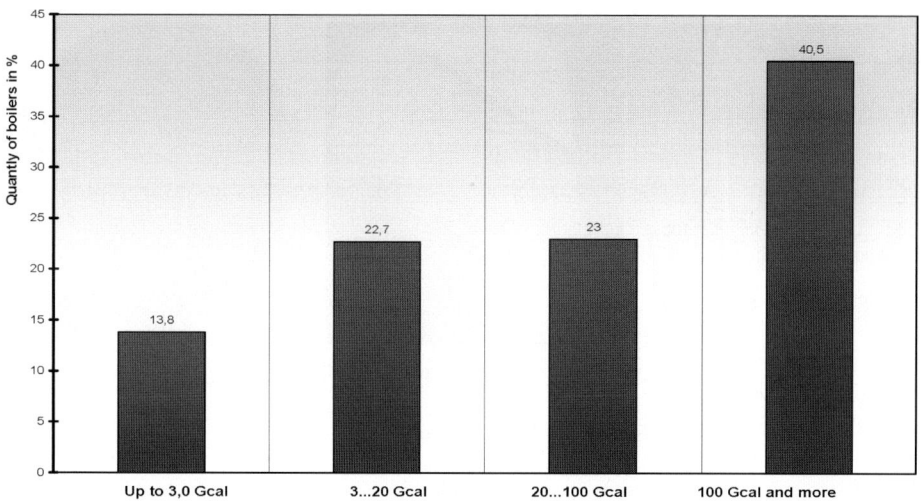

Figure 1. Capacity of boilers used in municipal services of Ukraine as of 31.12.2005

2. Modeling the Processes Arising in the Chamber of Combustion After the Installation of Secondary Radiators

Various authors have presented works directed on the increase of efficiency of boilers and reducing emissions of harmful gases by the development and installations of secondary radiators in the chamber of combustion of the boiler. It is necessary to note, that water-heating boilers can be divided into two groups by the construction method used: boilers with a "hot" chamber of combustion and boilers with a cooled or "cold" chamber of combustion.

The distinctive features of the boilers considered in the given work, is the compact sizes of the chamber of combustion (100, 200 dm^3) and the high degree of thermal pressure.

Methods for the intensification of the heat exchange process can be developed by changing the geometry of the chamber of combustion, taking into account processes of aerodynamics, the redistribution of temperature gradients, gas velocities and the completeness of chemical reactions. This can result in a stabilization and an intensification of the burning process and as a consequence a reduction in the harmful emissions of CO and NO$_x$.

The most widespread method of installation of secondary radiators is in the zone of burning, requiring the use of high-temperature materials, for example, stainless steel or ceramics (Figure 2).

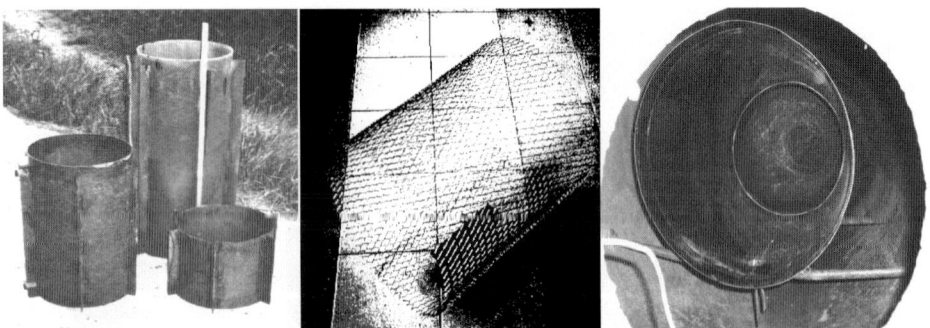

Figure 2. From left to right:
- Secondary radiators from stainless steel
- Secondary radiator from a metal grid
- Installation of a Secondary radiator in boiler "BK-22"

Taking into account, that recirculation results in the heating of the air and fuel effecting the burning, the efficiency of the chamber of combustion of boilers of low power can be determined using the formula:

$$\acute{\eta} = (Q_H - V_B \lambda C_{sg} T_{sg})/Q_H \times 100\% \tag{1}$$

- For the case without heating of the air and fuel (1)

$$\acute{\eta}_1 = (Q_H - V_B \lambda C_{sg} T_{sg} + V_T \lambda C_{вр} T_{вр})/Q_H \times 100\% \tag{2}$$

- For the case with heating (2)

where Q_H – the lowest calorific value of the fuel

V_T, V_B – the heated up and theoretical volumes of air going on burning

λ – factor of surplus of air

$C_{вр}$, C_{sg} – thermal capacity of gases of recirculation and leaving smoke gases

$T_{вр}$, T_{sg} – temperature of gases of recirculation and leaving smoke gases.

The numerical calculations carried out by authors have shown that a rise in the reaction temperature of the burning of 100°C, results in an increase of efficiency of the chamber of combustion of 9.5%.

Simultaneously, it has been shown by the working normative method of calculation for boilers that the installation of secondary radiators in the chamber of combustion of the boiler has an adverse effect on efficiency.

It is obvious, that the described design procedures were created for large-scale industrial and power generation boilers working with solid, gaseous and liquid fuels, and do not take into account the features and the processes occurring within boilers of low power.

2.1 MATHEMATICAL AND CFD MODEL

To define the influence of the stand degree of recirculation on the overall performance of the boiler, the parameters of the specially converted laboratory boiler "Victor-100" have been taken for a basis (Figure 3).

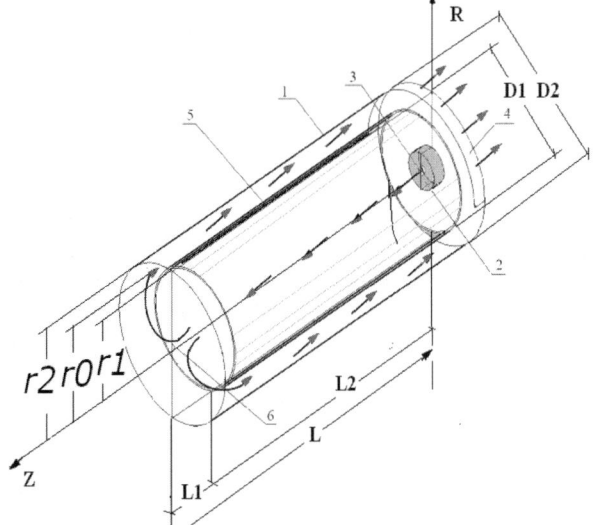

1 – The chamber of combustion;
2 – Submission of air;
3 – Submission of methane;
4 – An output of smoke gases;
5 – A secondary radiator;
6 – The central section of space of the boiler.
R, Z– The central axes,
r_0, r_1, r_2 – Radiuses of bifurcation, secondary radiator and chamber of combustion, accordingly.
$L = 1,030$, $L_1 = 880$, $L_2 = 150$, $D1 = 320$, $D2 = 420$ – overall dimensions in mm

Figure 3. The geometrical sizes of the chamber of combustion of boiler "Victor-100" and settlement area

In the standard boiler the burning of fuel occurs within the chamber of combustion (first course of gases). Further products of combustion are formed within the convective group of smoke-pipes (second course of gases) and from there the gases proceed through a modular box to a chimney.

To gain a better understanding of the combustion processes occurring in the chamber after the installation of a secondary radiator and the effect of the recirculating gases, it is necessary to define a factor of recirculation. This can be defined from knowing the distribution and speeds within the recirculation zone of the boiler and coordinates of the bifurcation points.

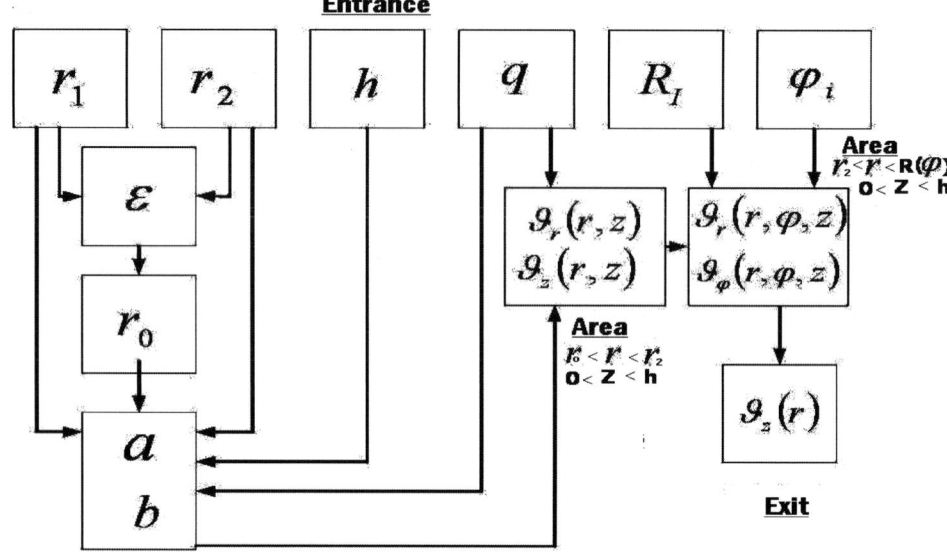

h – distance between a forward edges secondary radiator and internal brick-work wall of the boiler; r_0 – bifurcations Radius; a, b – unknown factors, $\acute{\varepsilon}$ – recirculation's factor, r_1 – radius of a secondary radiator; r_2 – radius chamber of combustion; q – charge of natural gas; $R_1, \varphi_i, v_z(r)$ – coordinates of an axis smoke-pipes

Figure 4. The block diagram of calculation of aerodynamics of ring and slot-hole areas of the combustion chamber of the boiler

The resulting block diagram of calculations for the aerodynamics of ring and slot-hole areas of the combustion chamber of the boiler is shown in Figure 4.

On output from the ring channel the products of combustion are divided into two streams: one acts in a convective bunch and the second returns into the chamber of combustion. The velocity fields in this area can be accepted as axisymmetric, and it is enough to limit the form of average trajectories of particles of products of combustion by curves of the second order. The equation of indissolubility of a stream was solved using the following simplifications: the distribution of the density of gas, ρ_1, ρ_2, ρ_3 were averaged for three areas: on an output of the ring channel, in space between a wall of the chamber of combustion and a trumpet bunch, inside a trumpet bunch; and the boundary conditions reflect an absence of radial speed at the walls and on a line of the unit of streams.

$$\vartheta_r = b \cdot \frac{(z - z_0)^2}{r\left(a(z - z_0^2)^2 - (r - r_0)^2\right)}$$

$$a = \left(\frac{r_2 - r_0}{z_0} tg\left(\frac{r_2}{r_2 - r_0}\left(\frac{z_0}{r_2} - \ln\frac{r_2}{r_0} \right) \right) \right)^2 \qquad b = \frac{aq}{\pi\left(r_2^2 - r_0^2\right)\ln\dfrac{r_2}{r_0}}$$

The dependent relationship of the component of velocities of a gas stream on an input in a smoke pipe is found, and is defined as:

$$\vartheta_z(r,\varphi,z) = \vartheta_k\left(\frac{z}{z_0} - \frac{z_0}{z}\frac{1}{r_m^2}\left(r^2 + r_k^2 - 2r_k r\cos(\varphi - \varphi_k)\right) \right)$$

It has been found using the following boundary conditions: all components of velocities for the walls are equal to zero; the stream of radial speed through a cylindrical surface $r = r_2$ (internal radius of the chamber of combustion) is equal to a stream of axial speed through a face section of all smoke pipes; the stream of radial speed through a lateral surface of a cone is equal to a stream of axial speed through its basis.

$$\varepsilon = \int_{r_1}^{r_0} \rho\vartheta_z rdr \Big/ \int_{r_2}^{r_1} \rho\vartheta_z rdr \ ,$$

The above formula defines the factor of recirculation of smoke gases in terms of the radial coordinate r_0. Which shows a share of the mass charge of gas in the ring channel is returned in a secondary radiator. The given relationship is determined under the assumption that there is no influence of the variations of density and speeds of a gas stream. Having entered an hypothetical radius of the unit of a gas stream, r_0 – on which occurs the division of a stream on circulating and transit (going in a bunch smoke pipe) it is possible to define the relationship for the factor of recirculation as an indicative – sedate function

$$\varepsilon = \Phi w h^\alpha e^{-\beta h}.$$

Thus, it was possible to calculate the factor of recirculation from the geometrical size of the chamber of combustion and the coordinates of the secondary radiator location and bifurcation points.

After the installation of a secondary radiator, the process of burning occurs with some of the products of combustion once in the fire-box, changing direction and returning through a gap formed by the walls of the chamber of combustion and the secondary radiator to the front of the boiler, forming a convective loop. It is necessary to note, that at the front the boiler there is a division of a stream of smoke gases which in part return to the secondary radiator. So is formed an internal recirculation zone in the chamber of combustion and an additional path for the gases.

Tests were carried out using diesel fuel in the laboratory and natural gas under industrial conditions. In addition, a CFD model was created to simulate the heat exchange in the chamber of the combustion, using the software product FLUENT 6.3.

The following boundary conditions were used for the CFD simulation: the burner produces a structure of uniform longitudinal speed containing a methane and air mixture of mass flow rate 0.004 and 0.044 kg, respectively, the initial turbulent kinetic energy was set to be 1 m^2/s, the walls are assumed to be absolutely smooth, with the normal velocity component equal to zero and the output from the boiler is defined as atmospheric pressure.

The simulation was solved in three-dimensional stationary statement. The boiler is of a double-thread welded type, with a fire-tube, and is gas-tight. The boiler produces low-temperatures with the reversive furnace, and is intended for use with gas or diesel fuel and has a capacity of 100 kW. The fuel is mixed with air by the burner to factors of surplus $\lambda = 1.1$ and higher. The chamber of combustion of the boiler is formed by a fire-pipe.

Figure 5 shows the velocity vector field in the chamber of combustion, and the distribution of temperatures for the boiler "Victor" with and without a secondary radiator.

It can be seen that for the boiler without the installation of a secondary radiator, there is a basic stream going from a torch surrounded by two steady vortices which are larger than the torch (Figure 5.1a). Besides due to a difference in density of the environment of the reagents and the products of reaction, the burning flame is displaced in the top part of the chamber of combustion; resulting in its nonuniform heating. The installation of a secondary radiator promotes the occurrence of zones of laminar return streams directed in a ring backlash to the front of the chamber of combustion (Figure 5.1b). The recirculation allows the repeated burning of the products of the combustion and a more uniform distribution of temperatures in the chamber of combustion (Figure 5.2). The recirculation of the products of combustion (basically – NO_x, CO_2, H_2O, and SO_2) results in an increase of the degree of blackness of the chamber of combustion and an intensification of the heat exchange process. Changes in the distribution of pressure and temperatures inside the chamber of combustion result in an intensification of the chemical processes of burning,

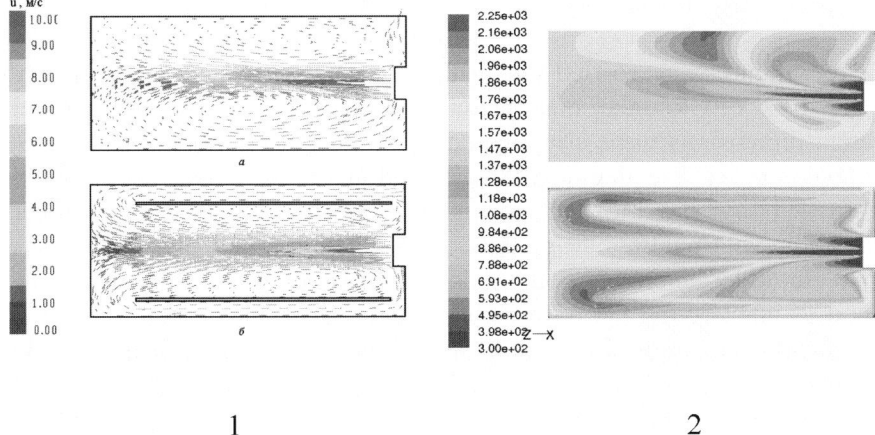

1   2

Figure 5. At the left on the right: change of (1) speeds of a stream, (2) temperatures in the chamber of combustion

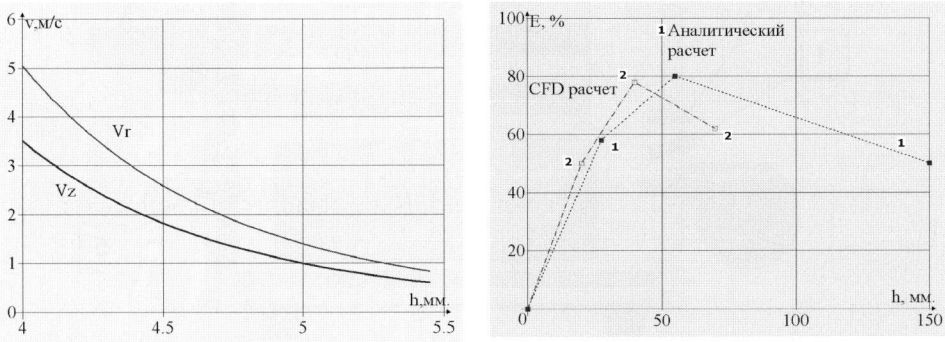

Figure 6. Left – distribution of speeds in the ring channel formed by the chamber of combustion and a secondary radiator on axes R and Z. Right – values of factor of recirculation depending on size of the gap between a secondary radiator and brick-work a forward door of the boiler. (1) Analytical model, (2) CFD models.

increasing their speed and reducing the duration of the products of reaction in a zone of the maximal temperatures, resulting in a reduction of the harmful emissions to atmosphere.

Calculations of the values of the recirculation factor carried out on analytical and CFD models differ by only 5%, which testifies to the adequacy of the two models of calculation (Figure 6).

Checking the results of the models against real operating conditions is complicated, due to the design constraints of the boiler, as well as carrying out all of the experimental tests required.

3. Experiment

All experiments were carried out in the laboratory using the modified boiler "Victor–100" of Ukrainian manufacture (see Figure 7). Secondary radiators of different dimensions were investigated: $L = 2.5D$, $1.84D$, $0.9D$, and $Ssr/Scc = 0.65$, 0.43, 0.21. The degree of recirculation of the smoke gases was also investigated by moving the secondary radiator inside the chamber of combustion within limits from 0.04 up to 0.16 m. The investigations were carried out under laboratory conditions to confirm the correctness of the chosen method of intensifying the heat exchange process in the combustion chamber, and reducting the harmful emissions to the atmosphere (see Figure 7 for boiler diagram).

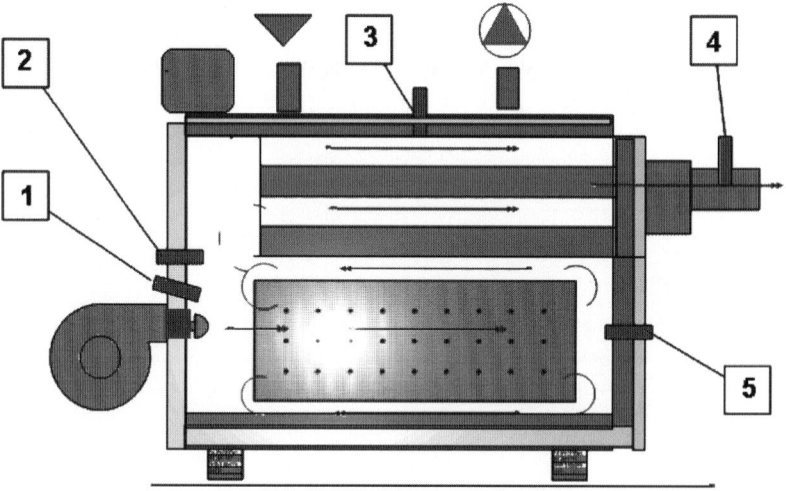

Figure 7. The device of the laboratory stand, points of visual supervision over process of burning, gauging of temperatures and chemical compound of smoke gases

- Apertures with quartz glasses for visual supervision and replaceable elements for introduction inside of the chamber of combustion of the boiler of probes and thermocouples (1, 5)

- Additional aperture for introduction inside of the boiler of probes and thermocouples (2)

- Point of gauging of temperature of water in the boiler (3)

- Point of definition of temperature and chemical compound of smoke gases in a chimney pipe (4)

The technique established of carrying out the investigations has allowed the industrial testing of secondary radiators in boilers of various capacities and from different manufacturers.

It is obvious, that the installation in the chamber of combustion of a secondary radiator provides, due to recirculation, a method to return the products of combustion from a zone in which the reaction of oxidation is nearing completion to a zone where it is beginning. There is therefore a heating of the reagents, causing a reduction in the values of activation energy, which in turn causes an increase in the burning of the oxides of carbon and sulfur on the surface of the secondary radiator.

It is important to note that the shielding a surface of the chamber of combustion from the burner through the use of a secondary radiator, is completely compensated by the heating of the radiator to temperatures ranging from 600°C up to 900°C. The average values of the temperatures at the front and inside the chamber of combustion without a radiator were found to be 605°C and 823°C, with a radiator they were 576°C and 1,163°C accordingly.

Temperature in degrees C   **Temperature in degrees C**

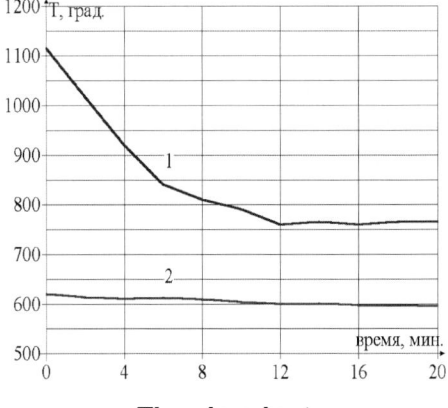
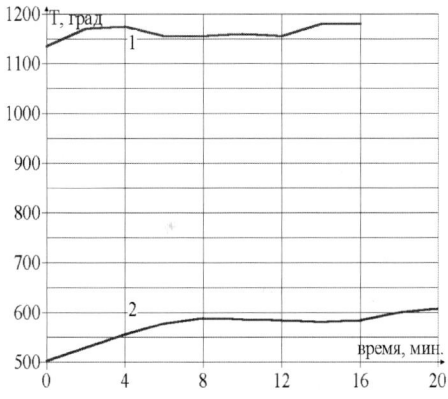

Time in minutes   **Time in minutes**

Figure 8. Change of temperature (1) in the chamber of combustion and (2) at the front the boiler. At the left – without a radiator, from the right – with a secondary radiator

Results of laboratory measurements are shown in Figure 8. The installation of a radiator has increased the ratio of the combustion chamber temperature to the front of the boiler temperature to a factor of 2, in comparison with the same ratio of 1.36 times for the case without a radiator. There is also a noticeable increase in length of the flame with the installation of a secondary radiator in the chamber of combustion of the boiler, which has been proved from measurements and visual supervision. It is recorded that the length of a flame without a radiator occupied about half of length of the chamber, and after his installation about three fourth of the length.

By analyzing the speed of the heating up of the water in the boiler, it can be concluded that the installation of a secondary radiator accelerates the heating process and improves boiler efficiency.

Changes in the temperature values testify that there is an increase in the overall performance of the chamber of combustion after the installation of a secondary radiator, and the linear trend indicates a stabilization of the complex processes occurring in the boiler. This will be compared to the results of the numerical analysis based on the mathematical models. The increase in efficiency corresponds with the results of the chemical gas analyses carried out using the computer gas analyzers: rbr-Ecom KD M, Testo-350 and Ecolaine-4000.

4. Results

The analysis of the conditions, by theory and practical testing, that cause the intensification of the heat transfer in the combustion chambers of boilers has revealed the practicality of the use of secondary radiators.

A design of a secondary radiator of cylindrical form has been developed, constructed of stainless steel with the edges providing its symmetrical accommodation in chambers of combustion. Such design provides the occurrence of an additional layer of gases and the occurrence of recirculation of gases to the root of the flame. (The Application for the invention «Water-heating boiler» No and 200511414 from 01.12.2005).

The developed mathematical model of the aerodynamics of smoke gases inside the boiler allows the definition of the magnitude of the circulating and transit gas streams.

This result has also been confirmed by CFD calculations which have been developed on a personal computer. The agreement of the two different results testify to the adequacy of the developed analytical model and CFD simulations, it also indirectly proves the results of the experimental data.

The improved distribution of the radiant thermal stream after the installation of a secondary radiator has been experimentally established. Visual supervision of the burning process has revealed a lengthening of the flame of approximately one third due to the introduction of a secondary radiator in the chamber of combustion of the boiler, this corresponds with the results of analytical modeling.

The experimental method has established the efficiency gains of applying secondary radiators in boilers of various designs, and highlights the features of movement in them of the aerodynamic streams.

The installation of secondary radiators in boilers of "Victor" type, in the gas boiler-house of the "KSW" enterprise has provided an increase of efficiency of up to 2.0% and a reduction of the emissions of NO_x to atmosphere of 5%. The installation of a secondary radiator in boiler VK-21, of capacity of 2,000 kW in the gas boiler-house of "Zhitomir-warmly communes-energy" has increased the efficiency of the boiler by 0.4%, and lowered the emissions of CO by 77.5%, and NO_x by 52.4%.

4.1 INFLUENCE OF SECONDARY RADIATORS ON EFFICIENCY OF GAS BOILERS

The initial investigations have established that in the water-heating smoke pipes of the boiler "Victor–100", the larger the size of an installed secondary radiator, the greater the increase in efficiency of the boiler and simultaneously the greater the reduction in emissions of NO_x. It was found that the optimum results are through the use of a secondary radiator of size parameters (0.9 D) and a ratio of radiation area to total area of the chamber of combustion of 0.21, with this it is possible to increase the efficiency of the boiler by 1.4% while decreasing the concentration of NO_x by 24 mg/m^3. The secondary radiators were also tested with boilers of foreign manufacture, for example the boiler of RTQ RIELLO/Italy with a capacity of 105 kW, equipped with a gas burner.

During tests the maximum increase in efficiency of the boiler was found to be 1.5%, with a secondary radiator length of 0.9 D and a ratio of its area to the total area of the chamber of combustion of 0.32. A simultaneous reduction in the concentration of NO_x of 34% (52.4 mg/m^3) was recorded with the installation of a secondary radiator of parameters L/D =1.84. From the point of view of increasing the efficiency, the optimum installation position of the secondary radiator of size 0.9 D is 140 mm from the edge of a face-to-face trumpet board within the combustion chamber.

The maximum reduction of NO_x was 30%, and the maximum increase of efficiency was 1.2% for the boiler of RTQ (105 kW capacity). This was achieved with tests of a secondary mesh radiator made of net steel construction and thickness 3.0 mm with cell sizes 40 × 40 mm^2, overall parameters were 0.9 D.

These initial investigations have revealed the dependence of the performance of secondary radiators on the design features of various boilers. The performance of the secondary radiator depends upon the initial aerodynamics of the product gases, the distribution of fields of pressure and temperature, and the varying turbulence intensity and radiating heat exchange in boilers of different designs.

To check the chosen method on boilers of medium capacity, initial tests of a secondary radiator were carried out in the boiler-house of enterprise "Zhitomir-warmly communes-energy" in the city of Zhitomir, on boiler "BK-21" whose capacity is 2.0 MWt. The secondary radiator was of stainless steel construction of size: length 1.0 m; diameter 0.5 m, radiator wall thickness 1.5 mm. The radiator was located in the center of the burner. The results of these industrial tests are shown in Table 1.

From the data shown in the table it is visible that the installation in the chamber of combustion as a secondary radiator has halved the concentration of NO_x.

TABLE 1. Tests of a secondary radiator for boiler «BK-21», capacity 2.0 MWt

Point of gauging	λ	CO	CO_2	NO_x	η
		(mg/m^3)	(%)	(mg/m^3)	(%)
Work of the boiler without a secondary radiator					
Output from the chamber	1.2	130.2	13.3	102	72
Output from the boiler	1.2	42.8	9.08	101.8	94.6
Work of the boiler with a secondary radiator					
Output from the chamber	1.3	169	12	59	76
Output from the boiler	1.26	57	9.1	88	95

4.2 INFLUENCE OF SECONDARY RADIATORS ON EFFICIENCY OF DIESEL BOILERS

For the investigations that were carried out with diesel fuel, a burner used during the study at Golling/Germany increased its capacity from 57 to 142 kW for a increase in fuel usage of 4.8–12 kg/h.

The burner was equipped with a fuel pump of type – Danfoss BFP 20/21 size 3, which provides the flow of diesel fuel from the tank to an atomizer at a set pressure. The size of the charge of diesel fuel to be burnt was defined by the pressure in the pump. During the study the charge was kept at 10 kg/h. The charge of fuel was monitored through the use of a graduated laboratory flask, of capacity 2.0 l, and a stop watch.

Results of the gas analysis of the boiler for a selected output and constant temperature conditions are tabulated in Figure 9. The photos illustrate the burning processes described. Analyzing the data shows that after the installation of a secondary radiator the concentration of NO_x in the smoke gases has decreased by more than in 1.2 times, this shows the value of applying secondary radiators as a means of suppressing the formation of NO_x without reducing the efficiency of the boiler.

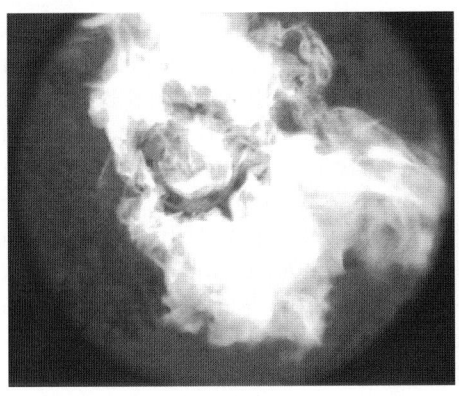 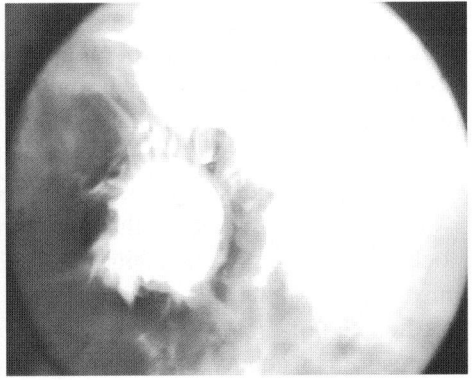

Point of gauging	Λ	CO	CO_2	NO_X	H (%)	Point of gauging	Λ	CO	CO_2	NO_X	H (%)
Output from the chamber of combustion	1.2	130	13,3	102	63	Output from the chamber of combustion	1.3	169	12	59	72
Output from the boiler	1.26	43	9.1	101.8	94.6	Output from the boiler	1.26	57	9.8	88	95

Figure 9. The data the works of the boiler received at tests on diesel fuel without a radiator – at the left, and with a secondary radiator – on the right

The photos of the burner in Figure 9 testify to the brighter luminescence of the flame with the secondary radiator. This is promoted by the heated surface of the secondary radiator which emits radiant energy. The installation of a secondary radiator has considerably extended the length of the flame.

5. Conclusion

Directions for the modernization and intensification of the heat exchange process in chambers of combustion are considered and determined. It is theoretically proven, that the proposed method of modernization of the chamber of combustion raises the efficiency due to increase in convection and radiating heat exchange. There is also a reduction in the harmful emissions to atmosphere due to change in the aerodynamics, thermodynamics and kinetic processes that take place in the boiler.

A mathematical model was developed to determine the volume of recycled gases and the velocities of the aerodynamic streams in the chamber of combustion and fire tube of the boiler. Calculations of the material balance of aerodynamic streams and thermal calculations for the chamber of combustion with a secondary screen-radiator installed were carried out. Experimental investigations under laboratory conditions were also carried out.

The results confirmed the ability of the chosen method to intensify the process of heat exchange within a boiler. The optimum geometrical sizes and constructional materials for secondary screens-radiators were established. The coordinates of installation of the screens-radiators in the chamber of combustion to achieve the maximum gains were determined. Changes in the structure of the flame were investigated; and a sharp decrease in the emissions of NO_x and the time required to heat the boiler's water recorded. CFD modeling has allowed detailed studying of the processes of aerodynamics, thermo-dynamics and chemical kinetics that occur in the chambers of combustion with a secondary screen-radiator installed.

The initial calculations by the analytical and CFD models agreed to within 5%, which testifies to their adequacy.

Initial tests under industrial conditions of secondary screens-radiators in boilers of 100 kW (Victor-100) and 2,000 kW (BK-22) of Ukrainian manu-facture, and other boilers of capacity 575 kW (Vitoplex-100 Viessmann/ Germany) and 105 kW (RTQ Riello/Italy) were carried out for natural gas and diesel fuels. These investigations have proved the chosen method intensifies the heat exchange process in the chamber of combustion The initial studies also allowed the development of the testing technique, whose results showed an increase in aerodynamic resistance of the boiler of up to 17%, an increase in efficiency of up to 1.7%, changes of temperature, and reductions in the emis-sions of CO of 77.5% and NO_x of 52%. The humidity of the leaving smoke gases were reduced by 15%, and changes in O_2, CO_2, NO, and SO_2 concen-trations measured.

Economic calculations have proven the practicality of the application of secondary radiators, determining an overall economic benefit as the decrease in fuel expenses quickly repays the initial capital outlay for the modernization of the boilers.

References

1. Demchenko V.G. Influence a fire pipe established in flue boilers, on their efficiency. Ecotechnologies and Resource saving, No. 5, 2005.
2. Demchenko V.G. Decrease of emissions NO_x by installation in chamber of combustion of the boiler of secondary screen-radiator. Pressing questions of thermo physics and physical hydrodynamics. Aluschta 2005.
3. Basok B.I., Demchenko V.G., Martinenko M.P. Numerical modeling of processes of aerodynamics in chamber of combustion the water-heating boiler with a secondary radiator. Industrial the heating engineer, No. 1, 2006.
4. Demchenko V.G., Serebrjanskiy D.A. Analyze of efficiency of an intensification chamber of combustion heat exchange at work of boilers on diesel fuel. Industrial the heating engineer, No. 6, 2005.
5. Demchenko V.G., Water-heating boilers of low power. Opportunities of their modernization and manufacture in view of criteria of efficiency. Materials of conference of the Problem of ecology and operation of objects of power, Sevastopol, 2005.

ENGINE EMISSION CONTROL USING OPTIMIZED COOLING AIR DISTRIBUTION BETWEEN COMBUSTOR AND TURBINE HOT SECTION

B. GLEZER

Optimized Turbine Solutions, San Diego, USA

Abstract. Presented paper is focused on turbine hot section cooling system design and the effects that it has on overall engine emission characteristics. Much consideration is given to industrial gas turbines due to their continuous operation as prime drivers for electric power generators and mechanical driven equipment like gas compressors and pumps. Although aircraft engines present a major part of the installed power in the gas turbine market, they play a less significant role in pollution due to a relatively short operating mission.

The paper provides analyses of compressed air distribution in a typical advanced gas turbine engine. The air distribution in a combustor section, role of air/fuel ratio in a primary zone and its effect on NO_x and CO formation are reviewed in detail. Means of emission improvement with the application of advanced combustor liner cooling techniques and optimized configurations for cooling of the combustor–turbine vane endwall transition are analyzed using practical design examples. The impact of overall engine thermal efficiency on emissions is discussed in comparison with the conventional characteristic that is based on parts of pollutants per million in the engine exhaust.

Keywords: emission, combustion, cooling technology, optimized cooling distribution

1. Introduction

The ever increasing demand for electric power generation and sources of mechanical drive for pipeline equipment and other applications was filled over the last two decades with gas turbines ranging from 5 to 200 MW output power. In many regions of the world gas turbines became a major source of fossil

N. Syred and A. Khalatov (eds.), Advanced Combustion and Aerothermal Technologies, 317–329.
© 2007 *Springer.*

energy conversion devices resulting in their combustors contributing a
noticeable portion of the man-made pollution. To limit emissions of harmful to
the environment pollutants, the combustors on both new and upgraded
industrial gas turbines in the recent years gravitated toward dry low NO_x
combustion systems. These systems employ lean premixed combustion to
achieve low NO_x and CO emissions. The cooler flame temperatures of the lean
premixed flames are the primary mechanism for producing lower NO_x levels.
This approach required significant rearrangement in the traditional distribution
of the air available for combustion, employing much greater amounts of air to
quench the stoichiometric flame temperature in the primary combustion zone
from over 1,700°C down to 1,450–1,550°C. At the same time it was undesirable for CO formation to reduce gas temperatures anywhere within the
combustor primary zone to below 1,200°C. These limitations combined with
continuously rising operating temperatures (Figure 1) and increasing demand
for turbine and combustor cooling flows create significant challenges for gas
turbine designers dealing with an extremely tight budget of the available
compressed air.

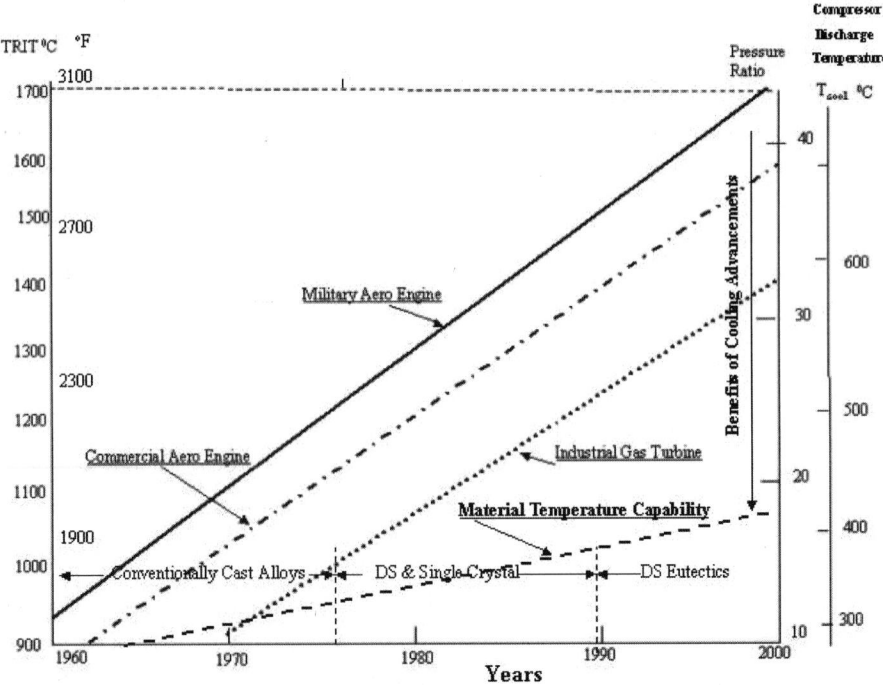

Figure 1. Trend of growth for key engine operating parameters

2. Combustor Cooling Strategies and their Effect on Engine Emission

Figure 2 presents a generic simplified turbine hot section with the combustor being the main consumer of the available compressed air. As it is shown later, every design possibility to save this air can become essential for achieving emission targets without sacrificing engine efficiency and durability.

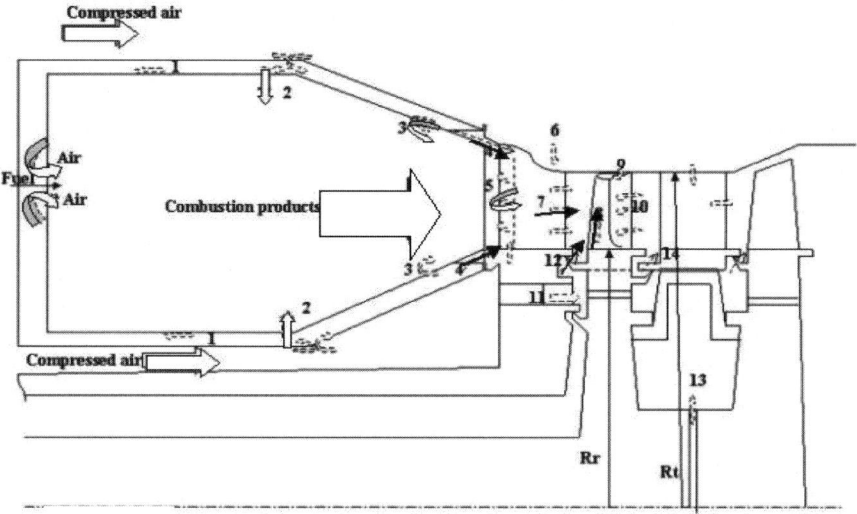

Figure 2. Generic turbine hot section cooling system

2.1 DEVELOPMENT OF COOLING TECHNIQUES WITH EVOLUTION OF COMBUSTOR DESIGN

The cooling requirements for the combustor liners vary with a number of parameters, which have continually changed in the last 20 years of engine development. The major parameters are:

- Hot gas temperature and type of fuel
- Cooling air temperature and allowable combustor pressure drop
- Allowable material temperatures, expected life, and durability
- Weight, cost, and complexity constraints

Considering that liner heat load is driven primarily by the flame radiation, the calculation of the flame temperature and the heat flux through the liner wall requires a good understanding of the combustion process. Figure 3 presents

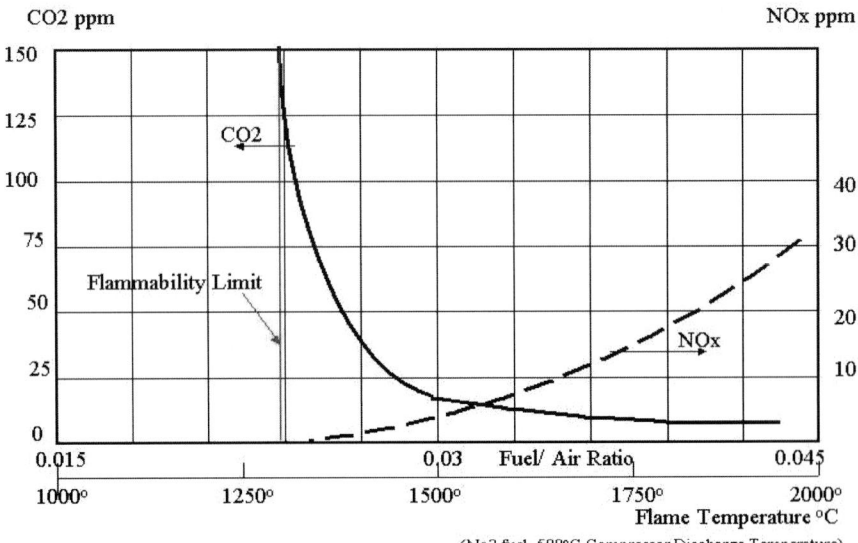

Figure 3. Effect of fuel to air ratio on engine emission

correlations between NO_x and CO vs combustor flame temperature defined by a fuel to air mass ratio.

A comprehensive monograph by Lefebre [1]. provides an excellent review of various gas turbine combustor systems. It also details the calculation procedures for the main factors that affect liner cooling. Figure 4 depicts some of the conventional and advanced cooling configurations. Many early gas turbine combustors were of a single or multiple can design and required a transition between combustor exit and turbine inlet. Liners for many of these combustors were assembled from a group of cylindrical shells that formed a series of annular passages at the shell intersection points. These passages created the louvers permitting a film of cooling air to be injected along the hot side of the liner wall to provide a protective thermal barrier. The annular gap heights were maintained by simple "wiggle-strip" louvers. Air metering was a major problem with this technique. The application of splash-cooling devices provided control of the cooling air entering the liner through a row of small-diameter holes with air jets impinging on a cooling skirt, which deflected the air along the inside of the liner wall. Annular combustors, which were introduced later, originally also employed wiggle-strip and splash-cooling configurations. Since then, the "machined-ring" or "rolled-ring" approach, which features accurately machined holes instead of louvers and combines accurate airflow metering with good mechanical strength, has been widely adopted in one form

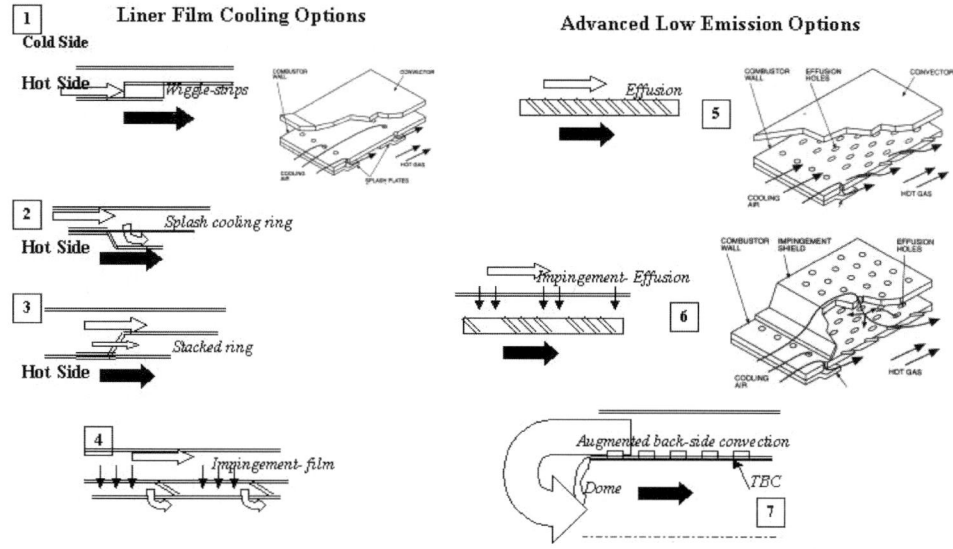

Figure 4. Combustor liner cooling techniques

or another. Introduction of highly penetrating combustion jets or significant film cooling flows can negatively affect CO emissions.

Modern cooling techniques include angled effusion cooling (EC) using multiple rows of small holes drilled through the liner wall at a shallow angle to its surface. With this scheme, the cooling air flows through the liner wall, first removing heat from the wall by convection–conduction, and then providing a thermal film barrier between the wall and the hot combustion gases. From the stand point of cooling EC presently is considered to be the most promising option among advanced combustor cooling techniques that are being actively developed for the new generation of industrial and aeroengines. For some advanced aeroengines it has reduced the conventional cooling air requirement by 30%. The main drawback of EC is an increase in liner weight of around 20%, which stems from the need for a thicker wall to achieve the required hole length and provide buckling strength.

A common alternative to increasing the efficiency of cooling techniques is to spray protective and thermal barrier coatings (TBCs) on the inner liner wall. As it has for the last 60 years, the search continues for new liner materials that will allow operation at higher temperatures. Current production liners are typically fabricated from nickel-based alloys such as Haynes 230.

The application of combined impingement – film or impingement – effusion techniques are often considered when a higher cooling effectiveness is required. These techniques require a double-walled liner design where the outer (in relation to the gas path) wall in the double-walled region is perforated. The advantage of the method derives from its use of cooling air to serve a dual purpose. First, the air is shaped into multiple small jets, which provide impingement cooling to the front (primary zone) section of the liner wall, and then the jets merge to form an annular sheet, which operates in a conventional film cooling mode to cool a downstream section of the inner liner wall. Another advantage of impingement cooling is that the impingement jets can be positioned to provide extra cooling on liner hot spots. The higher cooling effectiveness of these techniques comes with certain penalties in terms of cost, weight and higher pressure losses that affect overall engine efficiency. Another concern stems from the significant difference in temperature between the two walls leading to a differential expansion that might result in buckling of the inner wall if the local hot spots become too severe. Also, the high heat-transfer coefficients that are normally associated with impingement cooling cannot be realized fully in the downstream section, because the film of air discharged from the upstream protects the downstream section reducing the inner wall metal temperature and thus lowering the effectiveness of impingement cooling.

Similar to other cooled turbine components the combustor air inlet and hot gas outlet temperatures can be combined together with the maximum allowable liner wall temperature into a single parameter called cooling effectiveness, reflecting the thermal load on the wall at given heat transfer conditions. The larger this parameter, the more cooling air is needed, or the more effective the cooling method must be (Figure 5). Advanced large engines are clearly approaching the limit above which pure film cooling is no longer sufficient. In the past, therefore, an iterative solution was found in reducing the surface in need of cooling by shortening the flame tube. This was made possible by the adoption of improved fuel preparation systems. Over the years the ratio of the liner length to its height dropped from about 4 for old engines to about 2 for new engines.

Cooling configurations were extensively studied in model rigs, where the cooling action is determined at defined constraints [2–5]. When applying the cooling configuration to real combustors, the exact constraints are unfortunately not known. This applies in particular to constraints on the hot gas side. The flow pattern and the local temperature in the flame tube, familiarity with which is a prerequisite for determining the constraints, cannot normally be measured. For this reason, these quantities must be computed using 3D CFD computations for the combustor flow, where the local release of heat is to be determined accurately enough to derive local temperatures, velocities and radiation load.

Codes for this purpose already exist, although the confidence in the results they provide is still limited. However, the recent improvements in accuracy of the analytical predictions suggest that future designs of combustor cooling configurations should be made in combination with CFD computations.

Many details of practical combustor designs and liner cooling features can be found in references [6–10].

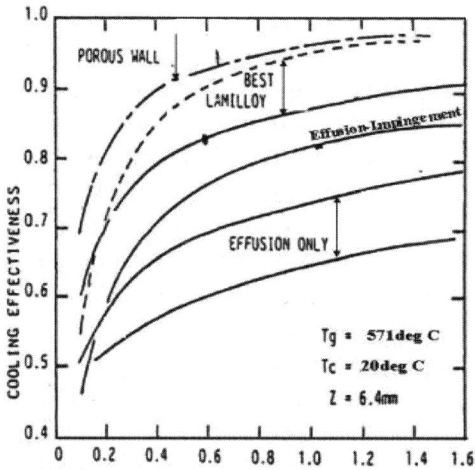

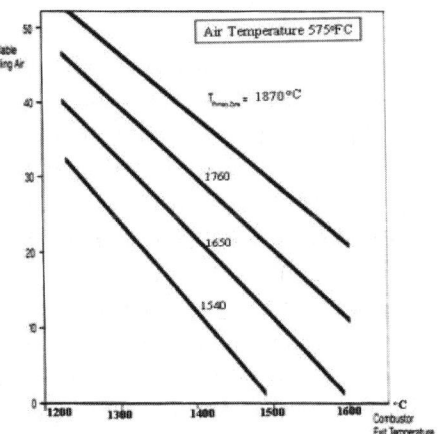

Figure 5. Effectiveness of liner cooling methods

Figure 6. Influence of combustor exit and primary zone temperature on the available amount cooling air

2.2 THERMAL BARRIER COATING

One attractive approach to the problem of achieving satisfactory liner life is to coat the inside of the liner with a thin layer of a very low thermal conductivity material, which is often called a thermal barrier coating (TBC). A suitable material of low emissivity and low thermal conductivity could reduce the wall temperature in two ways: by reflecting a significant portion of the radiation heat flux from the flame and by providing a layer of thermal insulation between the hot gas and the wall of a base metal. The steep temperature drop through the TBC varies with the thermal conductivity and thickness of the layer, and the heat flux through the layer that is greatly affected by the heat transfer from the hot gas and to the cooling air. If the TBC coated wall is not provided with sufficient back side cooling, the barrier helps little in lowering the temperature.

A further benefit may be gained if an oxidation-resistant base coat is applied because it reduces the oxidation constraint on the choice of liner-wall material.

An ideal TBC would be chemically inert and have good mechanical strength, resilience to thermal shock, and resistance to wear and erosion. Above all, it would have a low thermal conductivity and a thermal expansion coefficient that is similar to that of the base metal. A typical thermal spray-deposited TBC comprises a metallic base coat (e.g., 0.1 mm of Ni Cr AL Y), plus one or two layers of ceramic (e.g., yttrium-stabilized zirconium oxide ZrO_2).

Recent developments in the strain tolerance of the TBC have reduced the necessity for an intermediate coat, and two-layer coatings are now sometimes specified for improved mechanical integrity.

Plasma flame spraying is often used to apply the ceramic and base coat layers because it is found to provide durable and reproducible coatings. A typical overall coating thickness is around 0.4–0.5 mm, which gives metal temperature reductions of the order of 55–90°C, depending on the heat flux through the liner wall. In this context it was noted earlier that for a TBC to be fully effective there must be adequate heat removal from the "cold" side of the liner wall. The most favorable temperature reduction in the base material that results from implementation of a TBC is obtained at highest convection heat transfer coefficients on both hot gas side and air cooled back side of the liner wall. Inevitably, this means that liner geometries will become more complex as various features (such as fins and ribs) are added to augment the convective heat transfer from the cooled side of the wall in order to derive the full benefit from the TBC coating on the inner wall.

The reduction in wall temperature obtained from using a TBC can be calculated by adding a term of TBC resistance $R_{TBC} = (k/t)\,(T_h - T_i)$ and solving one-dimensional heat transfer equations for the composite liner wall, where:

- k – TBC conductivity
- t – TBC thickness
- T_h – hot-side surface temperature of TBC
- T_i – temperature at interface between TBC and liner wall

2.3 TRANSPIRATION COOLING

For the cooling air to be utilized much more effectively than in traditional film cooling-based designs, the single layer perforated sheet must be replaced with a multiple layer sheet structure where the cooling air is routed through a winding path between and through the layers. This arrangement is referred to as transpiration cooling. This method approaches an ideal wall-cooling system which can maintain the entire liner at the maximum temperature of the material avoiding cooler regions that would represent a wasteful use of cooling air. The transpiration cooled liner wall is constructed from a porous material that provides a large contact area for heat transfer to the air passing through it. Because the pores are uniformly dispersed over the surface of the wall, the tiny air jets emerging from each pore rapidly form a protective layer of a relatively cold air over the entire inner surface of the liner. While passing through the pores, the cooling air removes a significant amount of heat from the wall. When this combined convection–conduction wall heat transfer is coupled with the protective layer of discharged film, the overall cooling effectiveness is sufficient to counterbalance very high heat loads that include radiation from the flame. This means that, in addition to acting as a porous medium, the wall must also have good heat transfer properties and be of adequate thickness. A problem this poses is that, in order to form a stable boundary layer on the inner surface of the wall, the coolant flow should emerge with as a low velocity as possible, whereas for maximum heat transfer within the wall a high velocity is required. Although transpiration cooling is potentially the most efficient method of liner cooling, its practical implementation has been very limited due to availability of required porous materials. The porous materials developed to date have failed to demonstrate the required tolerance to oxidation, which has led to the small passages becoming blocked. These passages are also sensitive to blockage by foreign particles in the air.

2.4   EFFUSION COOLING

The simplest approach to a practical form of transpiration cooling is a wall perforated by a large number of small holes. Ideally, the holes should be large enough to remain free from blockage by impurities, but small enough to prevent excessive penetration of the air jets into the mainstream. Provided that the jet penetration is small, it is possible to produce along the inner surface of the liner a fairly uniform film of cooling air. If, however, the penetration is too high, the air jets rapidly mix with the hot gases and provide little cooling of the wall downstream. EC can be applied to all or any portion of the liner wall but due to the high rate of cooling flow required, it is best used for treating local hot spots in the liner wall. Another useful role of EC is in improving the effectiveness of a conventional film-cooling slot. As the film of air from this slot moves downstream, its temperature gradually rises due to the entrainment of the surrounding combustion gases. Eventually, it becomes so hot that it starts to heat the liner wall instead of cooling it. If EC is applied before this point is reached, the injection of cold air into the film enables it to maintain its cooling effectiveness for a longer distance downstream.

In conventional EC, the holes are drilled normal to the liner wall. The advantages to be gained from angled EC with the holes drilled at a shallower angle are twofold:

- An increase in the internal surface area available for heat removal. This area is inversely proportional to the square of the hole diameter and the sine of the hole angle. Thus, for example, a hole drilled at 20° to the liner wall has almost three times the surface area of a hole drilled normal to the wall.

- Jets emerging from the wall at a shallow angle have low penetration and are better able to form a film along the surface of the wall. The cooling effectiveness of this film also improves as the hole size and angle are decreased.

Some studies showed that the cooling effectiveness can be increased by 60% at a realistic pressure ratio of 1.03, if the hole is made at an angle of 20° vs being normal to the wall surface. It is clear that the practical implementation of angled EC is highly dependent on an ability to accurately, consistently, and economically manufacture large numbers of oblique holes of very small diameter. Advances in laser drilling have made this possible, and this cooling method is now regarded as a viable and economically acceptable technique. At the present time, the lower limit on a hole diameter is about 0.4 mm, whereas the lowest attainable hole angle is just below 20°. Andrews [2–5] presented a number of papers that are widely used for advanced liner cooling design. Full coverage discrete hole impingement cooling and effusion film cooling are extensively used in gas turbine blade and combustor wall cooling. However, most applications and most experimental investigations are for these cooling techniques used separately. The combination of impingement and EC offers a good means of improving the overall cooling effectiveness of both turbine blades and combustor walls and of minimizing the coolant flow required to achieve the desired cooling effectiveness. Combined impingement/EC with equal numbers of holes, but the main pressure loss at the impingement holes, have very good internal wall heat transfer characteristics with increases of 45% and 30% found for two designs, relative to the impingement only situation. Studies of combined impingement – EC yielded the following conclusions:

- The combined impingement/EC heat transfer was not greatly influenced by the effusion wall design for the effusion/impingement hole diameter ratios.

- The measured combined impingement/effusion heat transfer coefficients were lower than the sum of the separate impingement and effusion wall heat transfer by approximately 15–20% for the two designs tested. This indicates that there was an interaction between the two heat transfer modes, which reduced the net heat transfer.

- The overall cooling effectiveness results demonstrated the large benefits to be obtained from the addition of impingement cooling to EC. However, the film cooling part of the process, which was strongly dependent on the effusion hole size, had a strong influence on the wall impingement/effusion heat transfer coefficient.

Cost, increased weight, durability, and ability to repair the angled effusion cooled liners are the main concerns that limit their application. These issues can only be fully resolved by extensive service experience.

Future developments in angled EC will tend to focus on the optimization of hole geometry. A diffuser-shaped expansion at the exit portion of the hole has been shown to improve cooling effectiveness due to lower exit velocity and reduced penetration of the air jet into the hot gas stream. However, a cost-effective method of producing the shaped holes has yet to be developed.

2.5   AUGMENTED BACKSIDE CONVECTION

The introduction of low emission combustors has resulted in partly changed challenges compared to those associated with conventional combustion chambers. Low-emission combustors for advanced industrial engines target very low NO_x emission using the lean-combustion principle. In the lean-combustion concept the objective is to use a large fraction of the air for combustion in the primary combustor zone to achieve reduction in combustion temperature and NO_x emissions. The resultant effects on the available cooling air flow are shown in Figure 6. As expected, the amount of air required increases for lean combustion. As illustrated in Figure 6, the amount of cooling air available decreases proportionately. Consequently, with a primary zone temperature of 1,700°F and a combustion chamber exit temperature of over 1,500°C the portion of air available for cooling is only about 20%. Therefore, it is necessary to find methods of cooling, which require less air or methods that can use the air in sequence for liner cooling and then for the primary zone "quenching".

This application of in-series cooling can make most of the combustor air available for cooling the liner backside. The convective heat transfer rate on the back side of the liner can be increased by the application of fins, pedestals, ribs or any other form of secondary surface that augments convective heat transfer and increases the effective area for heat exchange. Augmented back-side convection (ABC) cooling method can be particularly beneficial when combined with TBC applied to the inner surface of combustor. Usually, this cooling technique will require an additional "cold" wall to control the air passage. Such dual-wall cooling structures are not necessarily easy to translate into practical designs. The inner wall being hot and the outer wall cold result in excessively high differential temperatures and hence differential expansions between inner and outer walls such that a fixed joint is made impossible.

Therefore, the outer wall is normally designed as a supporting structure, while the inner wall is designed as shingle-type individual plates with sufficient clearance between them to accommodate the difference in thermal growth. Some of the design concepts, however, can be based on a reversed scheme where the hot wall is a continuous structure and outer wall is sectioned and spring loaded against the liner [11]

Specific information on the use and performance of various extended-surface configurations for the liners, including ribs, fins, and pedestals, may be found in [12]. One of the critical parameters that can limit the application of these techniques is the pressure loss of the cooling system, which should not exceed 1.5–2% of the total compressor discharge pressure, otherwise causing unacceptable engine performance penalties. Among the various heat transfer augmentation techniques that might be attractive for this application is a surface with periodic concavities often called dimples, which has been recently introduced also to combustor cooling. The application of the dimpled back side cooled surface, when optimized for certain geometry applicable to combustors, showed significant improvement in heat transfer at remarkably small pressure loss [13]. Fabrication of a liner wall that is smooth on the gas side and dimpled on the cold side can result in certain manufacturing challenges, however. Combination of this cooling method with a TBC can provide promising design alternatives to more complicated low emission combustor systems.

Existing practice comparing the emission of different engines usually focuses on parts per million (ppm). This practice does not take in account overall engine thermal efficiency that has reverse function of the consumed fuel (also known as SFC). Meantime, it is directly related to the amount of produced pollutants and should be taken in consideration as important contributor to emission.

The significant effect of cooling air discharge downstream of the combustor on engine efficiency is well known. The cooling of the transition between combustor and turbine as well as the stage 1 nozzle endwalls presents a particular interest in optimizing cooling budget and improving turbine efficiency. A number of studies and some advanced engine designs [14] have demonstrated benefits of treating combustor exit transition and nozzle endwalls together by utilizing some of the spent combustor cooling air for effective cooling of the endwalls as well. The introduction of this air upstream of the nozzle leading edge, in the region of a low Mach number, helps to keep cooling air film next to the endwall without it being washed away by a horseshoe vortex. A triple positive effect can be achieved with this arrangement: saving some of the cooling air flow, improved endwall-cooling effectiveness and improved turbine efficiency. All of these benefits can assist in overall engine emission reduction.

References

1. Lefebvre, A.H., 1998, Gas Turbine Combustion, 2nd Edition, Tailor & Francis, Philadelphia, London.
2. Andrews, G.E., Asere, A.A., Hussain, C.I., and Mkpadi, M.C., Transpiration and impingement/effusion cooling of gas turbine combustion chambers. Seventh International Symposium on Air Breathing Engines, Beijing, China, pp. 794–803, AIAA/ISABE 85-7095, September 1985.
3. Andrews, G.E. and Kim, M.N., Influence of Film Cooling on Emission for a Low NOx Radial Swirler Gas Turbine Combustor, ASME/IGTI Paper 2001-GT-0071, New Orleans, June 2001. International Gas Turbine Conference, pp. 67–74.
4. Andrews, G.E., Asere, A.A., Mkpadi, M.C., and Tirmahi, A., Transpiration cooling: Contribution of film cooling to the overall cooling effectiveness. ASME Paper No. 86-GT-136. International Journal of Turbo and Jet Engines, **3**, 245–256, 1986.
5. Andrews, G.E. and Hussain, C.I., Full coverage impingement heat transfer: The influence of cross flow. Presented at the AIAA/ASME/sAE/ASEE 23rd Joint Propulsion Conference, San Diego, 1987, AIAA Paper 87-2010.
6. Burkhardt, S., 1992 Advanced Gas Turbine Combustor Cooling Configurations. Proceedings of International Symposium on Heat Transfer in Turbomachinery, pp. 261–285, Begell House Bahr, D.W., Technology for Design of High Temperature Rise Combustors, AIAA-Paper No.85-1292, 1985.
7. Glezer, B., Combustor Cooling, Handbook of Turbomachinery, Marcel Dekker, New York, 2003, pp. 215–224
8. Nealy, D.A., Reider, S.B., and Mongia, H.C., Alternate Cooling Configuration for Gas Turbine Combustion Systems, AGARD PEP 65th Meeting, Bergen, 1985.
9. Bahr, D.W., 1999, Gas Turbine Combustion and Emission Abatement Technology Current and Projected Status, Proceedings of International Gas Turbine Congress, Kobe, Japan, pp. 15–25.
10. Arellano, L., Smith, K., and Fahme, A., Combined Back-side Cooled Combustor Liner and Variable Geometry Injector Technology, ASME/IGTI Paper 2001-GT-0086, New Orleans, June 2001.
11. Glezer, B., Greenwood, S., Dutta, P., and Moon, H.-K., 2000, Combustor for a low-Emissions Gas Turbine Engine, US Patent 6,098,397.
12. Gardner, K.A., 1945, The Efficiency of Extended Surfaces, Transactions of ASME, Vol. 67, pp. 621.
13. Moon, H.-K., O'Connell, and T., Glezer, B., 1999, Channel Height Effect on Heat Transfer and Friction in a Dimpled Passage. ASME Paper 99-GT-163.
14. Harrogate, I. and Rolls-Royce W., Cooled Turbine Nozzle Assembly, US Patent 5,417,545, 1998.

THE INFLUENCE OF MOISTURE IN AIR ON THE WORKING EFFICIENCY OF BOILERS IN THE INDUSTRIAL AND MUNICIPAL ENERGY SECTORS

A. I. SIGAL[*] A. A. DOLINSKY
The Institute for Engineering Thermal Physics
National Academy of Science of Ukraine

Abstract. The necessity of taking into account and compensation of influence atmospheric air parameters that is used for combustion, for the dynamic of changes of heat-and-power characteristics process, is showed on the basis the review of situation which happened in the municipal energy sector of Ukraine. The double role of moisture in the hydrocarbon fuel oxidizing process is proved. The preliminary quantity estimation of water steam volume, which can take part in the combustion process chemically intensifying it, is given.

Keywords: combustion process, moisture, surplus moisture, nitrogen oxides

1. Results

A great number of studies have been devoted to the influence of moisture on the combustion process. Among the most significant of these studies are those conducted by Kormilitsyn (of the Krzhyzhanovskiy Institute of Energy, at the E.P Volkov Russian Scientific Academy) and Tsyrulnikov (of SredAsNIOgas, at the Moscow Institute of Energy)[1–6]. However, the deterioration of the fuel combustion process in dry air under increasing temperature remains unexplained. Similarly, the improvement of the combustion process with increasing temperature during the delivery of recirculation gases to the combustion zone, mentioned by various authors in the literature, also remains unexplained. For

_____
[*]To whom correspondence should be addressed. Alexander I. Sigal, Apt.115, 103-a, Gorkogo street, Kyiv, 01003. Work Tel.: +38044 456 92 62; Home Tel.: +38044268 18 22.

N. Syred and A. Khalatov (eds.), Advanced Combustion and Aerothermal Technologies, 331–339.

example, ref[7] mentions that after switching off the recirculation smoke exhausters at two boilers, the furnace regimes deteriorated significantly; the density and luminosity of the torch increased and the faintly blue color was changed to yellow.

In recent years, research conducted by the Institute of Thermal Physics Engineering at the Ukrainian National Academy of Sciences and in industrial conditions, has demonstrated that it is necessary to account for the role of moisture, delivered as both water and vapor, in the calculations of the combustion process, as well as the generation and decomposition of toxic substances (nitrogen oxides, carbon oxides, etc.).

In both the municipal and the industrial energy sectors, the air for combustion is taken directly from the atmosphere. The content of oxygen available for combustion in 1 m^3 of air changes with the temperature and moisture of the air. Changeable boiler input parameters can lead to the absence of stationary process parameters completely, i.e., an in-effective combustion supply at the constructive level, the necessity on the complicated basic automatics and at the end fuel overrun.

Calculations and experimental research have been performed where it was possible to define the volume of moisture in the input components in the process of hydrocarbon fuel combustion.

Until recently, it was considered that the influence of moisture on the combustion process was negligible, only contributing to the consumption of heat in the combustion zone as ballast through heating and evaporation. This idea was facilitated by the comparison of the moisture content in the fuel-oxidizer mixture with the moisture content in the combustion products. According to the equation of methane oxidization, the complete combustion of 1 kg of methane produces 2.25 kg of steam. In the introduction of oxidizer – air with a moisture content of 20 g/1 kg of dry air – 300 g of water per 1 kg of methane may be additionally inserted when $\alpha = 1.1$. Although, the starting moisture of the fuel–air mixture will form 8–9% of the combustion products. This is the basis for the idea that the moisture content of the input components has a negligible impact on the combustion process.

Experimental studies have shown that moisture in the initial components of the combustion process can be divided into two sections. The first is the moisture that takes part in the chemical process of fuel oxidation, i.e., the moisture decomposed during high-temperature dissociation to oxygen and hydrogen before combustion followed by the formation of hydroxyl OH and peroxide radical HO_2, which are strong oxidizers that accelerate the hydrocarbon combustion process. For example, the final oxidation of CO in the reaction:

$$CO + OH \rightarrow CO_2 + H_2O, \tag{1}$$

The second is the moisture that does not take part in the chemical combustion process, i.e., surplus moisture that plays the role of ballast and carries off the heat from the combustion zone.

The variety of data in the literature and the subsequent disagreement over the influence of moisture on the combustion process is essentially explained by the dual role of moisture[8-10]. Opinions differ from the intensification of the combustion process to the increase of ineffective ballast volume. It may be assumed that for such widespread processes as recirculation of the combustion products into the combustion zone and flame turbulization in diffuser burners, the former is the case.

Obviously, with increasing moisture content, the oxygen content in the air decreases when the temperature increases. This explains the initiation of the chemical under combustion and the yellow flame in small boilers during precipitation, as well as the necessity to increase the working excess air factor in contact with the installed air-heaters.

Using the experimental rig, the effect of adding a drop of distilled water of known volume to a laminar flame on the process of methane combustion was investigated. The drop "hovered" on the injection needle, where the flow rate was balanced by the speed of evaporation. In this way, the surface of evaporation was constant throughout the experiment.

It was demonstrated that both increasing the drop volume by almost a factor of 10 (Figure 1) and adding an additional drop of the same volume to the laminar flame boundary (Figure 2) did not significantly reduce the temperature

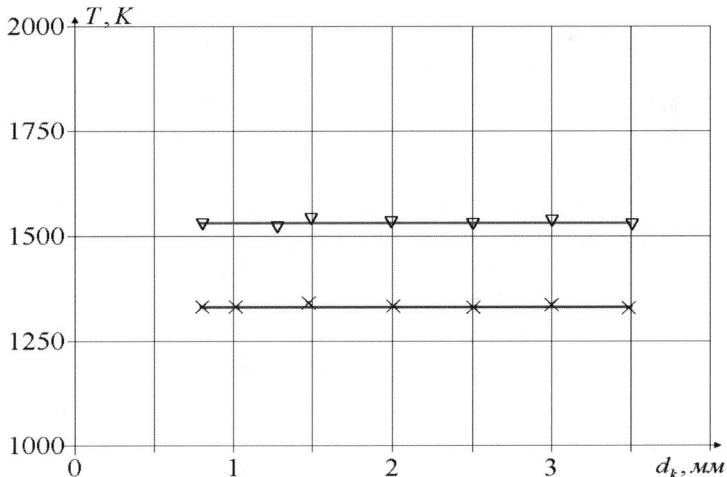

Figure 1. The "step" in temperature of laminar flame boundary ($\alpha = 1.1$) caused by the addition of one drop ∇ – Without moisture introduction x – With introduction of one drop

or the concentration of nitrogen oxide in the combustion products. In fact, comparing the influence of the effectiveness of the evaporation of the first and second drops as demonstrated by Figure 2, enables the comparison of the chemical and clearly thermodynamic impact of the added moisture in equal amounts.

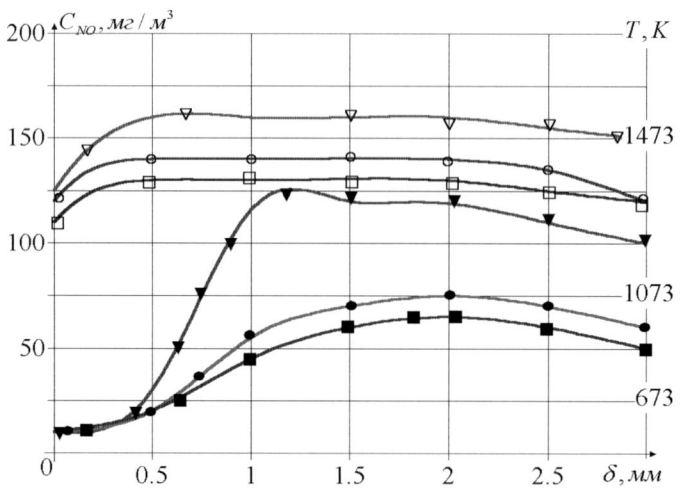

Figure 2. NO content and temperature by flame boundary width of laminar methane flame ▽ – Without moisture introduction ▼ – without moisture introduction. O – With introduction of one drop. ● – with introduction of one drop. □ – With introduction of two drops. ■ – with introduction of two drops

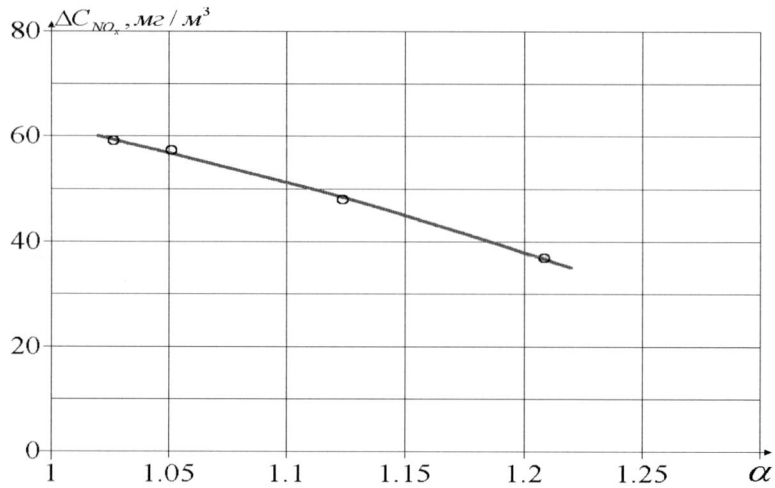

Figure 3. Ecological efficiency of moisture introduction in the combustion zone

During the introduction of additional oxidizer (i.e., increasing the excess factor by 20%), the efficiency of the repression of NO formation by moisture inserted into the combustion zone decreased by almost half (Figure 3).

In order to investigate the influence of peroxide radical and hydroxyl groups OH on the intensification of the combustion process, experiments were performed where hydrogen peroxide solution was added to the methane flame using the method described above. The results presented by Figure 4 clearly demonstrate that higher local temperatures develop with increasing H_2O_2 concentration in the flame.

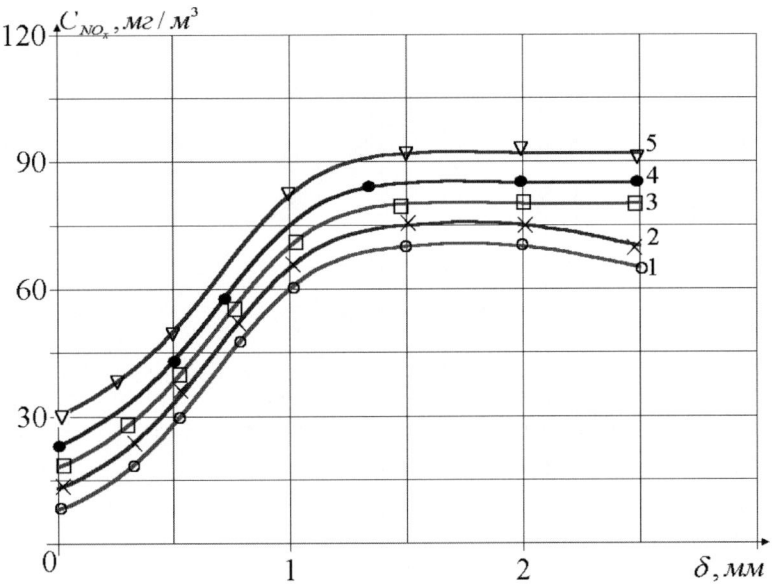

Figure 4. NO_x content by flame boundary thickness of laminar methane flame

- 1 – With introduction of one drop H_2O_2
- 2 – With introduction of one drop of 3.0% solution H_2O_2
- 3 – With introduction of one drop of 7.5% solution H_2O_2
- 4 – With introduction of one drop of 15.0% solution H_2O_2
- 5 – With introduction of one drop of 30.0% solution H_2O_2

The results indicate that the effect of introducing additional components to the combustion process cannot be referred to in thermodynamic terms alone. However, if we compare, for example, the most frequently introduced ballasts such as surplus air and steam, then we find an interesting result as demonstrated by Figure 5. As the maximum local temperature of the process increases, the efficiency of introduction of thermodynamically different ballasts draws together. This can be explained by the domination of kinetics in the field of maximum temperatures.

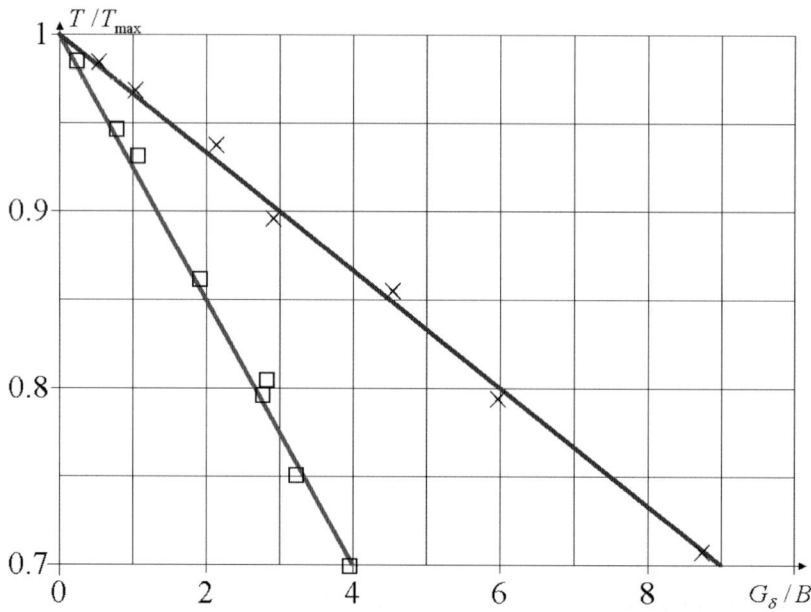

Figure 5. Thermodynamic efficiency of ballasts with different specific heat □ – Steam, $T = 373$ K, $\alpha = 1.0$ x – Air, $T = 373$ K

Reference[11] stated that choosing the correct amount of moisture to introduce, its dosage, characteristics and the content of the additions, which are the sources of radicals, define the combustion process. The author attempted to characterize the amount of moisture necessary for the required performance.

The amount of moisture that can be chemically used in the process of combustion is different for different fuel types and depends upon the excess factor.

The ratio between the moisture contained in the primary fuel–air mixture and the moisture that takes part in the chemical combustion process (stehiometry quantity) can be called the excess moisture factor using the analogy of the excess factor. Accordingly, if this ratio is equal to one, the combustion zone is over ballasted with moisture which causes energy overexpenditure in the process on its evaporation. Accordingly, if the excess moisture factor is less than one, slowing down of the combustion process and increasing of the flame length will occur due to the absence of the necessary amount of the hydroxyl group, other atoms and radicals necessary for quick oxidation. This is of practical importance in the regulation of gas combustion, primarily in the furnaces of boilers and stoves.

Fixing studies of the boiler may lead to the definition and fixing of the air–fuel ratio. In the absence of an air-heater or moisture control this ratio would be variable depending on the real temperature and atmospheric air moisture, which would be useful in the industrial (besides heavy and chemical industry) and municipal heat energy sector where the boilers are not equipped with air-heaters.

Calculations of the moisture content in the primary reactants necessary for combustion, the so-called stehiometry value, show that:

- The energy necessary for the evaporation of 1 kg of moisture is nearly 6,000 kcal/kg. The energy of water molecule dissociation (depending on how dissociation occurs) is 114–122 kcal/gram-molecule; approximately 5,500 kJ/kg. In general, the energy consumed by evaporation and dissociation is nearly 25,000 kJ/kg.

- For natural gas under adiabatic conditions the dissociation of 1.3 kg of moisture consumes over 33,500 kJ/m^3. In practical conditions it is much more difficult to estimate the energy that may be consumed by dissociation, but by various estimations, it will form 10–20% of the amount consumed under adiabatic conditions, i.e., 3–5 MJ/m^3 of dry gas will be enough for water dissociation. To estimate the moisture content of the primary reactants, with 280 g/m^3 natural gas, 20 g/m^3 moisture content of air and $\alpha =$ 1.3, then the ratio for practical conditions, i.e., the excess moisture content factor, will be nearly 1.4 (280/200). From these figures, the optimal moisture content in the air used for combustion will have a value between 13–15 g/m^3 of dry air.

2. Conclusions

1. The total amount of steam that takes part in the combustion process can be divided into "necessary" and "excess".

2. The ratio of steam in the primary products (fuel, oxidizer) to the theoretical amount of steam necessary in the combustion process, can be called the excess moisture factor.

3. Steam formed as a the result of the combustion process usually does not take part in the combustion process other than through the recirculation of

the combustion products or the insertion of water and steam into the combustion zone.

4. Insertion of an insufficient amount of steam to the combustion zone, i.e., less than is theoretically required, slows down oxidization and hastens the formation of the products of incomplete combustion.

5. For quick disbalance processes the role of water steam in the reactions of combustion increases with the speed increase and disbalance process.

6. For the purpose of normalizing the primary parameters of boilers and preventing fuel overexpenditure, it would be appropriate to fit boilers used in industrial and municipal power sector with air-heaters, thereby facilitating air moisture control.

References

1. Э.П. Волков, В.И. Кормилицын, И.Г. Збрайлов, Т.А. Тишина, Экспериментальное исследование влияния режимных и конструктивных факторов на концепции оксидов азота в дымовых газах паровых котлов при сжигании газа и мазута //Оксиды азота в продуктах сгорания и их преобразование в атмосфере. Киев: Наукова думка, 1987. С. 20–27.
2. Э.П. Волков, Н.Ю. Кудрявцев, Моделирование образования оксидов азота в турбулентном диффузионном факеле // ИФЖ. 1989. Т. 56. No. 6. С. 885–894.
3. В.И. Кормилицын, М.Г. Лысков, А.А. Румынский, Влияние добавки влаги в топку на интенсивность лучистого теплообмена //Теплоэнергетика. 1992. No. 1. С. 41–44.
4. В.И. Кормилицын, М.Г. Лысков, Ю.М. Третьяков, Экономичность работы парового котла при управлении процессом сжигания топлива вводом влаги в зону горения // Теплоэнергетика. 1988. No. 8. С. 13–15.
5. Л.М. Цирульников, К.Т. Баубеков, Особенности образования оксидов азота при ступенчатом сжигании природного газа в топке с многоярусной однофронтовой компоновкой горелок // Топливоиспользование и охрана окружающей среды. Моск. энерг.ин-т. 1989. Вып. 209. С. 15–22.
6. Л.М. Цирульников, Подавление токсичных продуктов сгорания природного газа и мазута в котельных агрегатах. М.: ВНИИЭГазпром, 1977. 60 С.
7. А.Л. Коваленко, В.Г. Козлов, А.П. Уткин, В.Н. Пермяков, Результаты испытаний горелок с малотоксичными выбросами ЗАО «ЭКОТОП» и фирмы «ТОД Combustion» США на котлах ТГ-104 и ТГМЕ-205 при сжигании попутного и природного газа. //Журн. Теплоэнергетика No. 4.-2003г. С. 41–45.

8. Г. Тартон, Влияние конструкции камеры сгорания газовой турбины и условий ее работы на эффективность снижения выбросов NO_x путем впрыска воды или пара // Энергетические машины и установки. 1985. No. 3. С. 118–126.

9. А.Г. Тумановский, В.Ф. Тульский, Влияние впрыска воды на образование окислов азота за камерой сгорания с последовательным вводом воздуха в зону горения // Теплоэнергетика. 1982. No. 6. С. 34–36.

10. В.А. Корягин, Сжигание водотопливных эмульсий и снижение вредных выбросов. С.-Петербург. Недра. 1995. С.304.

11. В.И. Кормилицын, Экологические аспекты сжигания топлива в паровых котлах. М.: МЭИ. 1998. С. 335.

NEW AND NOVEL TECHNIQUES
FOR POWER SYSTEMS

FLAMELESS OXIDATION TECHNOLOGY

A. MILANI*
WS Wärmeprozesstechnik – Dornierstr 14 – 71272 Germany WS

J. G. WÜNNING
WS Wärmeprozesstechnik – Dornierstr 14 – 71272 Germany WS

Abstract. Flameless combustion is the most significant recent advancement in high-temperature combustion technology and has been applied to industrial furnaces with well proven, very low NO_x performance and high energy savings. This experience has produced spinoffs in power-generating equipment, from innovative gas turbine combustors to small reformers for decentralized H_2 production, and R&TD of *flameless oxidation* techniques is quite promising for new advanced process design.

Keywords: flameless combustion, flameless oxidation, self-ignition temperature, flue gas recirculation, low NO_x firing in furnaces, energy savings, low NO_x gas turbine combustors, small-scale reformers, solid fuel pressurized combustion

1. Introduction

Concern for the environmental burden caused by the combustion of fossil fuels is a primary issue in the design of energy intensive processes for both large plants and distributed fuel fired devices The *flameless combustion* technology applied to high-temperature industrial processes stems from systematic investigations carried out at laboratory scale and from their application to large plants in the steel industry. Results are very satisfactory both for abatement of NO_x emissions and for energy savings; spinoff and ongoing R&TD in the field of power generation is very promising. All this started from looking again at basic principles. A conventional flame is based upon a mechanism as old as fire discovered in nature many centuries ago: a stable flame develops from a

---

* To whom correspondence should be addressed. Ambrogio Milani, WS Wärmeprozesstechnik – Dornierstr 14 – 71272 Germany WS; e-mail: Ambrogio.milani@fastwebnet.it

N. Syred and A. Khalatov (eds.), Advanced Combustion and Aerothermal Technologies, 343–352.

stationary flame front that is a few millimeter thick. Burner design is primarily
concerned with the problem of stabilizing the flame front by means of fluid
dynamic devices. Typically, a bluff body drives back hot reaction products that
heat up the fresh fuel–air mixture, thereby triggering a stable chain reaction.
High gradients of temperature and species concentration in a confined space are
required to obtain a stationary flame.

2. Flameless Combustion

In a stabilized flame burner most reactions occur within the flame front, where
local temperature approaches adiabatic temperature. In a flameless burner, the
flame front is deliberately avoided and combustion reactions occur as fuel and
air mix together with entrained recirculated combustion products. For the
process to occur, the combustion products must be above the self-ignition
temperature (>850°C for safety). The reaction rate is determined by the mixing
pattern between *three partners*: fuel, air, and combustion products entrained
before combustion. In the flameless *mode* the temperature profile is determined
by the mixing pattern with the recirculated combustion products and cannot
depart much from the temperature of these entrained combustion products or
flue gases.

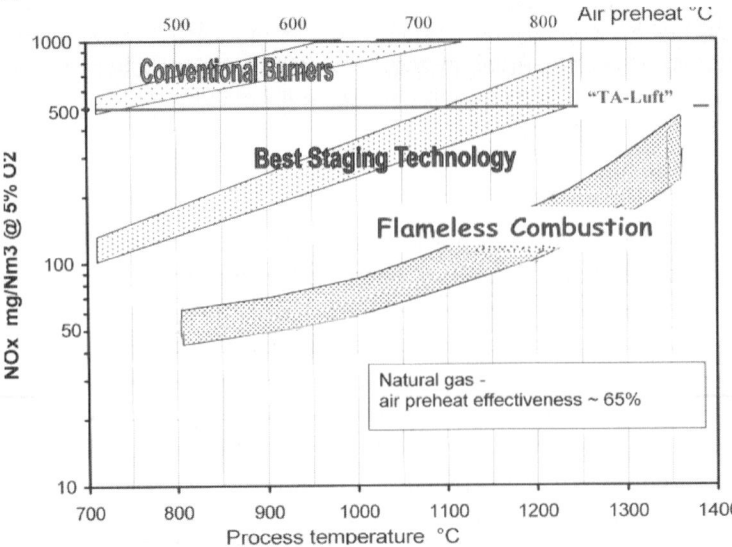

Figure 1. NO$_x$ emissions from steel furnaces vs process temperature

In the flame *mode*, the temperature profile peaks in the flame front close to
the burner and decreases downstream as mixing and reactions proceed. This

temperature peak is conducive to enhanced thermal NO formation as described by the Zeldovich mechanism (Wünning, 1991). To abate temperature peaks means to abate thermal NO and in fact flameless combustion does abate NO_x emissions by one order of magnitude. Figure 1 reports the accumulated data, on a logarithmic scale, relevant to many natural gas fired furnaces in the steel industry: the advantage of flameless technology for temperatures >850°C with respect to the best *low*-NO_x burner designs is quite evident.

Figure 2 shows how the flame and flameless *modes* are implemented in high velocity burners, common in heat treatment furnaces for steel products.

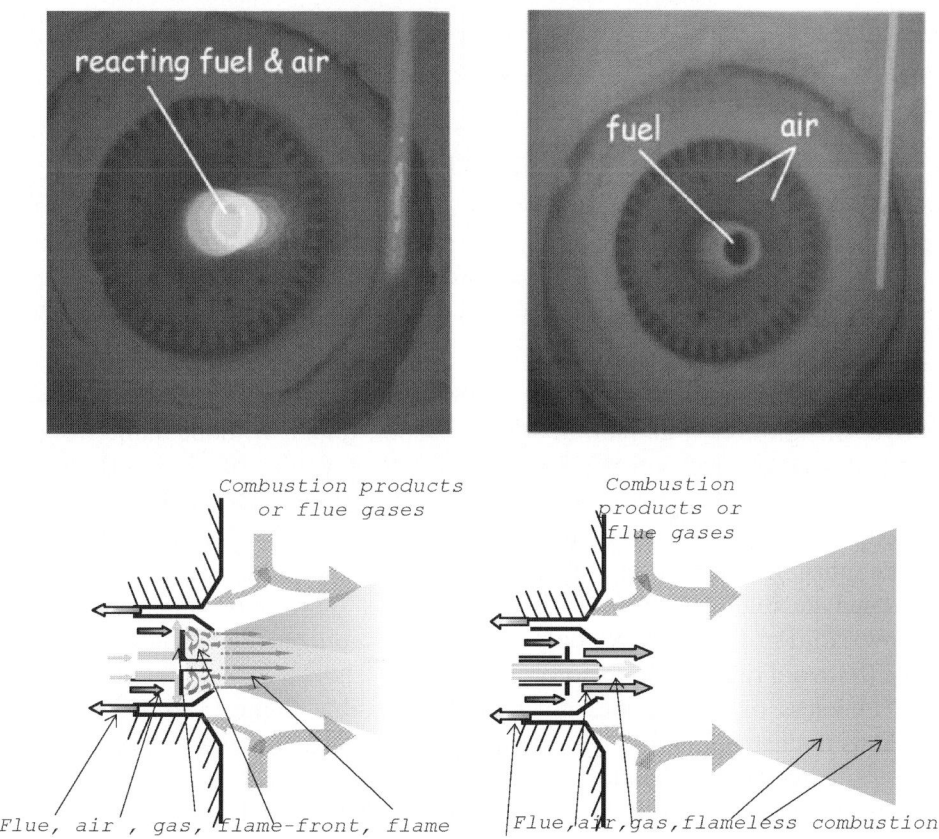

Figure 2. Flame mode and flameless oxidation mode

The domain of flameless combustion has been investigated on a test furnace as a function of the *recirculation ratio* K_v defined as the ratio of recirculated mass flow of combustion products (before reaction) with respect to the driving flow rate of reactants (Wünning, 1991, 1997):

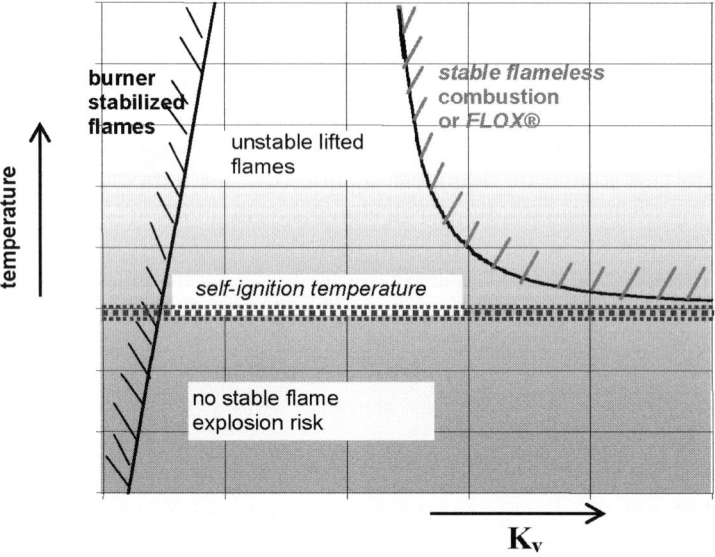

Figure 3. Domain of flameless oxidation (or FLOX vs K_v factor)

$$K_v = M_{rec} / (M_{air} + M_{fuel})$$

Figure 3 presents a schematic of the results: for temperatures $> \sim 850°C$, above self-ignition, a domain of stable reaction region without flame front can be established, corresponding to large K_v values (order of $K_v > \sim 3$), that are obtained with high momentum of the injected fluids. This domain has been called *flameless oxidation* or by the trademark name *FLOX*. It is not possible to establish a conventional flame front for K_v values $> \sim 0.3–0.5$ and the intermediate region is typical of "lifted flames" and of unstable combustion. Below ignition temperature, only the burner stabilized flame mode is admissible (to avoid risks of explosion).

Flameless oxidation does not produce a visible flame (Figure 2) and this explains the name *flameless*; furthermore, this combustion *mode* is almost silent and abatement in combustion noise (~ 15 dBA) is at least as impressive as the disappearance of a visible flame, proving that the turbulent flame front accounts for most of the typical combustion roar of high velocity burners (Wünning, 2005). Flameless oxidation has been thoroughly investigated by WS (EP Patents): *FLOX* has been shown to work for rich, near stoichiometric and for very lean combustion conditions; it works with and without air or fuel preheat. It also works for diffusion, partial premixed and premixed combustion. A well-known advantage concerns low-NO_x burners operated at very high air preheat: unlike the conventional flame mode, the flameless mode is insensitive to air preheat temperature as far as NO_x is concerned, and this is very important for application to high-temperature industrial processes or furnaces.

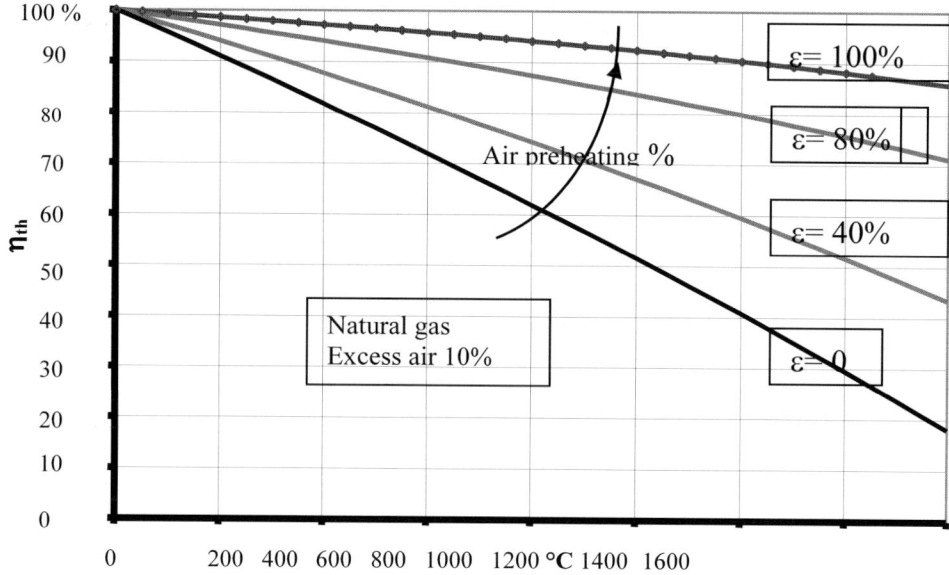

Figure 4. Thermal efficiency vs process temperature

3. Energy Savings in Steel Furnaces

The thermal efficiency of high-temperature furnaces can be greatly increased by means of efficient air preheating (Figure 4) (preheat temperature is defined as a percentage ε of the process temperature): high efficiency ηth is equivalent to reduction in fuel consumption and to a corresponding saving in greenhouse gas emissions. High preheat, like air at 800–1000°C, is only technically feasible if special combustion techniques are adopted in order to prevent unacceptable NO_x emissions and local overheating. Flameless oxidation fits this requirement perfectly and can be considered a prerequisite for such applications.

A preferred burner design is based on burner integrated heat recovery (Figure 5): flue gases are extracted through the burner itself and combustion air is preheated in the countercurrent while cooling the flue gases. This is a convenient solution for furnaces equipped with several burners: cold combustion air is distributed to the burners while almost cold flue gases are extracted from a common manifold. This design offers effective preheating efficiency: in a well-stirred furnace, centralized heat recovery allows thermal efficiency ~60% (~40% for no preheating at all), burner integrated recovery scores ~75–85%, which is a good step forward, corresponding to a fuel saving of 15–25% with respect to the state of the art (centralized heat recovery).

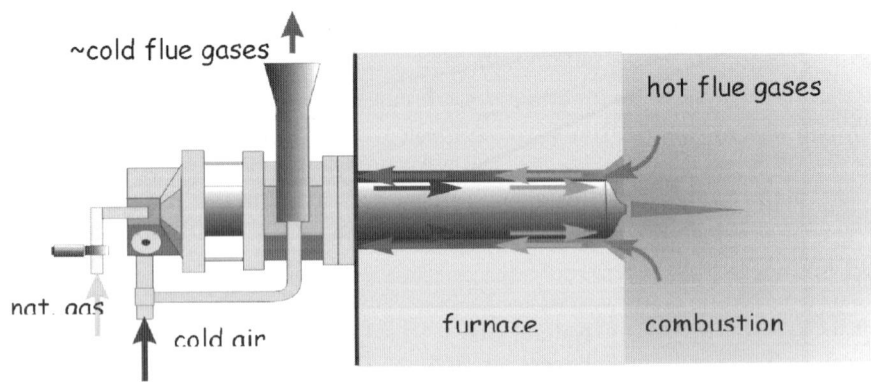

~cold flue gases

hot flue gases

nat. gas

cold air

furnace   combustion

Figure 5. Burner-integrated heat recovery

Thousands of *FLOX* burners have been installed in continuous industrial plants and perform satisfactorily. In addition, regenerative burners firing in *FLOX* mode have been adopted in several large annealing lines for stainless steel strips and in batch furnaces; *regenerative* air preheating is certainly most efficient and allows energy savings in the order of 30–50% (Milani, 2002).

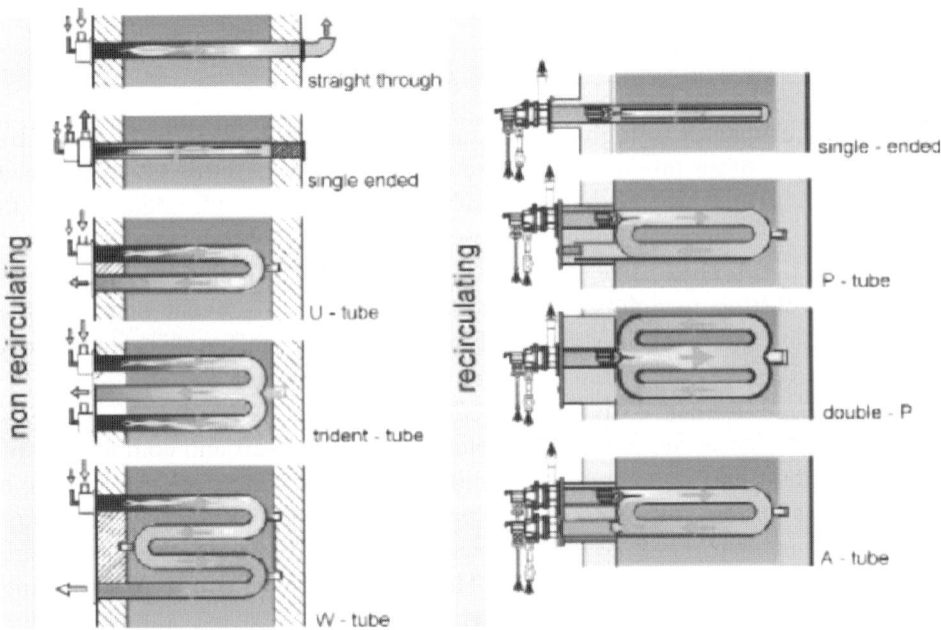

non recirculating

straight through

single ended

U - tube

trident - tube

W - tube

recirculating

single - ended

P - tube

double - P

A - tube

Figure 6. Recirculating and nonrecirculating geometries in radiant tubes

The *radiant tube* is a device used in large heat treatment furnaces for steel products: it radiates to the stock without permitting contact with the flue gases and combustion is developed inside a long tubular chamber, which makes combustion control difficult. Experience has demonstrated that internal recirculation of combustion products is the key to good performance: "recirculating geometries" (RHS in Figure 6) allow low-NO_x performance and uniform temperature of the radiant tube thanks to flameless oxidation. Temperature uniformity has beneficial consequences for the strength of the radiant tube and for the average allowable heat flux, which implies a better exploitation of the radiating surface: in other words, a saving in installation costs. Good examples are the annealing furnaces equipped with ceramic "single-end" tubes in SiSiC: the cost of ceramic radiant tubes has been largely overridden by excellent performance.

4. Power-generating Equipment

The *FLOX* principle is not limited to steel furnaces and can be applied to several high-temperature processes (Wünning, 2005). Examples are the Stirling engines, where heat is made available at high temperature with high efficiency, for the purpose of providing combined heat and power in small power generating units. A very promising application of flameless combustion to combustors for gas turbines is presently being developed and successfully tested: a specially designed *FLOX* prototype burner (Figure 7) ensures very low-NO_x, emissions and overcomes the nasty problem of fluctuations or "humming" that affects premix-based GT combustors, where the flame front stabilization is a critical issue. R&TD is ongoing with the participation of several academic and industrial partners in Europe.

Flameless oxidation has been investigated for gaseous fuels and in particular for natural gas. However, the basic principle holds good for any fuel, at least any fuel as soon as it is made available in fluid form (like evaporation of liquid droplets or release of volatile matter from pulverized solid fuel). Trials are being carried out together with German universities to test the effects of flameless oxidation (Figure 9): flameless mode occurs with any fuel and consistent NO_x reduction has been observed. Encouraging tests have been carried out under pressure, which might be applicable to envisaged future processes aimed at CO_2 sequestration.

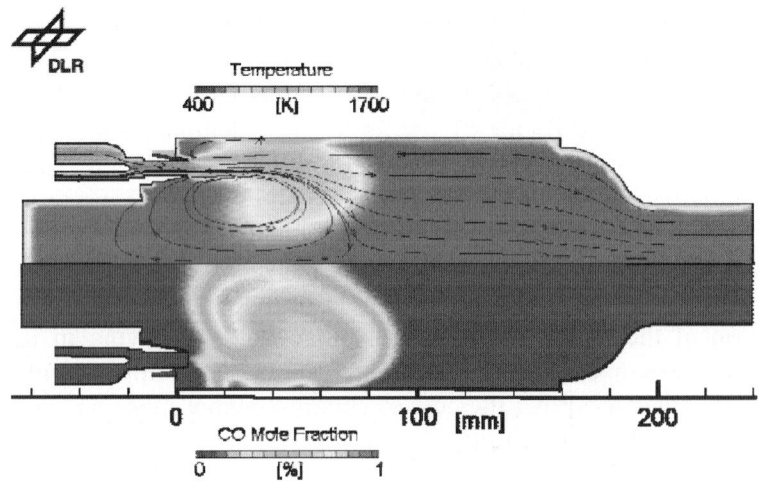

Figure 7. CFD computations concerning the FLOX® prototype GT combustor

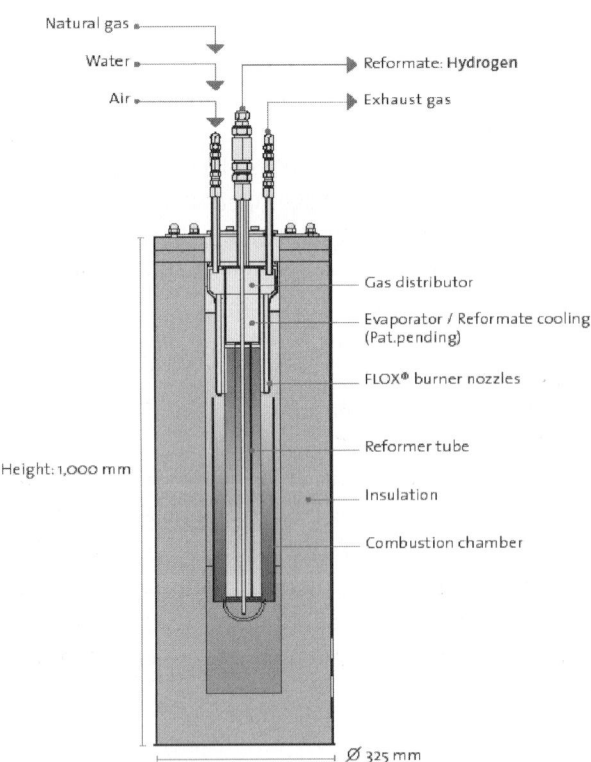

Figure 8. The minireformer based on the FLOX burner

The inherent temperature uniformity obtained with flameless combustion finds an ideal application in steam reformers for hydrogen production: reforming reactions take place inside vertical tubes filled with a catalyst and reheated from the outside. The uniform temperature distribution is essential for high productivity, reduced stress on the reaction tubes and better control. The experience of the *WS* company in steel process furnaces has been used to found a daughter company specialized in "mini-reformers" for producing small amounts of hydrogen (order of 5–200 Nm^3/h) for decentralized fueling stations for future H_2 powered vehicles. Figure 8 shows the scheme of the WS mini-reformer: such plants have been installed in the airports of Munich and Madrid to provide the H_2 used by local buses for passenger service.

Oil *flame - mode*   Oil *FLOX®-mode*   Coal *FLOX®-mode*

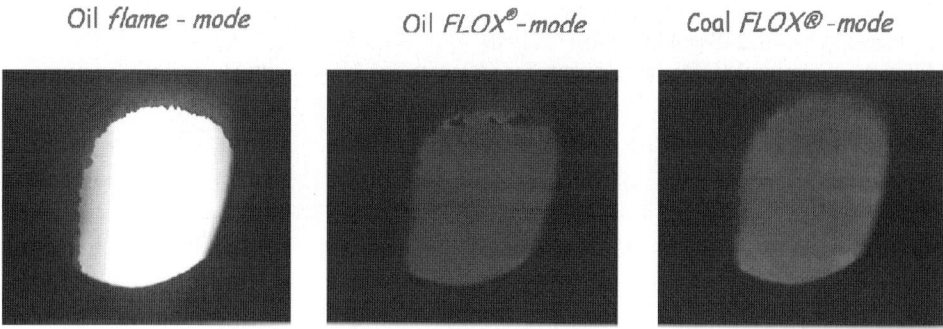

Figure 9. FLOX performance with different fuels

5. Conclusions

Referring to the case of high-temperature furnaces, the industrial application has demonstrated that flameless technology can greatly renew and improve the design and the performance of traditional plants/processes; advantages like "downsizing" (reduction of the furnace length), NO_x minimization, temperature uniformity, better control and improved product quality also make investment for revamping old plants advantageous. Similar arguments hold true for the R&TD applications to power generating devices as quoted in Section 4 above. We can conclude that the principle of flameless oxidation still has great potential for further development in equipment where combustion plays the important role.

The tendency is to tighten regulations concerning pollutant emissions and to limit specific emissions of greenhouse gases, which implies reducing specific fossil fuel consumption. This is based upon steady grounds: not the available or future fossil fuel resources put an effective limit to economic and abundant energy, but the available *clean air*. Clean air for combustion is a limited global resource that cannot be wasted or corrupted beyond a sustainable threshold. In former times California had promoted use of catalytic converters and had thereby stimulated the competitive production of cleaner engines. A similar, virtuous pattern should be followed in other domains related to fossil energy conversion as awareness of the worldwide "environmental challenge" proceeds.

References

Wünning, J.A. Flammenlose Oxidation von Brennstoff mit hochvorgewärmter Luft, Chem. Ing. Tech., **63** (12) 1243–1245, 1991

Wünning J.A., Wünning J.G. Flameless Oxidation to reduce thermal NO formation, Prog. Energy Combust. Sci., **23**, 81–94, 1997

Wünning J.G. Flameless Oxidation, 6th int Symposium HTACG – Essen, 17–19 October 2005

WS Patents EP 0463218 and EP 0685683, 1990

Milani A., Wünning J.G. Design concepts for radiant tubes, Millennium Steel, 2002

PULSE DETONATION ENGINES: ADVANTAGES AND LIMITATIONS

N. SMIRNOV*
Moscow M.V. Lomonosov State University, Leninskie Gory, 1, Moscow 119992, Russia

Abstract. This paper reviews the efforts made over the years in adapting detonations for propulsion applications, and highlights new challenges in studying detonation dynamics.

Keywords: combustion, detonation, pulse, wave, ignition, turbulence, cavity

1. Introduction

Developments in modern propulsion technology have created the need for more and more powerful energy converters. The conversion of chemical energy in a fuel into propulsion energy supplied to a vehicle is limited by the rate of release of chemical energy in combustion. The rates of energy release in detonation modes of gas combustion are three orders of magnitude higher than in deflagration combustion modes. This could make the use of detonation combustion modes more efficient for creating high power energy converters. However, the rate of combustible mixture supply is usually lower; thus necessitating the pulsed operation mode for such an energy converter.

In terms of thermodynamic efficiency and the reduction CO emissions, the advantages of the constant volume combustion cycle over constant pressure combustion have focused advanced propulsion research on detonation engines. One of the schemes for producing enhanced thrust at both static and dynamic

---

*To whom correspondence should be addressed. Nickolay Smirnov, Moscow M.V. Lomonosov State University, Faculty Mech & Math, Leninskie Gory, 1, Moscow 119992, Russia, e-mail: ebifsun1@mech.math.msu.su

N. Syred and A. Khalatov (eds.), Advanced Combustion and Aerothermal Technologies, 353–363.

conditions is pulse detonation. The high thermodynamic efficiency of Chapmen-Jouget detonation compared to other combustion modes is due to the minimal entropy of the exhaust jet. Based on this, efforts have been made over several decades to show that proper utilization of the operation cycle does result in improved performance. However, there are several issues in developing this technology which represent scientific and technological challenges. The success in resolving these problems will determine the implementation of pulse detonation propulsion.

Control of the onset of detonation is of major importance in pulse detonating devices. The advantages of detonation over constant pressure combustion necessitate the promotion of the deflagration to detonation transition (DDT) and shortening of the predetonation length. With most fuel–air mixtures being heterogeneous the problem of liquid droplet interaction with the surrounding gas flow is of key interest with particular reference to atomization and heat and mass transfer.

The DDT has turned out to be the key factor that characterizes the pulse detonation engine (PDE) operating cycle. Thus, the problem of DDT control in gaseous and polydispersed fuel–air mixtures has become very acute.

This paper contains the results of theoretical and experimental investigations into DDT processes in combustible gaseous mixtures. In particular, this paper investigates the effect of cavities incorporated in PDD at the onset of detonation in gases.

2. Experimental Investigations of DDT

Propagation of waves in metastable systems is sustained by the energy release triggered by the wave front. Combustion waves in chemically reacting systems, and boiling waves in superheated fluids could be considered as typical examples of such self-sustained waves, which can have two modes of propagation: subsonic and supersonic – due to different mechanisms. The process of transition from one mode of propagation to the other is the most intriguing issue. Investigations of DDT in gases have been carried out for pulse detonating devices.

Investigations of DDT in hydrogen–oxygen mixtures[1-5] and later in hydrocarbons–air mixtures[6-8] have demonstrated the multiplicity of the transition processes scenario. The various modes of detonation onset were shown to depend on particular flow patterns created by the accelerating flame, thus making the transition process nonreproducible in its detailed sequence of events. Currently, there exist different points of view with regard to the DDT mechanism: the "explosion in explosion" mechanism proposed by Oppenheim[3,5] and the gradient mechanism of "spontaneous flame" proposed by

Zeldovich[9]. The latter theoretical analysis showed that microscale nonuniform-mities (temperature and concentration gradients) arising in local exothermic centers ("hot spots") ahead of the flame zone could be sufficient for the onset of detonation or normal deflagration[9–16].

The precision of results from experimental investigations into the sensitivity of DDT processes to variations in the mixture parameters is naturally limited because various modes of detonation onset depend on stochastic flow patterns created by accelerating turbulent flames thus making the transition processes nonreproducible in its detailed sequence of events. In any case, it is hardly possible to vary the different parameters independently in physical experiments. This empha-sizes the importance of numerical modeling for investigating the detonation initiation sensitivity to variations in the govern-ing parameters. The numerical investigation of the transition processes provides a unique possibility to vary each parameter independently and incrementally.

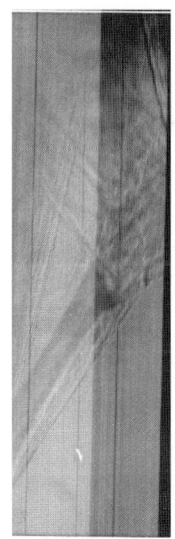

Figures 1–4 present schlieren images of the detonation onset, illustrating different scenarios of DDT in hydrocarbon–air gaseous mixtures. They illustrate the types of flow structure that exist at various distances from the initiating section in tubes filled with stoichiometric hydrocarbon–air mixtures. The flame is propagating from the left to the right, time increasing from bottom to top. Thus the schlieren pictures give the x–t diagrams of the process. The x-axis gives the actual coordinate along the axis of the tube. The t-axis provides only the timescale, but not the actual point (the zero point is adjusted to the beginning of the registration).

Figure 1. Onset of detonation in the flame

Figure 2 shows the flow structure before the onset of detonation. The presence of turbulizing chambers contributes to flow irregularity ahead of the flame, which could promote the onset of detonation. The flame velocity is 950 m/s. The later shock waves overtake the primary ones until a strong shock wave supported by the flame induced compression waves is formed ahead of the flame (Figures 3 and 4).

The detonation wave occurs after ignition in local exothermic centers ("hot spots") either in the near proximity (Figure 1) or ahead of the flame (Figs. 3–4). The transition scenario illustrated in Figure 3 is characterized by hot spot formation in the high enthalpy zone on the contact surface resulting from the interaction of two primary shock waves. Figure 4a illustrates the transition scenario characterized by the formation of the secondary combustion zone between the flame and the leading shock due to autoignition in a local exothermic center. The combustion zone expands in all directions and the onset of detonation takes place 180 µs later. Figure 4b illustrates the transition scenario under which ignition takes place subsequently in a number of hot spots ahead of the flame. These ignitions do not lead directly to the formation of detonation waves. Flames propagating in all directions from the ignition centers expand in both directions leading to the formation of volume combustion and further compression of the mixture behind the leading shock. The detonation wave arises in one of the subsequent exothermic centers closer to the leading shock beyond the limits of the photographic zone. The retonation wave moving backward at a speed of 1,350 m/s in the upper part of Figure 4b testifies to this. Analysis of the experiments shows that detonation onset takes place in one of the exothermic centers ("hot spots") originating stochastically in the compressed gas between the leading shock and the flame zone. Depending on the hot spot local structure, combustion can give rise to either a detonation or deflagration wave propagating from the hot spot.

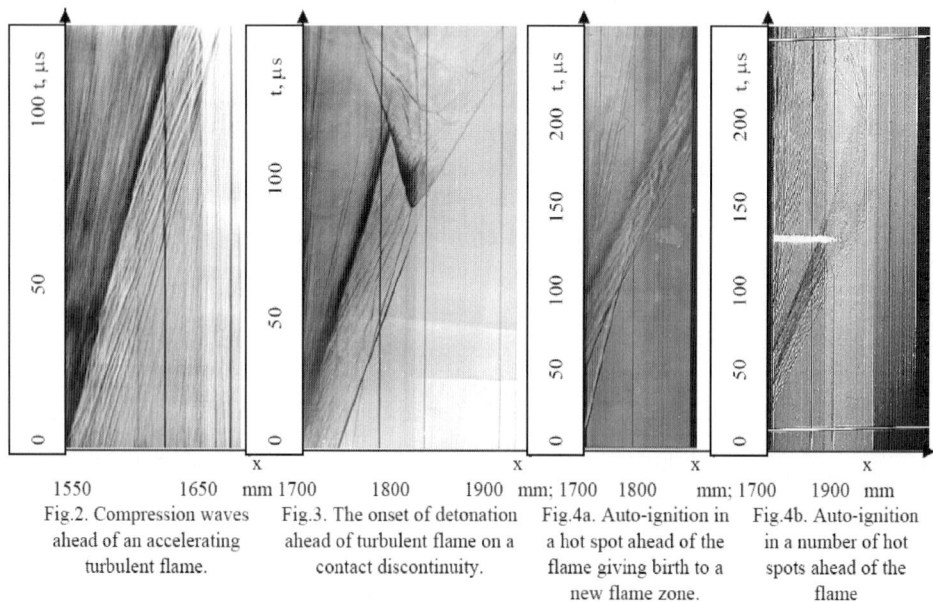

Fig.2. Compression waves ahead of an accelerating turbulent flame.

Fig.3. The onset of detonation ahead of turbulent flame on a contact discontinuity.

Fig.4a. Auto-ignition in a hot spot ahead of the flame giving birth to a new flame zone.

Fig.4b. Auto-ignition in a number of hot spots ahead of the flame

3. Mathematical Model

Numerical investigations of the DDT processes have been performed using the system of equations for the gaseous phase obtained by Favre-averaging of the system of equations for multicomponent multiphase media. The modified *k-epsilon* model was used. To model temperature fluctuations the third equation was added to the *k-epsilon* model to determine the mean-squared deviation of temperature. The production and kinetic terms were modeled using the Gaussian quadrature technique. With the term responsible for chemical transformations, $\dot{\omega}_k$ being very sensitive to temperature variations, as it is usually the Arrhenius law type function for the rates of reactions, the third equation for mean temperature deviation was added to the *k-epsilon* model.

The mathematical models for simulating chemically reacting turbulent flows in heterogeneous mixtures were described in detail in[17–19].

Averaging by Favre with the $\alpha\rho$ weight, where α is the volumetric fraction of the gas phase ($\alpha = 0$ for homogeneous gaseous mixtures) and ρ is the gas density, we obtain the following system for the gas phase in a multiphase flow (the averaging bars are removed for simplicity):

$$\partial_t(\alpha\rho) + \nabla \cdot (\alpha\rho\vec{u}) = \dot{M} \tag{1}$$

$$\partial_t(\alpha\rho Y_k) + \nabla \cdot (\alpha\rho\vec{u}Y_k) = -\nabla \cdot \vec{I}_k + \dot{M}_k + \dot{\omega}_k \tag{2}$$

$$\partial_t(\alpha\rho\vec{u}) + \nabla \cdot (\alpha\rho\vec{u} \otimes \vec{u}) = \alpha\rho\vec{g} - \alpha\nabla p + \nabla \cdot \tau + \dot{\vec{K}} \tag{3}$$

$$\partial_t(\alpha\rho E) + \nabla \cdot (\alpha\rho\vec{u}E) = \alpha\rho\vec{u} \cdot \vec{g} - \nabla \cdot p\vec{u} - \nabla \cdot \vec{I}_q + \nabla \cdot (\tau \cdot \vec{u}) + \dot{E} \tag{4}$$

Equations (1–4) present the mass balance of the gas phase, the mass balance of the *k*th component, the momentum balance and the energy balance, respectively (p – pressure, $\vec{u}$ – fluid velocity vector, $\vec{g}$ – gravity acceleration vector, E – specific energy, $\dot{\vec{K}}$ – specific momentum flux to gas phase, $\dot{M}$ – specific mass flux, $\dot{E}$ – specific energy flux, τ – turbulent stress tensor). The following relationships define the difference between Eqs. (1) and (2):

The equations of state for the gaseous mixture are as follows:

$$p = R_g\rho T \sum_k Y_k W_k , \quad E = \sum_k Y_k(c_{vk}T + h_{0k}) + \frac{\vec{u}^2}{2} + k . \tag{5}$$

where k – turbulent kinetic energy, W_k – molar mass of kth gas component,

h_{0k} – specific chemical energy, c_{pk}, c_{vk} – specific heat capacity.

The gaseous phase was assumed to contain the following set of species: O_2, C_nH_m, CO, CO_2, H_2, H_2O, N_2.

4.   Results of Numerical Modeling

Numerical simulations have been undertaken for the purpose of comparing the roles of various turbulizing elements in the promotion of DDT and the location of those elements in the tube. The test vessel contained a detonation tube with a number of chambers of a wider cross section placed at various locations along the tube, filled with a combustible gaseous mixture at ambient pressure (Figure 5). Ignition of the mixture was performed by a concentrated energy release in either the center of the first chamber or the center of the tube itself near the closed end on the left-hand side. The number of chambers was varied from 1 to 20.

4.1   THE ROLE OF CHAMBERS IN THE IGNITION SECTION

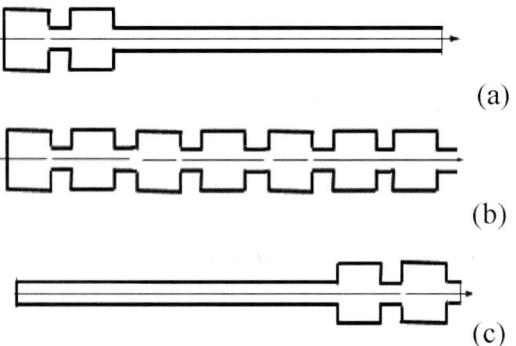

Numerical modeling of the onset of detonation was performed for a test vessel containing a detonation tube with two chambers of a wider cross section (Figure 5) filled with a combustible gaseous mixture at ambient pressure. Ignition of the mixture was performed by a concentrated energy release in the center of the first chamber. The results

Figure 5. Geometry of the computational domain

show that the flame accelerates and penetrates the bridge between the two chambers due to a flow of gas caused by the expansion of the reaction products. A high velocity jet penetrating the second chamber creates very fast flame propagation due to both additional flow turbulization and the piston effect of the expanding reaction products, supported by the continuing combustion in the first chamber. Fast combustion in the second chamber pushes the flame further into the tube. At some point detonation arises from a hot spot within the combustion zone, which gives rise to strong detonation and retonation waves.

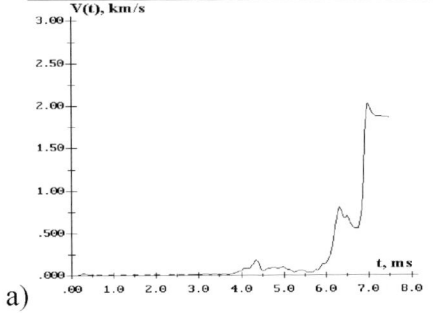
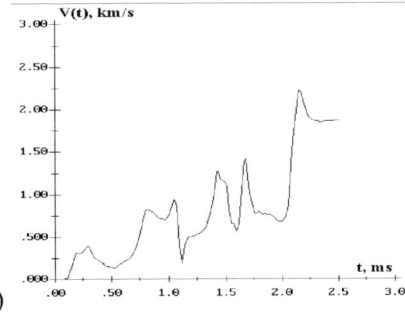

a)   b)

Figure 6. Variation of the mean flame front velocity with time for a fuel concentration C_{fuel}=0.012: (a) tube incorporating two chambers in the ignition section, (b) tube without any chambers

Figure 6 demonstrates the variation of mean flame front velocity with time for two test cases; a tube incorporating two chambers in the ignition section (Figure 6a), and a tube without any chambers (Figure 6b). It can be seen that the onset of detonation in a tube without chambers is an unstable stochastic process, and each pulsation of velocity, depending on some additional disturbance, could result in the onset of detonation. Increasing the number of chambers makes the DDT more stable and reduces the predetonation length.

4.2   THE INFLUENCE OF FUEL CONCENTRATION

Analysis of the results presented in Figure 7 indicates that by decreasing the fuel content the ability of the mixture to detonate decreases via DDT. The predetonation time increases (Figure 7a and b), but once the onset of the detonation takes place it propagates at a practically constant velocity. Decreasing the molar concentration of the fuel below $C_{fuel} = 0.011$ brings about the formation of galloping combustion regimes.

4.3   THE ROLE OF CHAMBERS AT THE END OF THE TUBE.

In order to provide comparative data, the role of two chambers of a wider cross section incorporated at the far end of the tube was investigated (Figure 5c). Numerical results showed that after ignition in the narrow tube (ignition energy was increased) acceleration of the flame zone, accompanied by a number of oscillations, created a detonation wave propagating with a mean velocity of 1,850 m/s. On entering the first chamber, decoupling of the shock wave and the reaction zone took place and the mean velocity of reaction zone propagation decreased to 200 m/s. Subsequently, the flame accelerated to 400 m/s in the narrow bridge, and slowed down in the second chamber to 100 m/s.

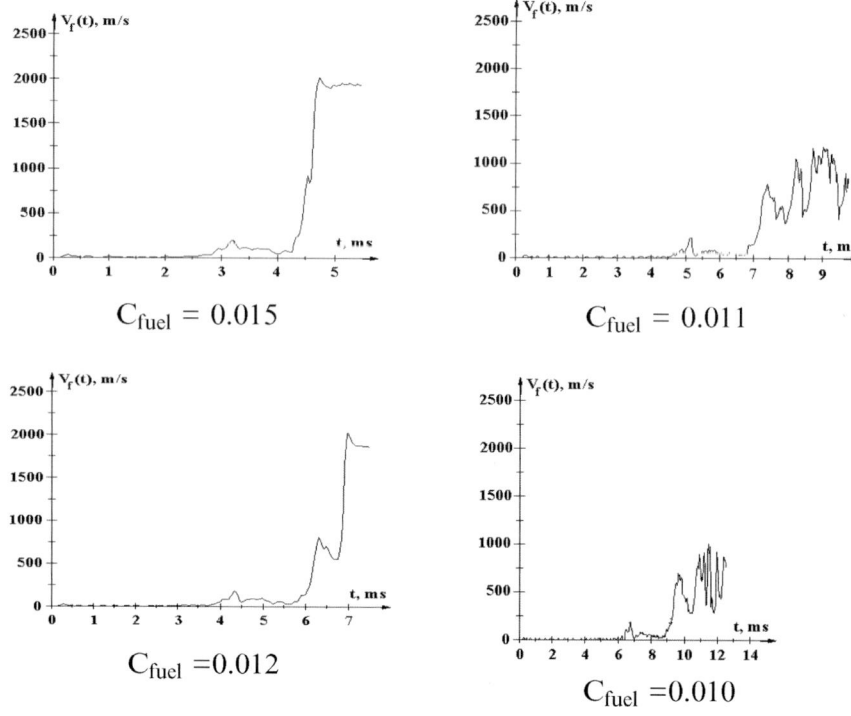

Figure 7. Reaction front velocity for different fuel concentrations in a two-chamber device

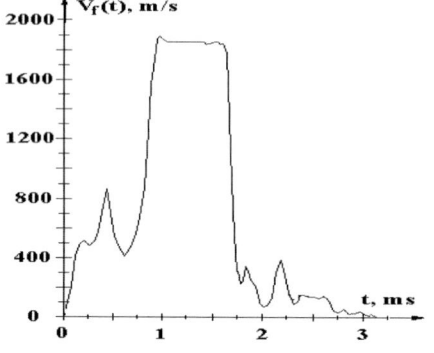

Figure 8. Flame velocity for fuel volumetric concentration 0.012

Figure 8 illustrates the trajectory of the reaction front and the variation of velocity with time for the onset of detonation and degeneration in a tube with two chambers at the end (fuel concentration was 0.012). Thus, similar chambers of wider cross section incorporated at the end of the detonation tube bring the detonation wave to a halt.

4.4   THE EFFECT OF CHAMBERS INCORPORATED IN THE TUBE ALONG THE WHOLE LENGTH

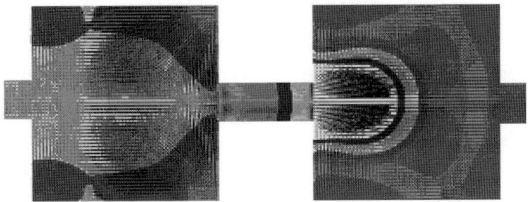

The detonation tube was 2.95 m in length, incorporating 20 similar turbulizing chambers uniformly distributed along the axis. The results of numerical simulations indicate that increasing the number of turbulizing chambers did not promote the DDT for the present configuration, but just the opposite; it

Figure 9. Density maps in the 6th–7th chambers. Expansion ratio β_{ER} = 0.96, fuel volume concentration 0.012

prevented the onset of detonation and brought about the galloping combustion mode. The effect was caused by very sharp jumps in the cross-sectional area of the detonation tube and the periodical deceleration of the flame caused by its expansion. (In the present numerical experiment the expansion ratio $\beta_{ER} = (S_{chamb} - S_{tube}) / S_{chamb}$ was equal to 0.96).

Investigations of the sensitivity of self-sustaining combustion modes to the expansion ratio indicate that at low expansion ratios low velocity galloping detonation was established and at high expansion ratios self-sustained galloping high speed combustion took place (Figure 11).

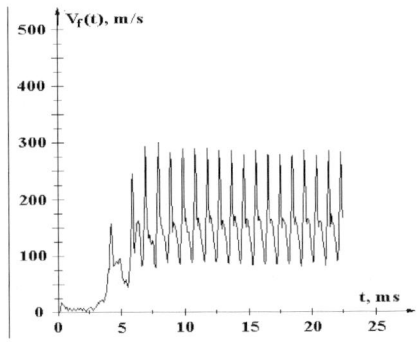

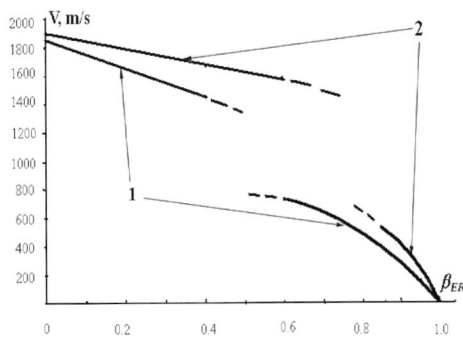

Figure 10. Reaction front position and velocity in a multichamber tube

Figure 11. Mean reaction front velocities in a multichamber tube: 1 – fuel content 0.012; 2 – fuel content 0.015

5. Conclusions

Extensive numerical simulations have permitted the investigation of the peculiarities of DDT in gases in tubes incorporating cavities of a wider cross section. The presence of cavities fundamentally affects the combustion modes being established in the device and their dependence on the governing parameters of the problem.

The influence of the confinement geometry and flow turbulization on the onset of detonation and the influence of the temperature and fuel concentration in the unburned mixture have been discussed. It has been demonstrated both experimentally and theoretically that the presence of cavities of a wider cross section in the ignition part of the tube promotes DDT and shortens the predetonation length. At the same time cavities incorporated along the whole length or in the far end section inhibit detonation and bring about the onset of low velocity galloping detonation or galloping combustion modes. The presence of cavities in the ignition section turns the increase of initial mixture temperature into the DDT-promoting factor instead of the DDT-inhibiting factor.

Acknowledgments

The present investigation was supported by the Russian Foundation for Basic Research (05-03-32232) and US Office of Naval Research (N62558-03-M-0022 and N0001404RC20027).

References

1. G.D. Salamandra, On interaction of a flame with a shock wave. In: Physical Gasdynamics, USSR Academy of Sciences Press, Moscow, 1959, 163–167.
2. T.V. Bazhenova and R.I. Soloukhin, Gas ignition behind the shock waves. Internatinal VII Symposium on Combustion, Buttterworths, London, 1959.
3. A.K. Oppenheim and P.A. Urtiew, Experimental observations of the transition to detonation in an explosive gas, *Proceedings of the Royal Society*, **A295** (1966) 13.
4. R.I. Soloukhin, Methods of measure and main results of experiments in shock tubes. Novosibirsk State University Publ., Novosibirsk, 1969.
5. A.K. Oppenheim and R.I. Soloukhin, *Annual Review of Fluid Mechanics*, **5** (1973) 31.
6. N.N. Smirnov and A.P. Boichenko, Deflagration to detonation transition in gasoline-air mixtures, *Combustion, Explosion and Shock Waves*, **22**, 2 (1986) 65–68.

7. R.P. Lindstedt and H.J. Michels, Deflagration to detonation transition in mixtures of alkane LNG/LPG constituents with O_2 /, *Combustion and Flame,* **72**, 1 (1988) 63–72.

8. N.N. Smirnov and M.V. Tyurnikov, Experimental investigation of deflagration to detonation transition hydrocarbon-air gaseous mixtures, *Combustion and Flame*, **100** (1995) 661–668.

9. Ya.B. Zeldovich, V.B. Librovich, G.M. Makhviladze, and G.I. Sivashinsky, On the onset of detonation in a non-uniformly pre-heated gas. *Soviet Journal of Applied Mechanics and Technical Physics,* **2**, (1970) 76.

10. A.G. Merzhanov, *Combustion and Flame*, **10** (1966) 341–348.

11. A.A. Borisov, *Acta Astronautica*, **1** (1974) 909–920.

12. K. Kailasanath and E.S. Oran. Ignition of flamelets behind incident shock waves and the transition to detonation. *Combustion Science Technology*, **34** (1983) 345–362.

13. Ya.B. Zeldovich, B.E. Gelfand, S.A. Tsyganov, S.M. Frolov, A.N. Polenov. In: Dynamics of Explosions (A Kuhl et al. Eds.) AIAA Inc., New York, 114 (1988), p. 99.

14. P. Wolanski, *Archivum Combustions*, 3–4 (1991) 143–149.

15. N.N. Smirnov and I.I. Panfilov, Deflagration to detonation transition in combustible gas mixtures. *Combustion and Flame*, **101** (1995) 91–100.

16. N.N. Smirnov, et al. Gaseous and Heterogeneous Detonations: Science to Applications (G.D. Roy ct al. Eds.) ENAS Publ., Moscow (1999), pp. 65–94.

17. N.N. Smirnov, V.F. Nikitin, M.V. Tyurnikov, A.P. Boichenko, J.C. Legros, and V.M. Shevtsova. In: High Speed Deflagration and Detonation. (G.D. Roy et al. Eds.) Elex-KM Publ., Moscow (2001) pp. 3–30.

18. N.N. Smirnov and V.F. Nikitin, *Combustion and Flame*, **111** (1997) 222–256.

19. N.N. Smirnov, V.F. Nikitin, and J.C. Legros, *Combustion Flame*, **123** (1/2) (2000), 46–67.

THE TRAPPED VORTEX COMBUSTOR: AN ADVANCED COMBUSTION TECHNOLOGY FOR AEROSPACE AND GAS TURBINE APPLICATIONS

C. BRUNO*
*Department of Mechanics and Aeronautics (DMA),
Via Eudossiana 18, 00184 Rome, Italy*

M. LOSURDO
*Ph.D. student Techincal University of Delft, Mechanical
Engineering, Process and Energy Section, Leeghwaterstraat 44,
2628 CA, Delft*

Abstract. In the past, it has been demonstrated that combustion technology based on premixing reactants with combustion products may improve both combustion efficiency and emissions for some industrial applications. Work currently in progress in the European Union aims at developing applications to use this combustion strategy for gas turbines and aerospace applications. The challenge is to provide a new class of combustors performing at a high combustion efficiency and low emission indices, together with multifuel capability (gas and liquid fuels), and low pressure drop. The trapped vortex combustor (TVC) may be considered a very promising form of technology for both pollutant emissions and pressure drop reduction. This strategy is based on mixing hot combustion products and reactants at a high rate. Turbulence occurring in a TVC combustion chamber is "trapped" within a cavity where reactants are injected and efficiently mixed. Since part of the combustion occurs within the recirculation zone, a "typically" flameless regime can be achieved, while a trapped turbulent vortex may provide significant pressure drop reduction. The work presented in this paper is the result of having investigated all of these aspects.

Keywords: trapped Vortex, flameless combustion, Rich-Quench-Lean combustion, low pressure drop

---

* To whom correspondence should be addressed. Claudio Bruno, Department of Mechanics and Aeronautics (DMA), Via Eudossiana 18, 00184 Rome, Italy

N. Syred and A. Khalatov (eds.), Advanced Combustion and Aerothermal Technologies, 365–384.
© 2007 *Springer.*

1. Introduction

Combustion stability is often achieved using recirculation zones to provide continuous sources of ignition, by mixing the hot products and burning gases with the incoming fuel and air. Cavities have been extensively investigated in aerodynamics as a means of pressure recovery or drag reduction. Rules have been investigated[1] for cavities formed between two axisymmetric disks spaced along a central spindle. It has been shown that for the ratio of afterbody to forebody disk diameters of <1, an optimal separation distance existed reducing drag to its minimum. Furthermore, the minimum drag using two disks was reportedly lower than using only the forebody disk alone. On the other hand, there are separation distances raising drag dramatically above that of the forebody alone. Somewhat intuitively, the minimum drag corresponds to a stable recirculation zone in the cavity[12]. Determined by disk separation distance and the afterbody disk diameter, certain values of these parameters yield a stable and locked (trapped) vortex. Further studies[3] were performed on low and high drag regimes. Among these results, low-drag occurs when the cavity shear layer stagnates at the downstream corner of the cavity: when this happens, the vortex motion within the cavity is stable. Low drag conditions can be considered as the main desirable result that can be collected when a vortex rotates in a cavity: oscillations produced by an external (mainstream flow) self-sustain the cavity flow through a feedback mechanism that keeps the position of the stagnation zone locked to the downstream corner of the cavity. This behavior occurs even when extra fluid is injected directly into the cavity. Numerical simulations performed by Katta and Roquemore[3-5] showed that a vortex can be trapped within a cavity when the afterbody and the forebody disk diameter are consistent with the rules established from experimental correlations. Furthermore, according to Little and Whipkey[2], a stable vortex condition produces the minimum drag for a given geometry.

When a vortex is trapped in a cavity, very little mainstream fluid can enter into the cavity (~3% of the total vortex mass)[6]. When turbulent combustion occurs inside this cavity, a turbulent flame is anchored by means of recirculation zones where reactants can mix and eventually burn. In the TVC geometry, because of the low entrainment of fluid into the recirculation area, reactants must be directly injected into the cavity. This strategy provides both a very efficient turbulent mixing of reactants and compensates for the lack of oxidizer due to a low fluid exchange between the mainstream flow and vortex region. Furthermore, the direct injection of reactants can strengthen the vortex itself, stabilizing it in a wide range of inlet airflow and fuel rates. This suggests that some form of premixing between the fuel and air may take place inside the cavity to the effect that so-called flameless combustion can be the combustion mode. This indeed is what has been found in the following section, using different fuels. Since the mode of operation of a TVC is found to be conducive to a lower pressure drop than in conventional combustors, it seems possible it may become the next combustor technology. This conjecture is also borne out by research work at General Electric and NASA[11,12]. It is for this reason TVC in the flameless combustion mode is being examined here for gas turbine applications (GT): in principle, low pressure drop and short combustion chambers enabled by such TVC may result in compact overall engine designs with higher thermodynamic efficiency.

2. The Trapped Vortex Combustor

The original TVC geometry presented in this work was proposed by Katta and Roquemore[5,6] and its performance was studied by Hsu et al.[3,6–10]. As shown in Figure 1, the structure of the burner is very simple.

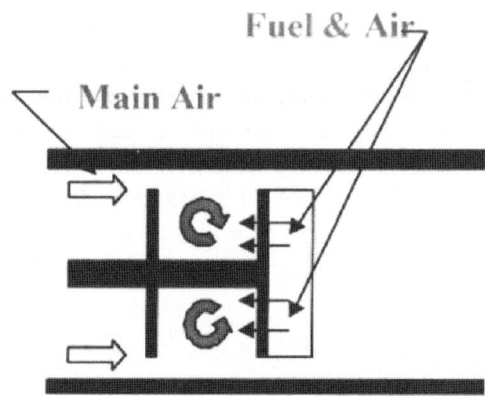

Figure 1. Sketch of the TVC strategy. (From[11,12].)

The desgin consists of a forebody disk and an afterbody disk connected along their centerline by means of a spindle (see Figure 1). Figure 2 shows an afterbody tested at WPAFB, with details of the fuel and secondary air holes[11,12].

The spindle is composed of two concentric tubes where air and fuel flow and are eventually delivered into the cavity from injectors within the afterbody disk facing the cavity itself. The forebody diameter in this example is 70 mm, while the afterbody diameter is 50.8 mm. The annular duct which contains the TVC has an 80 mm diameter. The blockage ratio, defined as the ratio of the area of the forebody to the duct frontal area is 76%. Two series of injectors are concentrically placed (8 injectors of 1.75 mm diameter and 24 injectors of 2.29 mm diameter).

Direct injection of reactants provides:

1. Direct control of the local equivalence ratio inside the recirculation zone (flame stability control)

2. Enhanced mixing by increasing the mixing region through distributed fuel and air jets

3. Some cooling of cavity walls

4. Reinforced vortex strength

Since good flame stability is very desirable, the length of the cavity (separation distance between the disks) has been chosen according to the numerical results already present in literature[1–3]: 0.59 $D_{forebody}$ is the length that provides the most stable vortex flame (with gas propane fuel) and the minimum pressure drop (see Figure 3). In order to increase combustion performance, a second cavity (hence a second afterbody disk) can be added (see Figure 4). This operation does not dramatically compromise pressure drop performance, if the second cavity is designed following the rules given by Little and Whipkey[2].

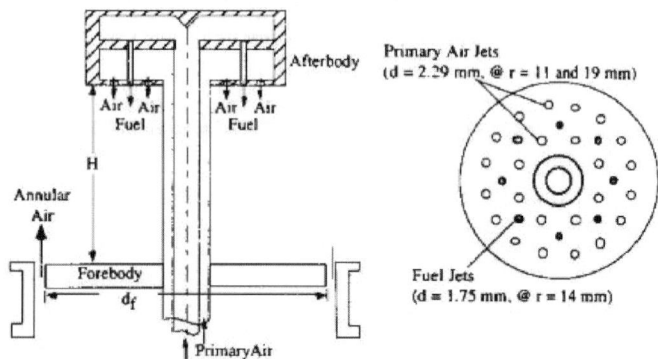

Figure 2. Schematic diagrams of the TVC structure (*left*) and the injection surface of the afterbody (*right*). (Picture taken from[3].)

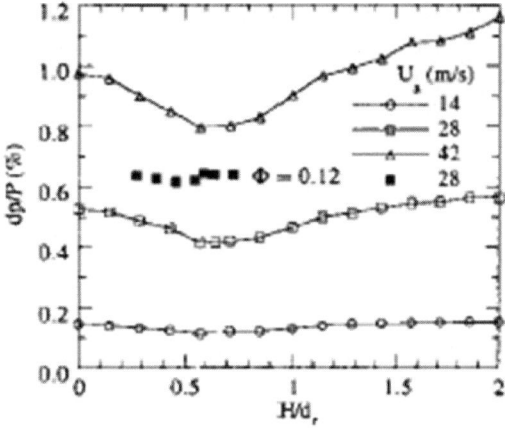

Figure 3. Impact of H/D$_f$ (cavity length *H* on forebody diameter) on pressure drop in cold and combustion flows[3]

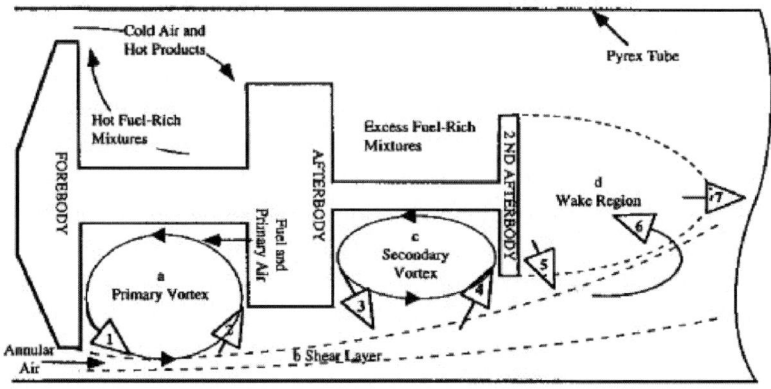

Figure 4. two cavities TVC sketch. (Picture taken from[3].)

3. The Trapped Vortex Theory

The numerical simulations presented in this work are aimed to show the main features of this low-pressure drop combustor. Literature has reported that the TVC combustor can implement the Rich-Quench-Lean (RQL) strategy, but, it is also true that its high hot combustion products recirculation enables, or may enable, prior mixing between fuel/air and combustion products.

Flameless combustion (FLOX)[13] is in fact known to be essentially based on a high recirculation level of hot products diluting the reactants, mixing them at high temperature and eventually burning them without any visible flame front. This strategy reduces the temperature peaks and lowers pollutant emission (NO$_x$ in particular) as it occurs at a low oxidizer concentration.

Similarly to a flameless burner, in a TVC the reactants are mixed at high temperatures by mean of a vortex, and burn at low oxidizer concentration regimes and at high recirculation factors (the vortex being locked within a cavity).

However, there are two main differences between these two strategies (as implemented so far):

1. In a flameless combustor, the entire streams injected into the combustion chamber takes part in the combustion process whereas in a TVC only a small amount of the mainstream flow is involved.

2. In a TVC, the combustion process is not completed within the cavity itself, but it may continue downstream of the afterbody due to the mainstream oxidizer. In a flameless burner the combustion process completes within its combustion zone. In a TVC the cavities cannot be considered combustion chambers as the TVC itself is inside the combustion chamber (see Figure 5).

Concerning the numerical simulations presented in this work, the results show that the TVC can be considered a "hybrid" combustor since it seems to work according to two different strategies at the same time. Actually, the TVC could be considered a radial–axial-staged combustor, as flameless combustion occurs within the cavity at low oxidizer levels and high recirculation factors, while hot combustion products are quenched along the axis.

The main advantage that such combined combustion can provide is that it can stably occur at high air speed and high pressure levels (at least up to 10 atm), reducing the pressure drop (because of its specific geometry) and also pollutant emissions. In order to achieve all goals desired, the geometry parameters, fluid dynamics and chemistry must be simultaneously coupled.

As reported in literature[13], in a flameless combustion process the exhaust gas recirculation plays an important role in the strategy. The recirculation parameter K_V is considered to be the key parameter to evaluate whether a flameless regime is theoretically reached (or reachable) within a combustor.

$$\text{It is defined as: } K_V = \frac{\dot{m}_e}{\dot{m}_e + \dot{m}_r},$$

where $\dot{m}_e$ is the mass flow rate of the exhausted gas and $\dot{m}_r$ is the mass flow rate of the reactants injected into the burner. When $K_v > 3.5$–4, the total amount of hot combustion products that recirculates represents about 75–80% of the total amount of the gas in the combustion chamber. If the temperature in the combustion zone is above the autoignition temperature, a stable and "frontless" flame condition is achieved (see Figure 6). In a TVC K_v is around 18–22 (95% of recirculation means $K_v = 20$), and the temperature within the cavity is above the threshold temperature, as the hot combustion products recirculate at high velocity rates.

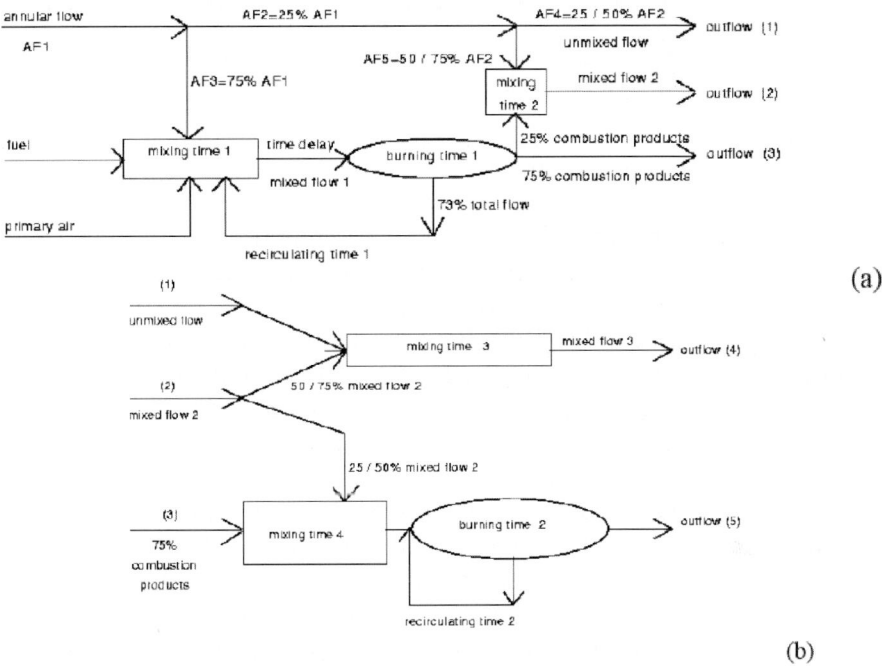

(a)

(b)

Figure 5. TVC working diagrams. (a) 1st cavity and (b) 2nd cavity. The quenching effect is provided by the 25% of the mainstream flow[15]

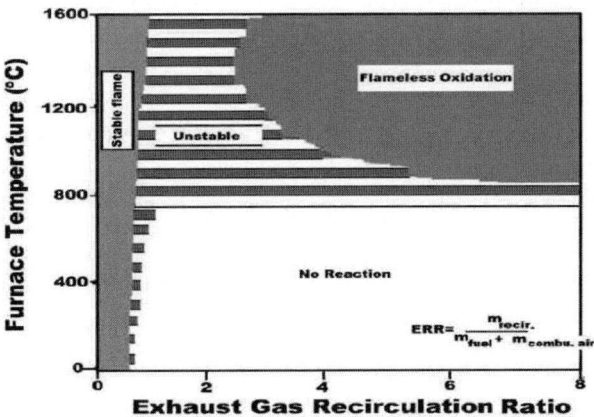

Figure 6. Combustion temperature and hot gas recirculation factor[14] K_V

3.1 PROPANE NUMERICAL SIMULATIONS

3.1.1 Katta and Roquemore Geometry

Since propane was used in the experiments at WPAFB[4-8], numerical simulations with propane were performed first to validate the numerical models[15,16]. Experimental results were at atmospheric pressure (1 atm) and different power settings (different equivalence ratio). The best combustion performance was at 35 kW, while pressure loss results seemed not to be strongly influenced by power. This result confirmed what is stated in literature[1], i.e., that the pressure drop is strongly influenced by the fluid dynamics and geometry of the cavity (aspect ratio, blockage ratio, and length). Due to heat release, the pressure loss worsens, but hot and cold cases maintain similar pressure trends. Simulations were validated using radial temperature profiles collected at three different axial locations, the pressure drop, combustion efficiency, unburned hydrocarbons and CO and NO_x emissions. Different combustion models (finite rate and/or eddy dissipation model) as well as different kinetic reaction mechanisms (one-step and two-step) were tested with FLUENT. With respect to computational costs, a coarse grid was used as well as a very fine mesh grid. Models were validated by means of RANS simulations. Using a fine mesh, the eddy dissipation model in conjunction with the two-step kinetic reaction mechanism provided the best results while, using a coarse mesh, the one-step mechanism yielded results in better agreement with experimental data (see Figure 7). Eventually, LES simulations were performed using a one-step kinetic model and the coarse grid.

The pressure drop results (see Tables 1 and 2) show that a properly dimensioned TVC geometry can significantly lower the predicted pressure

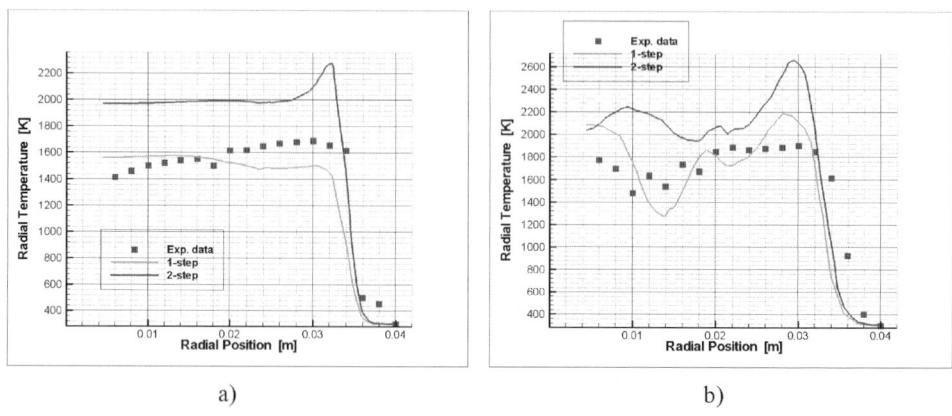

a)   b)

Figure 7. Temperature profile at axial location 5 mm (a), and 25 mm (b) from the forebody (coarse grid)

drop. It is instructive to highlight results when considering each cavities effect separately as well as those for the entire combustor: the second cavity enhances the combustion performance, reducing UHC and CO and increasing η_B, but also raises by about 10% the total pressure drop. It is expected that for unstable regimes (unlocked vortex or vortex fluctuations) the second cavity plays an important role in stabilizing the fluid dynamics: as there are no reactants further injected, the vortex trapped within takes its kinetic energy from the main stream. If the main stream is affected by fluctuations, the vortex in the second cavity dissipates them by transferring the turbulent kinetic energy collected down to smaller turbulent scales. This mechanism is supposed to stabilize the fluid motion, as the second vortex can be regarded as an *inflatable pillow* that changes its shape depending on the fluid "load" received.

Concerning the NO and CO concentrations (not in good agreement with experimental data), their behavior is justifiable based on the reduced kinetic mechanism. The one- and two-step mechanisms overestimate temperature and, since in the second cavity reactions occur and total residence time increases because of the vortex trapped within the second cavity, the NO concentration is overpredicted. Only the two-step mechanism can predict CO concentration, via the reaction $CO + \frac{1}{2}O_2 \rightarrow CO_2$. As several other reaction paths are missing, this further step is not sufficient to provide good estimates. For this reason, the CO prediction worsens using a coarser mesh. The RANS simulation carried out at 10 atm does not show any significant disagreement when compared to the 1 atm simulations. This means that at pressures up to 10 atm, the TVC keeps its performance without any change (see Figure 8a and b). Briefly results can be summarized as follows:

1. Even though hot temperature zones seem to be located downstream in respect of the first cavity, the temperature contours appear to be rather uniform within the cavity and above the temperature ignition threshold level.

2. Fuel is concentrated upstream of the cavity (in the area surrounding the forebody). This means that most of the UHC are produced within this region, since only a small fraction of the fuel is trapped within the second cavity. This conclusion is clearer when looking at radial concentration profiles as well as concentration contours (see Figure 9a–c).

3. Oxidizer is missing within the first cavity. This fact can be explained since only about 3% of the mainstream air is entrained into the cavity and mixes with the fuel. The amount of oxidizer directly injected is barely sufficient to let combustion occur completely within the cavity.

4. The largest proportion of combustion products can be found in the first cavity. High hot gases recirculation factors can be achieved within this cavity.

TABLE 1. Results of the RANS simulation with two-step reaction mechanisms and experimental data for a fine grid mesh. These results highlight the enhance in combustion performance achieved using a two cavities TVC geometry

C_3H_8 two-step (fine mesh)	Only 1st cavity	Only 2nd cavity	Entire combustor	Exp. data
η_B	0.824	0.315	0.88	0.92
$\frac{\Delta P_{Tot}}{P_{Tot}}$ %	0.683	0.0978	0.781	0.8
NO_{PPM}	$2 \cdot 10^3$	–	$3.86 \cdot 10^3$	<20
EI_{NOx}	$21.57 \frac{g}{kg}$	–	$40.18 \frac{g}{kg}$	
CO_{PPM}	$1.47 \cdot 10^4$	–	$1.125 \cdot 10^4$	<1,600
UHC_{PPM}	$1.456 \cdot 10^3$	–	$9.861 \cdot 10^2$	<1,400

In Figure 10 it can be noticed that the C_3H_8 outlet concentration has decreased by one fourth when compared to the concentration reported at the outlet of the first cavity. This result is in agreement with the TVC working diagrams (see Figure 5a and b): it is reported that about 25% of the mainstream flow quenches the combustion products burnt in the vortex.

TABLE 2. Comparison between one- and two-step reaction mechanisms andexperimental data for a coarse grid (two cavities TVC geometry)

C_3H_8 (coarse mesh)	One-step (1 atm)	Two-step (1 atm)	Exp. data
η_B	0.7	0.85	0.92
$\frac{\Delta P_{Tot}}{P_{Tot}}$ %	0.8	0.83	0.8
CO_{PPM}	–	$2.29 \cdot 10^4$	<1,600
UHC_{PPM}	$1.538 \cdot 10^3$	$7.07 \cdot 10^2$	<1,400

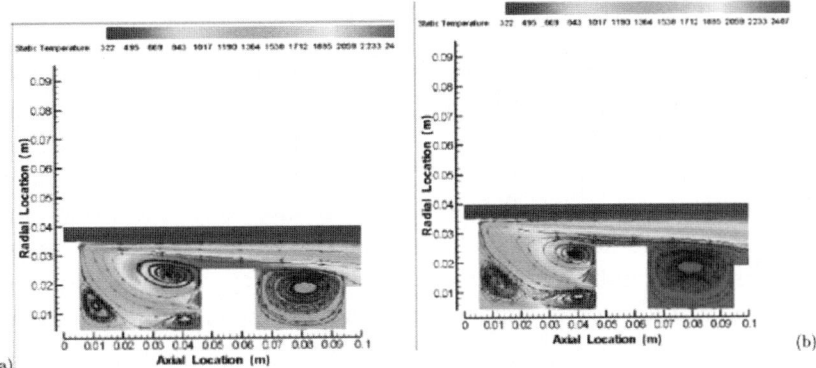

Figure 8. Temperature contours. RANS C_3H_8, two step, coarse mesh. (a) pressure = 1 atm; (b) pressure = 10 atm

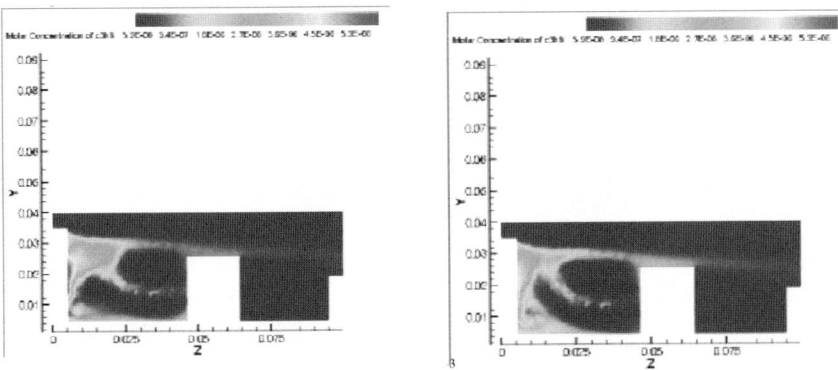

Figure 9. Contours of C_3H_8 concentration (kg m). LES C_3H_8, one step, 1 atm, coarse mesh, at three different instants

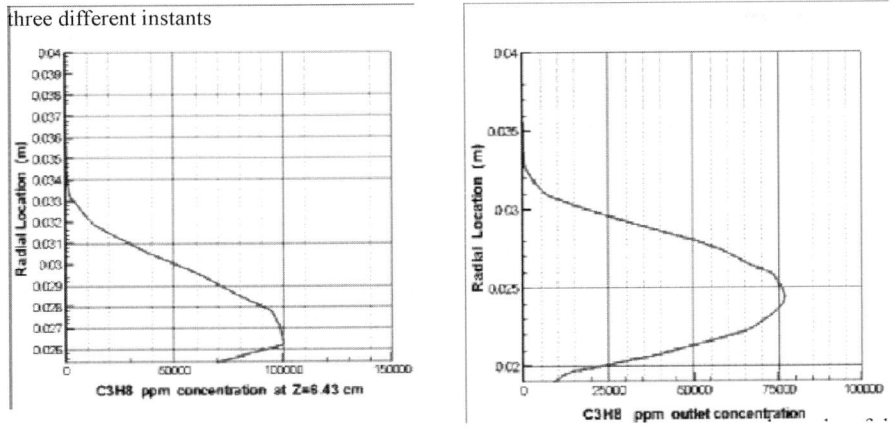

Figure 10. C_3H_8 concentration (ppm). LES C_3H_8, one step, 1 atm, coarse mesh at the outlet of the 1st cavity (a) and at the TVC outlet (b)

3.2 TVC IN A TYPICAL COMBUSTION CHAMBER

The aim of this section is to investigate a feasible integration of the TVC strategy in a combustion chamber for gas turbine applications. Particular importance is given to the integration in aircrafts, as the engine's frontal area and weight represent a critical issue. For this reason, in the present work a typical aircraft configuration is investigated numerically: the external case is given consideration and a new TVC geometry is set up within (see Figure 11). The combustion chamber is composed of a primary turbulent mixing zone, where reactants are injected, and the main reaction zone, within which the main vortex is trapped.

Two single-can geometries have been extensively studied and hence reported in this work. Works currently in progress at DMA show that it is possible to convert a single-can TVC geometry into an annular geometry with no significant effect on the combustor performance.

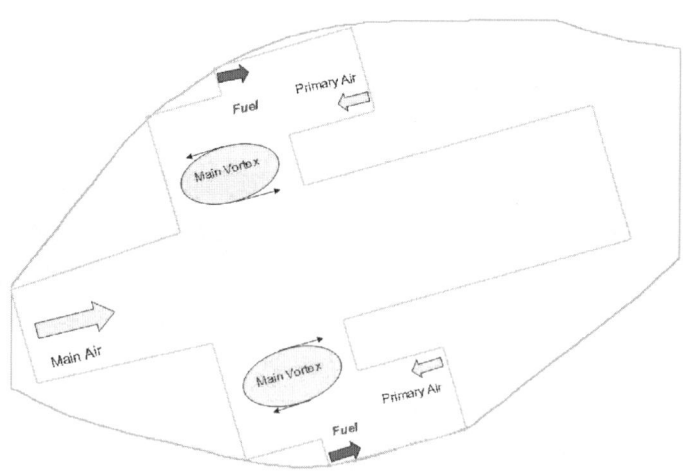

Figure 11. Sketch of the proposed TVC: first configuration

3.2.1 First Single-can Geometry

As showed in Figure 11, this geometry is composed of a main recirculation cavity, and a secondary region where reactants are injected by means of a set of single injectors.

In Figure 12 the computational domain is shown. Since it has been demonstrated that within the range of 1–10 atm the operating pressure does not influence the TVC performance, the numerical simulation has been carried out at 1 atm. Three different power conditions have been investigated: 32 kW (fuel velocity 10 m/s), 49 kW (fuel velocity 15 m/s) and 64 kW (fuel velocity 20 m/s), the primary air velocity is 25 m/s in the first and the second group while in the third it is 30 m/s. The main air velocity is constant at 50 m/s. In Table 3 the performance of the TVC are compared. Rich combustion conditions have been chosen as starting cases in order to evaluate the mixing capability of the TVC, aiming towards lean combustion conditions. According to the performance collected in Table 3, there are no relevant differences between the investigated cases. Since the combustion efficiency η_B is extremely low and the main vortex seems not to be trapped (see Figure 13), the geometrical ratios need to be modified. The geometrical ratios used by Katta and Roquemore cannot directly be applicable as the presence of a second cavity (turbulent premixing zone) must be taken into account. The ratio between the afterbody and the forebody diameter is hence decreased, as the surface where the vortex can really anchor itself is decreased (see Figure 13). The vortex motion seems to be driven by the reactants injected instead of by the main airflow. This effect can occur as the afterbody surface is not sufficient to trap a vortex according to the flow rates injected. As it appears in Figures 13a and 14a, the temperature contour is rather not homogeneous: combustion mainly occurs along the combustor length instead of in the main cavity, where the equivalence ratio reaches unity (see Figure 13b).

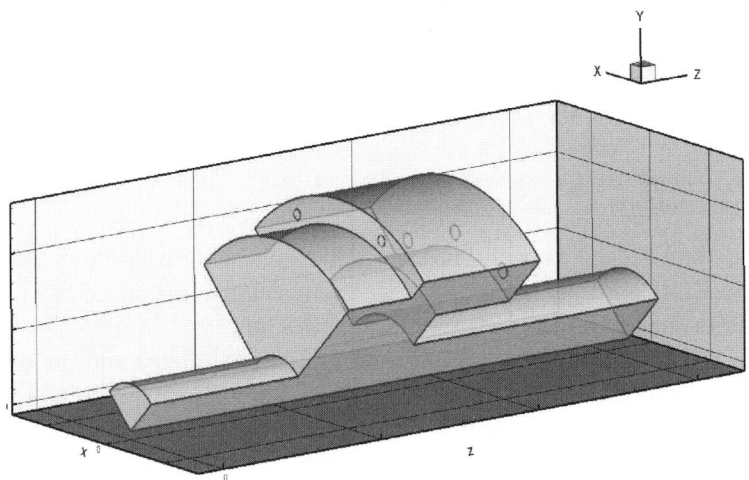

Figure 12. Sketch of the computational domain (90° 668,000 tetrahedral cells, 123,466 nodes)

TABLE 3. Performance of the TVC in rich combustion

Power (kW)	η_B (%)	$\dfrac{\Delta P_{Tot}}{P_{Tot}}\%$	EI_{NO}	$\Phi_{P.Air}$	$\Phi_{Overall}$
32	48	0.91	–	5.5	0.71
49	50	0.88	0.0418	8.25	1.06
64	50	0.87	0.04	9.16	1.35

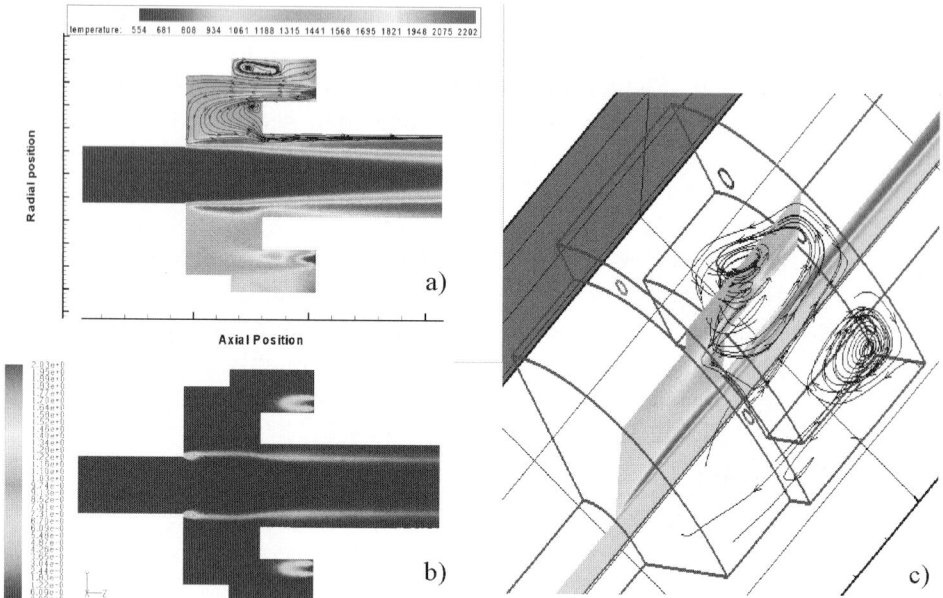

Figure 13. Temperature contour with streamlines (a) and reaction rate contour (b). RANS C_3H_8 one step, power 64 kW, 1 atm

According to the contour profiles, this combustion strategy is closer to a complete RQL configuration, that has been recently proposed by[21] (see Figure 14b), rather than to a radial staged flameless-RQL.

In Figure 14a and b the simulated single-can geometry and the experimental annular section of the TVC proposed by the American Air Force are compared. At first glance, these geometries show a similar behavior. This result was expected since both of the geometries are based on the same strategy. The aim of this research is to move the TVC strategy as close as possible to a fast flameless regime.

In Figure 13c, the secondary vortex structures are shown: these structures are tilted along the radius. This mechanism drives the fuel directly to the main chamber where it is mixed with the hot combustion products. The selected

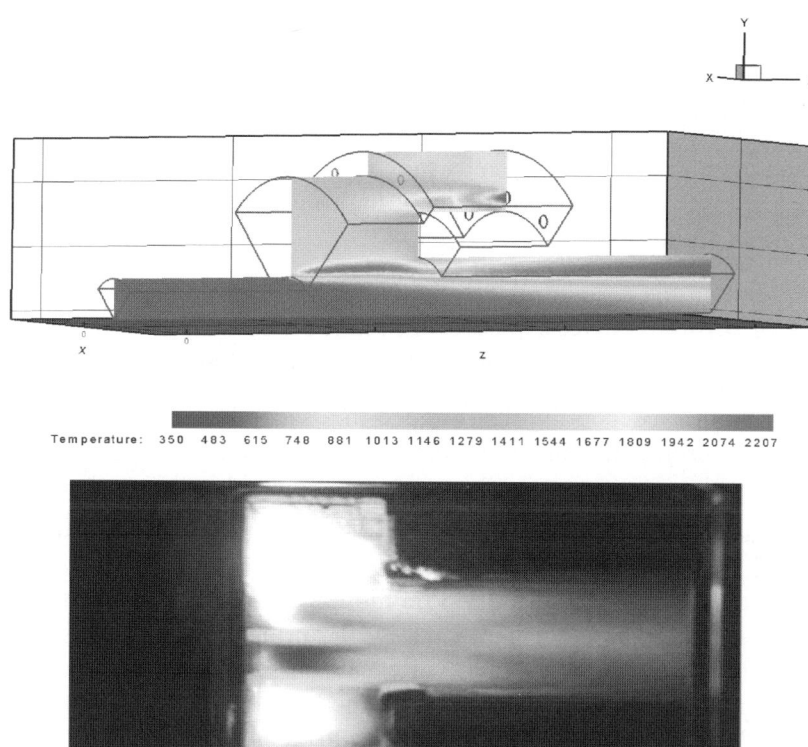

Figure 14. Temperature contour. RANS C_3H_8 one step, power 64 kW, 1 atm (a). TVC tripass diffuser[21] (b)

combustion regime results are too rich to assess whether these structures have a positive contribution in combustion and pressure drop performance.

3.2.2 Second Single-can Geometry

As shown in Figure 15, this configuration essentially differs from the previous one from the use of the second cavity: a self-standing vortex is created by means of a secondary stream, which replaces the primary air injection devices.

Two independent and coherent vortices increase the combustor performance, since the turbulent mixing rate is faster and more efficient compared to the single vortex strategy.

As —reported, the main airstream is only of 3–5% of the total. Because of this fact, neither the vortex momentum nor the oxidant concentrations are sufficient to provide an efficient mixing of reactants and a complete combustion process, unless primary air is correctly injected into the cavity. A critical issue highlighted in the previous works is that the residence time within the cavity must be increased as well as oxidant concentration. Three possible paths can and will be investigated:

1. An increase in the amount of oxidant by means of a primary air multiinjector. This strategy may lead to vortex instability, since the vortex motion may no longer be driven by the main stream if the amount of primary air injected into the cavity becomes of the same order of magnitude as that of the main airstream.

2. An increase in the amount of the main airstream entering into the cavity by means of diffuser devices, placed upstream of the cavity. This strategy is limited by the amount of air that can be turned into the main airstream.

3. The addition of a second vortex trapped in a secondary cavity: this strategy seems more promising, since two independent vortices reinforce each other, increasing the mixing rate and the residence time within the combustor.

In this section, some preliminary results for the third strategy above are presented.

Large eddy simulations performed on the first geometry (see Section 3.2.1) show that the vortex structure generated by the primary air injectors is intrinsically unstable: the motion of each vortex becomes unsteady and a vortex is periodically released from the cavity. This means that a single airstream does not have sufficient momentum to generate and keep a secondary vortex in a separate cavity. To prevent vortex shedding and motion instability, a secondary (annular) airstream is hence added.

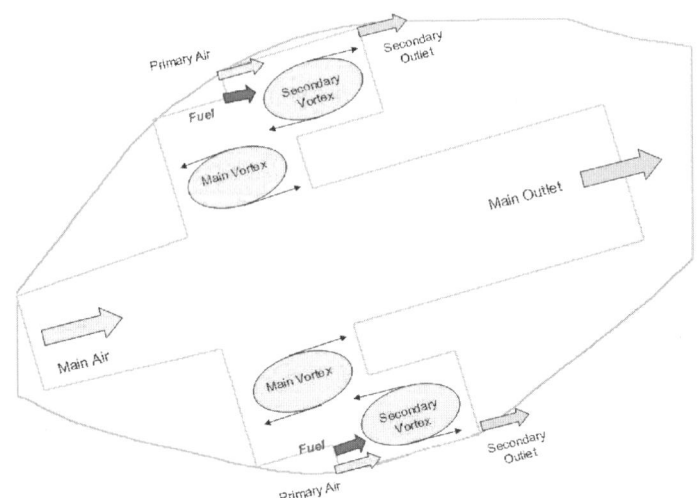

Figure 15. Sketch of the proposed TVC: second configuration

This configuration is, in essence, a combined staged combustion, since it uses two trapped vortices working "in parallel" but not oriented either axially or radially with respect to the combustion chamber. Such a geometry stabilizes the vortex motion and increases effectively both the amount of oxidant in the reaction zone (without increasing the total pressure drop) and the residence time.

Figures 16 and 17 show the contours of temperature and O_2 concentration. Most of the combustion process occurs within the main cavity, but because of insufficient oxidizer (see Figure 17) combustion is not complete. The unburned fuel is later oxidized along the path towards the combustor outlet, producing most of the (thermal) NO_x (see Figure 18).

As shown in Table 4, this TVC configuration increases both the combustion efficiency and total pressure drop. Since combustion takes place downstream of the cavity, the temperature peaks are found where the hot unburned fuel meets the main airstream. NO_x emission indices are about 100. Work currently in progress at DMA investigating possible solutions of this important issue.

TABLE 4. Performance of the second configuration. Pressure 30 atm

Power (kW)	η_B (%)	$\dfrac{\Delta P_{Tot}}{P_{Tot}}$ %	EI_{NO}	$\Phi_{Overall}$
340	99.15	0.7	99	0.6
450	99	0.7	105	0.7

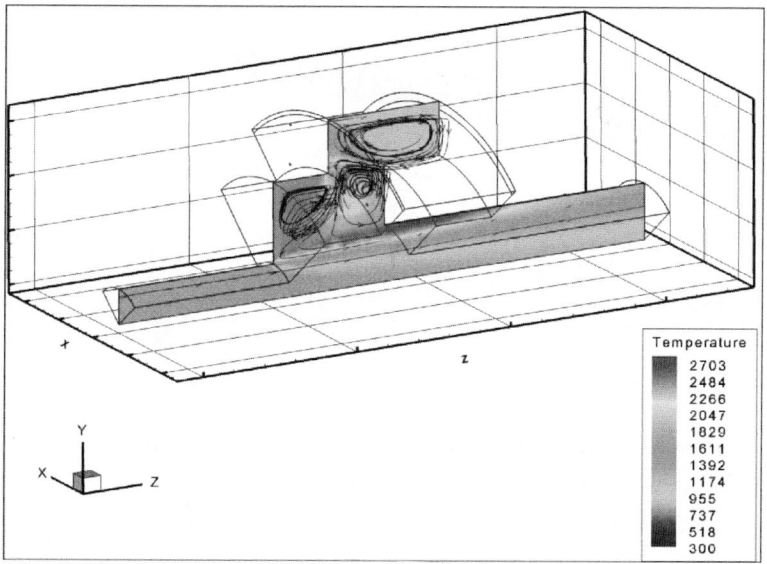

Figure 16. TVC geometry: second configuration. Temperature contour

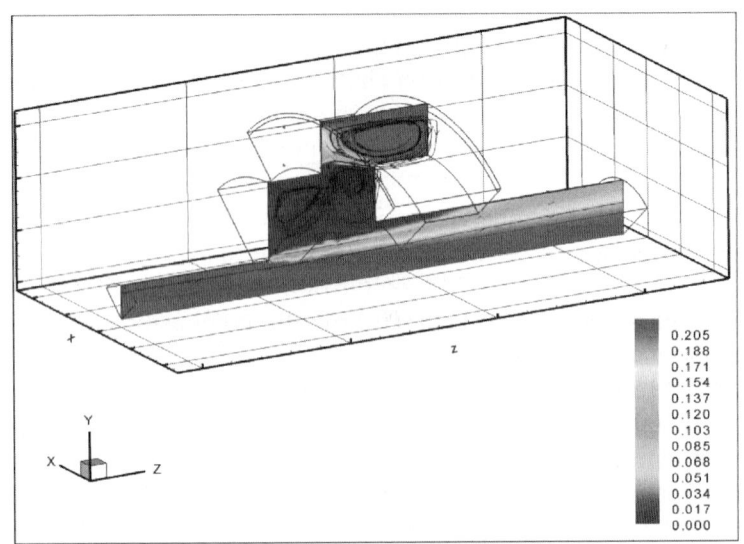

Figure 17. TVC geometry: second configuration. O_2 concentration contour (kg/m3)

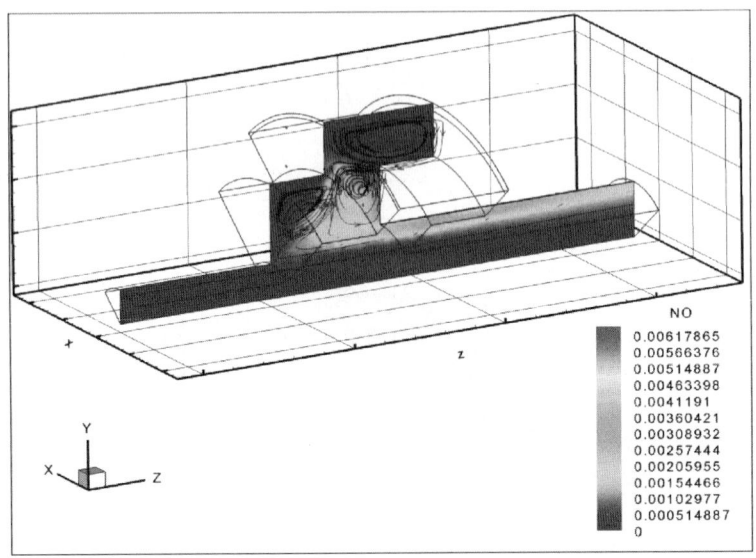

Figure 18. TVC geometry: second configuration. NO_x mass fraction contour

4. Conclusions

Two new TVC geometries have been numerically investigated, besides the base-line geometry proposed more than 10 years ago by Katta and Roquemore[5–7]. The results of the original TVC geometry highlight the intrinsic flameless combustion nature of the TVC, lowering the NO_x emission as well as yielding a very low pressure drop. These main features have driven a new TVC design.

The main task faced is to provide a set of geometrical ratios and fluid dynamic ranges enabling the TVC strategy to replace the conventional GT combustion strategy based on some type of pilot flame. In order to achieve this goal, a typical aircraft engine case has been used to preliminarily investigate the new TVC within. Preliminary results coming from new TVC configurations show that the radial dimension is critical, leading to further developments already under investigation. Using a premixing zone stages the combustion process and makes the TVC technology much more flexible, but on the other hand, could trigger combustion instability. This potential problem, found in the first new TVC configuration explored, is solved in the second configuration presented. The first geometry does not achieve complete combustion, as combustion efficiency is around 50%, while in the second configuration it is 99.1%. According to the numerical results, the NO emission index in the first case dramatically decreases with respect to the baseline, due to the short residence time; combustion is lean and at the same time somewhat incomplete, with part of the combustion taking place in the main airstream. On the other hand, the second configuration provides very good combustion efficiency, but also excessive NO_x emissions.

The pressure drop stays low, and seems not to be a critical issue in all of the configurations here investigated. A serious issue is the relative scarcity of air in the main combustion zone: increasing the primary air flow as well as the turbulent mixing can be considered the critical issue in this context. The second geometry investigated can only be considered as the first step toward TVC integration in already existing GT facilities, providing a good compromise among cost, emissions, and performance.

References

1. Mair, W. A., The Effect of Rear-Mounted Disc on the Drag of a Blunt-Based Body of Revolution, *Aeronautical Quarterly*, **10**(4), 1965, 350–360.
2. Little, B. H. jr. and Whipkey, R. R., Locked Vortex Afterbodies, *Journal of Aircraft*, **16**(5), 1979, 296–302.

3. Gharib, M. and Roshko, A., The Effect of Flow Oscillations on Cavity Drag, *Journal of Fluid Mechanics*, **177**, 1987, 501–530.
4. Hsu, K.-Y., Goss, L.P., Trump, D.D., and Roquemore, W.M., Performance of a Trapped-Vortex Combustor, *AIAA Paper 95-0810, January 1995.*
5. Katta, V. R. and Roquemore, W. M., Numerical Studies on Trapped-Vortex Combustor, *AIAA paper 96-2660, July 1996.*
6. Katta, V. R. and Roquemore, W. M., Study on Trapped-Vortex Combustor: Effect of Injection on Flow Dynamics, *Journal of Propulsion and Power*, **14**(3), 1998, 273–281.
7. Hsu, K-Y, Goss, L.P. and Roquemore, W.M., Characteristics of a Trapped-Vortex Combustor, *Journal of Propulsion and Power*, **14**(1), 1998, 57–65.
8. Sturgess, G. and Hsu, H.-Y., Entrainment of mainstream flow in a Trapped-Vortex Combustor, *AIAA paper 97-026, 1997.*
9. Stone, C. and Menon, S., Simulation of Fuel-Air Mixing and Combustion in a Trapped-Vortex Combustor, *AIAA paper 00-0478, 2000.*
10. Ramirez, M. and Bruno, C., Simulation of TVC without turbulence models, 20th Annual meeting of the Italian Section of The Combustion Institute (ISCI 1997), July 2–5, Paper. III 1.1, 1997, Naples.
11. Losurdo, M. and Bruno, C., Effects of Different Fuels in a Trapped Vortex Combustor, presented at the Italian Section of the Combustion Institute (ISCI 2002) as paper II 13, 3–5 June 2002, Rome.
12. Hendricks, R.C., Shouse, D.T., Roquemore, W.M., Burrus, D.L., Duncan, B.S., Ryder, R.C. Branckovic, A., Liu, N.-S., Gallagher, J.R., Hendricks, J. A., Experimental and Computational Study of Trapped-Vortex Combustion Sector Rig with Tri-pass Diffuser, *NASA/TM-2004-212507.January 2004.*
13. Shouse, D.T. Roquemore, W.M. Arana C.A., Burrus, D., Cooper, C., Trapped Vortex Combustion Technology Research, *Air Force and General Electric Aircraft article. Reference Document PR-01-11.*
14. Wuenning, J.A. and Wuenning, J.G, *Proc. Energy Combustion Sci.*, **23**, 1997, 81.
15. Milani, A., Saponaro, A., Diluted Combustion Technologies, *IFRF Combustion journal, Article 200101, February 2001.* Losurdo M., Numerical Simulation of a Trapped Vortex Burner using propane, methane and hydrogen, Technical report (not published) EC project No. NNES-1999-20226, ENEA report No. 1DA4A.
16. Losurdo, M., Numerical Studies on Vortex Stability of a Trapped Vortex Combustor, Cardiff University Technical Report (unpublished), No. 3019, 2003, Cardiff.
17. Mularz, J. (1981) (personal communication).
18. Straub, D.T., Sidwell, T.G., Maloney D.J., Casleton, K.H. and Richards, G.A., Simulations of a Rich Quench-Lean (RQL) Trapped-Vortex Combustor, presented at the American Flame Research Committee (AFRC) International Symposium, September 2000, Newport Beach, CA, USA.
19. Straub, D.L., Casleton, K.H. Lewis, R.E., Sidwell, T.G., Maloney, D.J. and Richards, G.A., Assessment of Rich-Burn Quick-Mix, Lean-Burn Trapped Vortex Combustion for Stationary Gas Turbines, *Transaction of the ASME, Vol. 127, pp. 36–41, January 2005.*
20. Hendricks, R.C., Shouse, D.T., Roquemore, W.M., Burrus, D.L., Duncan, B.S., Ryder, R.C., Brankovic, A., Liu, N.S., Gallagher, J.R., and Hendricks, J.A., Experimental and Computational Study of Trapped Vortex Combustor Sector Rig with Tri-Pass Diffuser, *NASA/TM-2004-212507, January 2004.*

THE ROLE OF FUEL CELLS IN GENERATING CLEAN POWER AND REDUCING GREENHOUSE GAS EMISSIONS

S. A. ALI[*]

Principal, Clean Energy Consulting Corporation

Abstract. The Kyoto Protocol and the Clear Skies Initiative of the USA are both aimed to achieve significant global reductions in greenhouse gas emissions, which is beneficial for the application of fuel cells. Fuel cell techno-logy provides high thermal efficiency and is environmentally benign. Current global fuel cell development is focused on mobile and stationary power applications; PEM fuel cells are used as hybrid power sources in transportation; auto manufacturers are offering PEM fuel cell-based hybrids combined with gas engines; and solid oxide fuel cell (SOFC) technology is being developed globally. The US Department of Energy (DOE) is sponsoring the Solid Oxide Energy Conversion Alliance (SECA) program, and a multiyear project to demonstrate coal-based SOFC. These systems aim to achieve 50% efficiency from coal, capture 90% of CO_2 emissions, and cost \$400/kw. The SOFC technology is driven towards commercial deployment because of its high thermal efficiency and environmentally superior characteristics.

Keywords: fuel cell, clean electro-chemical power, greenhouse emissions

1. Introduction

Fuel cells are electrochemical devices that convert chemical energy in fuels into electrical energy directly. Although a fuel cell is similar to a typical battery in many ways, it differs in several respects. The battery is an energy storage device in which all the energy available is stored within the battery itself (at least the reductant). The battery will cease to produce electrical energy when the chemical reactants are consumed (i.e., discharged). A fuel cell, on the other

---

[*] To whom correspondence. 7971 Black Oak Drive. Plainfield. IN. 46168. USA. Tel.: 317 839 6617; e-mail: sy.ali@cleanenergyconsulting.com

N. Syred and A. Khalatov (eds.), Advanced Combustion and Aerothermal Technologies, 385–403.

hand, is an energy conversion device to which fuel and oxidant are supplied continuously. In principle, the fuel cell produces power for as long as fuel is supplied.[1]

Fuel cells are applicable for power generation with high efficiency and low environmental impact. Since the intermediate steps of producing heat and mechanical work typical of most conventional power generation methods are avoided, fuel cells are not limited by the thermodynamic limitations of heat engines such as the Carnot efficiency. In addition, since there is no combustion in a fuel cell process, fuel cells produce power with minimal pollutants. Although fuel cells could, in principle, process a wide variety of fuels and oxidants, of most interest today are those fuel cells that use common fuels (or their derivatives) or hydrogen as a reductant, and ambient air as the oxidant. Most fuel cell power systems comprise a number of components:

- Unit cells, in which the electrochemical reactions take place
- Stacks, in which individual cells are modularly combined by electrically connecting the cells to form units with the desired output capacity
- Balance of plant (BOP) which comprises components that provide process conditioning (including a fuel processor if needed), thermal management, and electric power conditioning among other ancillary and interface functions

Fuel cell technology is described here with a brief review of the key potential applications of fuel cells.

2. Basic Structure of Unit Cells

Unit cells form the core of a fuel cell. These devices convert the chemical energy contained in a fuel electrochemically into electrical energy. The basic physical structure of a fuel cell consists of an electrolyte layer in contact with an anode and a cathode on either side. A schematic representation of a unit cell with the reactant/product gases and the flow directions of ion conduction through the cell is shown in Figure 1.

In a typical fuel cell, fuel is fed continuously to the anode (negative electrode) and an oxidant (often oxygen from air) is fed continuously to the cathode (positive electrode). The electrochemical reactions take place at the

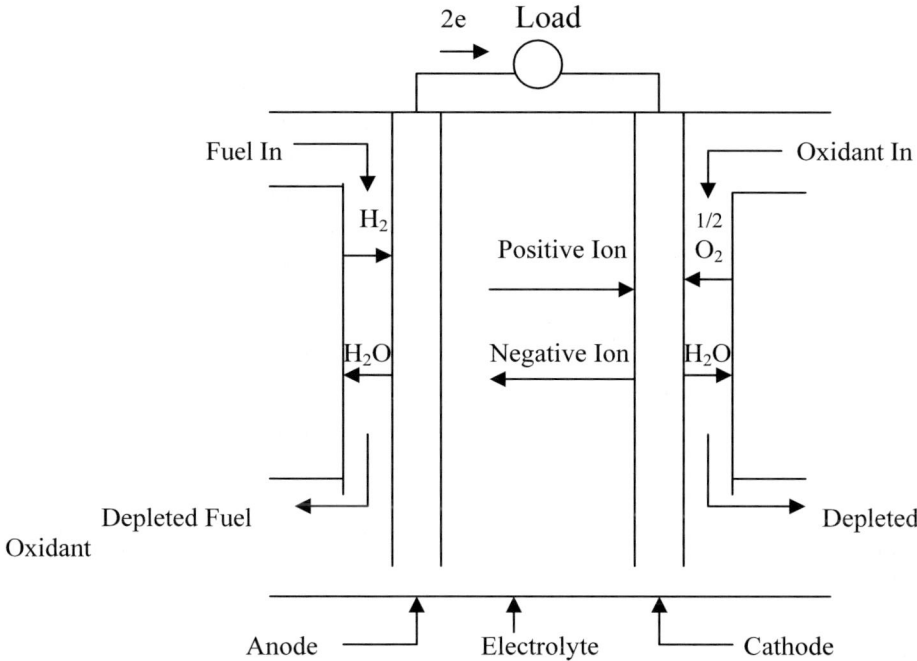

Figure 1. Schematic of an individual fuel cell

electrodes to produce an electric current through the electrolyte, while driving a complementary electric current that performs work on the load.

Fuel cells are classified according to the choice of electrolyte and fuel, which in turn determine the electrode reactions and the type of ions that carry the current across the electrolyte. Some scientists have noted that, in theory, any substance capable of chemical oxidation that can be supplied continuously (as a fluid) can be burned galvanically as fuel at the anode of a fuel cell. Similarly, the oxidant can be any fluid that can be reduced at a sufficient rate. Though the direct use of conventional fuels in fuel cells would be desirable, most fuel cells under development today use gaseous hydrogen, or a synthesis of gas rich in hydrogen, as a fuel.

Hydrogen reacts readily at the anode. It can be produced chemically from a wide range of fossil and renewable fuels, as well as through electrolysis. The most common oxidant is gaseous oxygen, which is readily available from air. For space applications, both hydrogen and oxygen are stored compactly in cryogenic form, and the reaction product is water.

3. Critical Functions of Cell Components

The electrochemical reactions take place where either electrode meets the electrolyte. For a site or area to be active, it must be in electrical contact with the electrode, in ionic contact with the electrolyte, and contain sufficient electro-catalyst for the reaction to proceed at the desired rate. The density of these regions and the nature of these interfaces play a critical role in the electrochemical performance of both liquid and solid electrolyte fuel cells.

3.1 LIQUID ELECTROLYTE FUEL CELLS

The reactant gases diffuse through a thin electrolyte film that wets portions of the porous electrode and react electrochemically on the surface of their respective electrode. If the porous electrode contains an excessive amount of electrolyte, the electrode may restrict the transport of gaseous species in the electrolyte phase to the reaction sites. This reduces electrochemical performance of the porous electrode. Thus, a delicate balance must be maintained among the electrode, electrolyte, and gaseous phases in the porous electrode structure.

3.2 SOLID ELECTROLYTE FUEL CELLS

The main challenge is bringing the catalyst sites into contact with the electric and ionic interface between the electrode and the electrolyte, while keeping them sufficiently exposed to the reactant gases. In most successful solid electrolyte fuel cells, a high-performance interface requires the use of an electrode which, in the zone near the catalyst, has mixed conductivity (i.e., it conducts both electrons and ions).

The continual development of fuel cell technologies has resulted in dramatic improvements in unit cell performance. Improvements in the three-phase boundary, reduced thickness of the electrolyte, and improved electrode and electrolyte materials have broadened the temperature range over which the cells can be operated.

In addition to facilitating electrochemical reactions, each of the unit cell components has other critical functions. For example, the electrolyte transports dissolved reactants to the electrode, and also conducts ionic charge between the electrodes, thereby completing the cell electric circuit as illustrated in Figure 1. It also provides a physical barrier to prevent the fuel and oxidant gas streams from directly mixing.

In addition to providing a surface for electrochemical reactions to take place, the porous electrodes in fuel cells also serve these three functions:

1. Conducting electrons away from or into the three-phase interface once they are formed (so an electrode must be made of materials that have good electrical conductance) and providing current collection and connection with either the load or other cells

2. Ensuring that reactant gases are equally distributed over the cell

3. Ensuring that reaction products are efficiently led away to the bulk gas phase

The electrodes are typically porous and made of an electrically conductive material. At low temperatures, only a few relatively rare and expensive materials provide sufficient electrocatalytic activity. Such catalysts are deposited in small quantities at the interface where they are needed. In high-temperature fuel cells, the electrocatalytic activity of the bulk electrode material is often sufficient. Most fuel cells currently under development are either planar (rectangular or circular) or tubular (either single ended or double ended and cylindrical or flattened).

4. Fuel Cell Stacking

Unit cells are combined into a cell stack to achieve the voltage and power output level required for the application. Generally, the stacking involves connecting multiple unit cells in series via electrically conductive interconnecttions. Various stacking arrangements have been developed, which are presented here.

4.1 PLANAR–BIPOLAR STACKING

The most common fuel cell stack design is the planar/bipolar arrangement. Individual unit cells are electrically connected with interconnections. Because of the configuration of a flat plate cell, the interconnection becomes a separator plate with two functions:

1. Provide an electrical series connection between adjacent cells, specifically for flat plate cells
2. Provide a gas barrier that separates the fuel and oxidant of adjacent cells

In many planar–bipolar designs, the interconnection also includes channels that distribute the gas flow over the cells. The planar–bipolar design is electrically simple and leads to short electronic current paths (which helps to minimize cell resistance).

Planar–bipolar stacks can be further characterized according to the arrangement of the gas flow:

- Cross-flow: air and fuel flow perpendicular to each other.

- Coflow: air and fuel flow parallel and in the same direction. In the case of circular cells, this means the gases flow radially outward.

- Counterflow: air and fuel flow parallel but in opposite directions. Again, in the case of circular cells this means radial flow.

- Serpentine flow: air or fuel follow a zigzag path.

- Spiral flow: applies to circular cells.

The choice of gas-flow arrangement depends on the type of fuel cell and the application. The channeling of gas streams to the cells in bipolar stacks can be achieved in various ways:

- Internally: the manifolds run through the unit cells

- Integrated: the manifolds do not penetrate the unit cells but are integrated in the interconnects

- External: the manifold is completely external to the cell, much like a wind-box

4.2 STACKS WITH TUBULAR CELLS

For high-temperature fuel cells, stacks with tubular cells have been developed. Tubular cells have significant advantages in terms of sealing and in their structural integrity. However, they present a particular geometric challenge to the stack designer when it comes to achieving high power density and short current paths. In one of the earliest tubular designs the current is conducted tangentially around the tube. Interconnections between the tubes are used to form rectangular arrays of tubes. Alternatively, the current can be conducted along the axis of the tube, in which case interconnection is done at the end of the tubes. To minimize the length of electronic conduction paths for individual cells, sequential series connected cells are being developed. The cell arrays can be connected in series or in parallel. To avoid the packing density limitations associated with cylindrical cells, some tubular stack designs use flattened tubes.

5. Fuel Cell Systems

Fuel cell systems require several subsystems and components. These are referred to as the BOP. Together with the stack, the BOP forms the fuel cell system. The precise arrangement of the BOP depends heavily on the fuel cell type, the fuel choice, and the application. Specific operating conditions and requirements of cells and stack designs determine the BOP requirements.

Most fuel cell systems contain:

- Fuel preparation: except for fuels such as hydrogen, fuel preparation requires the removal of impurities and thermal conditioning. Fuel reforming is usually required, allowing the fuel to react with steam or air to form a hydrogen-rich anode feed mixture.

- Air supply: including air compressors or blowers as well as air filters.

- Thermal management: all fuel cell systems require careful management of the fuel cell stack temperature.

- Water management: water is needed in some parts of the fuel cell, while overall water is also a reaction product. To avoid having to feed water in addition to fuel and to ensure smooth operation, water management systems are required.

- Power conditioning: fuel cell stacks provide a variable DC voltage output. Power conditioning is typically required.

The BOP represents a significant fraction of the weight, volume, and cost of most fuel cell systems.

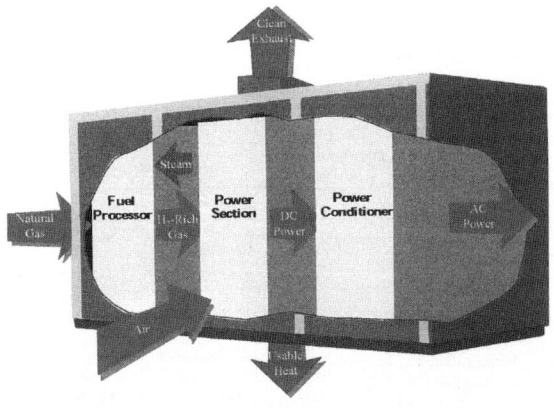

Figure 2. Fuel cell system components

6. Types of Fuel Cell

A variety of fuel cells are in different stages of development. The most common classification of fuel cells is by the type of electrolyte used in the cells and includes:

1.   Polymer electrolyte fuel cell (PEFC)
2.   Alkaline fuel cell (AFC)
3.   Phosphoric acid fuel cell (PAFC)
4.   Molten carbonate fuel cell (MCFC)
5.   Solid oxide fuel cell

TABLE 1. A summary of the major differences between the fuel cell types

	PEFC	AFC	PAFC	MCFC	SOFC
Electrolyte	Hydrated Polymeric Ion Exchange Membranes	Mobilized or Immobilized Potassium Hydroxide in asbestos matrix	Immobilized Liquid Phosphoric Acid in SiC	Immobilized Liquid Molten Carbonate in $LiAlO_2$	Perovskites (Ceramics)
Electrodes	Carbon	Transition metals	Carbon	Nickel and Nickel Oxide	Perovskite and perovskite / metal cermet
Catalyst	Platinum	Platinum	Platinum	Electrode material	Electrode material
Interconnect	Carbon or metal	Metal	Graphite	Stainless steel or Nickel	Nickel, ceramic, or steel
Operating Temperature	40 – 80 °C	65°C – 220 °C	205 °C	650 °C	600-1000 °C
Charge Carrier	H^+	OH^-	H^+	$CO3^=$	$O^=$
External Reformer for hydrocarbon fuels	Yes	Yes	Yes	No. for some fuels	No. for some fuels and cell designs
External shift conversion of CO to hydrogen	Yes, plus purification to remove trace CO	Yes, plus purification to remove CO and CO_2	Yes	No	No
Prime Cell Components	Carbon-based	Carbon-based	Graphite-based	Stainless-based	Ceramic
Product Water Management	Evaporative	Evaporative	Evaporative	Gaseous Product	Gaseous Product
Product Heat Management	Process Gas + Liquid Cooling Medium	Process Gas + Electrolyte Circulation	Process Gas + Liquid cooling medium or steam generation	Internal Reforming + Process Gas	Internal Reforming + Process Gas

The choice of electrolyte defines the operating temperature range of the fuel cell. Both the operating temperature and useful life of a fuel cell determine the physicochemical and thermomechanical properties of materials used in the cell components (i.e., electrodes, electrolyte, interconnection, and current collector). Aqueous electrolytes are limited to temperatures of about 200°C or lower because of their high vapor pressure and rapid degradation at higher temperatures.

The operating temperature defines the extent of fuel processing required. In low-temperature fuel cells, all the fuel must be converted to hydrogen prior to entering the fuel cell. In addition, the anode catalyst in low-temperature fuel cells (mainly platinum) is strongly poisoned by CO. In high-temperature fuel cells, CO and even CH_4 can be internally converted to hydrogen or even directly oxidized electrochemically. Table 1 provides an overview of the key characteristics of the main fuel cell types.

In parallel with the classification by electrolyte, some fuel cells are classified by the type of fuel used:

6.1 DIRECT ALCOHOL FUEL CELLS

Direct alcohol fuel cells (DAFC) use alcohol without reforming.

6.2 DIRECT CARBON FUEL CELLS

In direct carbon fuel cells (DCFC), solid carbon is used directly in the anode, without an intermediate gasification step. The thermodynamics of the reactions in a DCFC allow very high-efficiency conversion. With technology development into practical systems, it could have a significant impact on coal-based power generation.

6.3 POLYMER ELECTROLYTE FUEL CELL

Polymer electrolyte fuel cells (PEFCs) are being pursued for fuel cell vehicles (FCVs). As a consequence of the high interest in FCVs and hydrogen, the investment in PEFC over the past decade exceeds all other types of fuel cells combined.

The electrolyte in this fuel cell is an ion exchange membrane that is an excellent proton conductor. The only liquid in this fuel cell is water; thus, corrosion problems are minimal. Typically, carbon electrodes with platinum electrocatalyst are used for both anode and cathode, with either carbon or metal interconnections. Water management in the membrane is critical for efficient performance. The fuel cell must operate under conditions where the by-product water does not evaporate faster than it is produced because the membrane must be hydrated. Due to the limitation on the operating temperature imposed by the polymer, and problems with water balance, a H_2-rich gas with minimal or no CO is used. High catalyst loading is required for both the anode and cathode. Extensive fuel processing is required with other fuels, as the anode is easily poisoned by even trace levels of CO, sulfur species, and halogens.

The PEFC has a solid electrolyte which provides excellent resistance to gas crossover. The low operating temperature of the PEFC allows rapid start-up and, with the absence of corrosive cell constituents, avoids the need for exotic materials required in other fuel cell types. Test results have demonstrated that PEFCs are capable of high current densities over 2 kW/l. The PEFC lends itself particularly to situations where pure hydrogen can be used as a fuel.

However, the low and narrow operating temperature range makes thermal management problematic. Similarly, the use of the rejected heat for cogeneration or in bottoming cycles is also made difficult. Water management is another significant challenge. PEFCs are quite sensitive to poisoning by trace levels of contaminants including CO, sulfur species, and ammonia. If hydrocarbon fuels are used, the extensive fuel processing has negative impacts on system size, complexity, efficiency, and system cost. For hydrogen PEFCs the need for a hydrogen infrastructure to be developed poses a barrier to commercialization.

6.4 ALKALINE FUEL CELL

The AFC was one of the first modern fuel cells to be developed, beginning in 1960. The application at that time was to provide on-board electric power for the Apollo space vehicle. The AFC has enjoyed considerable success in space applications, but its terrestrial application has been challenged by its sensitivity to CO_2.

The electrolyte in this fuel cell is either concentrated KOH in fuel cells operated at high temperature ($\sim$250°C), or less concentrated KOH for lower temperature ($<$120°C) operation. The electrolyte is retained in a matrix, and a wide range of electrocatalysts can be used. The fuel supply is limited to nonreactive constituents except for hydrogen. CO is a poison, and CO_2 will react with the KOH to form K_2CO_3, thus altering the electrolyte. Even the small amount of CO_2 in air must be considered a potential poison for the alkaline cell. Generally, hydrogen is considered as the preferred fuel, although some direct carbon fuel cells use alkaline electrolytes.

Desirable attributes of the AFC include its excellent performance on hydrogen and oxygen compared to other fuel cells. The sensitivity of the electrolyte to CO_2 requires the use of pure H_2 as a fuel. As a consequence, the use of a reformer would require a highly effective CO and CO_2 removal system. In addition, if ambient air is used as the oxidant, the CO_2 in the air must be removed. This requirement imposes a significant impact on the size and cost of the system.

6.5 PHOSPHORIC ACID FUEL CELL

Concentrated phosphoric acid is the electrolyte in this fuel cell, which typically operates at 150–220°C. At lower temperatures, phosphoric acid is a poor ionic conductor, and CO poisoning of the Pt electrocatalyst in the anode becomes severe. The relative stability of concentrated phosphoric acid is high compared to other common acids; consequently, the PAFC is capable of operating at the high acid temperature range (100–220°C). In addition, the use of concentrated acid minimizes the water vapor pressure, making water management relatively simple. The matrix most commonly used to retain the acid is silicon carbide, and the electrocatalyst in both the anode and cathode is Pt.

PAFC is mostly developed for stationary applications. Both in the USA and Japan, hundreds of PAFC systems were produced, and used in field tests and demonstrations. It remains one of the few fuel cell systems that are available for purchase. Recently, however, development of PAFC had slowed down in favor of PEFC, which is thought to have better cost potential.

PAFCs are less sensitive to CO than PEFCs and AFCs. The operating temperature is still low enough to allow the use of common construction materials, at least in the BOP components. The operating temperature also provides considerable design flexibility for thermal management. PAFCs have demonstrated system efficiencies of 37–42% (based on LHV of natural gas fuel), which is higher than most PEFC systems, but lower than many of the SOFC and MCFC systems. The waste heat from PAFC can be used in most commercial and industrial cogeneration applications, and would allow a bottoming cycle. Cathode-side oxygen reduction is slow, and requires the use of a Platinum catalyst. PAFCs require extensive fuel processing, typically including a water gas shift reactor to achieve good performance. The highly corrosive nature of phosphoric acid requires the use of expensive materials in the stack.

6.6 MOLTEN CARBONATE FUEL CELL

The electrolyte in this fuel cell is usually a combination of alkali carbonates, which is retained in a ceramic matrix of $LiAlO_2$. The fuel cell operates at 600–700°C where the alkali carbonates form a highly conductive molten salt, with carbonate ions providing ionic conduction. At the high operating temperatures in MCFC, Ni (anode) and nickel oxide (cathode) are adequate to promote reaction. Noble metals are not required for operation, and many common hydrocarbon fuels can be reformed internally. The focus of MCFC development has been stationary and marine applications, where the relatively large size and weight of MCFC and slow start-up time are not an issue. MCFCs are under development for use with a wide range of conventional and renewable fuels. After the PAFC, MCFCs have been demonstrated most extensively in stationary applications, with many demonstration projects either under way or completed. While the number of MCFC developers and the amount of investment has declined over the last decade,, development and demonstrations continue.

The relatively high operating temperature of the MCFC (650°C) has several benefits. No expensive electrocatalysts are needed as the nickel electrodes provide sufficient activity, and both CO and certain hydrocarbons are fuels for the MCFC, as they are converted to hydrogen within the stack, simplifying the BOP and bringing the system efficiency up to the high 40s and low 50s. The high-temperature waste heat allows the use of a bottoming cycle to further boost the system efficiency.

The main challenge for MCFCs is the very corrosive and mobile electrolyte, which requires the use of nickel and high-grade stainless steel as the cell hardware. The higher temperatures promote material problems, impacting on mechanical stability and stack life.

6.7 SOLID OXIDE FUEL CELL

The electrolyte in this fuel cell is a solid, nonporous metal oxide, usually Y_2O_3-stabilized ZrO_2. The cell operates at 600–1000°C where ionic conduction by oxygen ions takes place. Typically, the anode is Co-ZrO2 or Ni-ZrO2 cermet, and the cathode is Sr-doped LaMnO3. The recent development of thin-electrolyte cells with improved cathodes has allowed a reduction in operating temperature to 650–850°C. Some developers are attempting to push SOFC operating temperatures even lower. Over the past decade, this has allowed the development of compact and high-performance SOFCs, which utilize relatively low-cost construction materials. Concerted stack development efforts through the US DOE Solid State Energy Conversion Alliance (SECA) program, have considerably advanced the knowledge and development of thin-electrolyte planar SOFCs. SOFCs are now considered for a wide range of applications, including stationary power generation, mobile power, auxiliary power for vehicles, and specialty applications.

The SOFC has been in development for a long time. Because the electrolyte is solid, the cell can be cast into various shapes, such as tubular, planar, or monolithic. The solid ceramic construction of the unit cell alleviates any corrosion problems in the cell. The solid electrolyte also allows precise engineering of the three-phase boundary and avoids electrolyte movement or flooding in the electrodes. The kinetics of the cell is relatively fast, and CO is a directly useable fuel as it is in the MCFC. There is no requirement for CO_2 at the cathode as with the MCFC. The materials used in SOFC are modest in cost. Thin-electrolyte planar SOFC unit cells have been demonstrated to be cable of power densities close to those achieved with PEFC. As with the MCFC, the high operating temperature allows use of most of the waste heat for cogeneration or in bottoming cycles. Efficiencies ranging from around 40% for a simple cycle, to over 50% for a hybrid system application have been demonstrated, and the potential for over 60% efficiency exists as it does for MCFC.

The high temperature of the SOFC has its drawbacks. There are thermal expansion mismatches among materials, and sealing between cells is difficult in the flat plate configurations. The high operating temperature places severe constraints on materials selection and results in difficult fabrication processes. Corrosion of metal stack components, such as the interconnections, is a challenge. These factors limit stack-level power density and thermal cycling and stack life.

7. Fuel Cell Performance

This section contains a brief description of the chemical and thermodynamic relations governing fuel cells and how operating conditions affect their performance. Understanding the impacts of variables such as temperature, pressure, and gas constituents on performance allows fuel cell developers to optimize their design of the modular units and to maximize systems application performance. To understand the operation of a fuel cell one needs to define its ideal performance. Once the ideal performance is determined, losses arising from nonideal behavior can be calculated and then deducted from the ideal performance to describe the actual operation.

7.1 THE ROLE OF GIBBS FREE ENERGY AND NERNST POTENTIAL

The maximum electrical work (W_{el}) obtainable in a fuel cell operating at constant temperature and pressure is given by the change in Gibbs free energy (ΔG) of the electrochemical reaction in Eq. (1):

$$W_{el} = \Delta G = -n\boldsymbol{F}\,E \tag{1}$$

where n is the number of electrons participating in the reaction, F is Faraday's constant (96,487 coulombs/g-mole electron), and E is the ideal potential of the cell.

The Gibbs free energy change is also given by the following state function:

$$\Delta G = \Delta H - T\Delta S \tag{2}$$

where ΔH is the enthalpy change and ΔS is the entropy change. The total thermal energy available is ΔH. The available free energy is equal to the enthalpy change less the quantity $T\Delta S$ which represents the unavailable energy resulting from the entropy change within the system.

The amount of heat that is produced by a fuel cell operating reversibly is $T\Delta S$. Reactions in fuel cells that have negative entropy change generate heat (such as hydrogen oxidation), while those with positive entropy change (such as direct solid carbon oxidation) may extract heat from their surroundings if the irreversible generation of heat is smaller than the reversible absorption of heat.

The standard state Gibbs free energy change of reaction is a function of partial molar Gibbs free energy at a given temperature. This potential can be computed from the heat capacities as a function of temperature and from values of changes in entropy and enthalpy at a reference temperature.

7.2   IDEAL PERFORMANCE

The Nernst potential E, gives the ideal open circuit cell potential. This potential sets the upper limit or maximum performance achievable by a fuel cell.

The Nernst equation for a SOFC is presented here as an illustration in Eqs. (3–6):

Anode reaction: $H_2 + O^= \rightarrow H_2O + 2e^-$ (3)

$CO + O^= \rightarrow CO_2 + 2e^-$ (4)

$CH_4 + 4O^= \rightarrow 2H_2O + CO_2 + 8e^-$ (5)

Cathode reaction: $\frac{1}{2} O_2 + 2e^- \rightarrow O^=$ (6)

The Nernst equation defines a relationship between the ideal standard potential for the cell reaction and the ideal equilibrium potential at other partial pressures of reactants and products. For the overall cell reaction, the cell potential increases with an increase in the partial pressure of reactants and a decrease in the partial pressure of products. For example, for the hydrogen reaction, the ideal cell potential at a given temperature can be increased by operating at higher reactant pressures.

7.3 CELL ENERGY BALANCE

The energy balance around a fuel cell is based on the energy absorbing/ releasing processes that occur in the cell. The energy balance varies for the different types of cells because of differences in reactions that occur according to the cell type. The cell energy balance states that the enthalpy flow of reactants entering the cell will equal the enthalpy flow of the products leaving the cell plus the sum of the net heat generated by physical and chemical processes within the cell, the DC power output from the cell and the heat loss from the cell to its surroundings.

7.4 CELL EFFICIENCY

The thermal efficiency of a fuel conversion device is defined as the amount of useful energy produced relative to the change in enthalpy between the product and feed streams. Fuel cells convert chemical energy directly into electrical energy. Theoretically, in a fuel cell the change in Gibbs free energy ΔG of the reaction is available as useful electric energy at the temperature of conversion. The ideal efficiency of a fuel cell is defined in Eq. (7):

$$\eta_{ideal} = \frac{\Delta G}{\Delta H} \qquad (7)$$

The thermal energy of an ideal fuel cell operating reversibly on hydrogen and oxygen at standard conditions is given in Eq. (8)

$$\eta_{ideal} = \frac{237.1}{285.8} = 0.83 \qquad (8)$$

The actual cell potential is decreased due to irreversible losses. These losses are caused by:

- Activation-related losses
- Ohmic losses
- Mass-transport related losses

7.5   FUEL CELL PERFORMANCE VARIABLES

The fuel cell performance is a function of temperature, pressure, gas
composition, reactant utilization, current density, cell design, impurities, and
cell life. Increasing current density influences activation, ohmic and concentration losses which occur as current is changed. At high current densities,
diffusion of enough reactants to the reaction site is limited, resulting in a drop
in the performance through reactant starvation. Ohmic losses govern normal
fuel cell operation. Ohmic loss and voltage change are a direct function of
current, which is given by the current density multiplied by the cell area.
Figure 3 contains a simplified illustration of the voltage/power relationship.

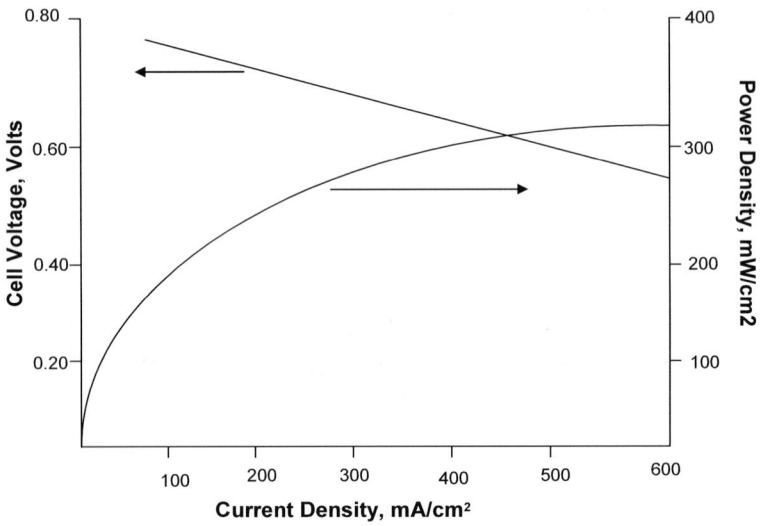

Figure 3. Voltage/power relationship

8.   Worldwide Fuel Cell Programs

Fuel cells are being developed to deliver efficiencies above those achievable
from a Carnot cycle system. Global interest in achieving higher performance
from power systems, anticipated shortages of fossil fuels, and concerns about
greenhouse gas emissions have encouraged the USA, Europe, Japan, and other
countries to accelerate the development of fuel cells. In the USA, the Department of Defense and National Aeronautics and Space Administration have
sponsored PFC and AFC fuel cells programs for space and classified applications. NASA is currently supporting the development of SOFC fuel cells for
auxiliary power unit applications[2].

The US Department of Energy (DOE) have initiated and sponsored MCFC and SOFC programs to enhance efficiency and reduce pollution for stationary and transportation applications. The SECA is currently in its 7th year. DOE held its 6th SECA annual workshop[3] in April, 2005. This program has given the go-ahead for selected industry projects in stationary and transportation applications. In addition, concurrent core technology research is also being sponsored by the DOE. In August 2005, DOE selected two industrial companies to demonstrate coal-based SOFC systems at approximately 100 MW for central power applications. These demonstrations are planned for 2012. The Hydrogen Fuel Initiative project of DOE is focused on accelerating the development of fuel cells utilizing hydrogen as the fuel for the transportation sector[4]. A simplified fuel cell hybrid system is presented in Figure 4[5,6].

Solid Oxide Fuel Cell Hybrid

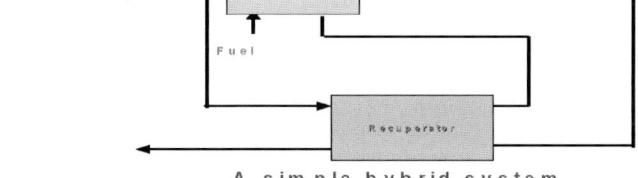

A simple hybrid system

Figure 4. A simplified solid oxide fuel cell hybrid system

Dr. Wolfgang Winkler of Germany provided the status information on fuel cell programs in Europe[7]. He graciously agreed to provide the information presented here. Dr. Winkler serves as the principal for the European Union (EU) helping to establish solid oxide and MCFC systems, operating on coal, biomass and other renewable energy derived fuels. The 3-year EU MCFC/MTG Hybrid Power Plant project aims to provide low cost production systems. The main objective of this project is to develop and demonstrate a small-sized hybrid system with a combination of MCFC technology and Micro Gas Turbines (MTG).

A similar 3-year demonstration project on pressurized IP-SOFC is also being sponsored by the EU. The project's main objectives are to develop a pressurized 10 kWe fuel cell (FC) block with integral internal reforming and to operate it with a hydrocarbon fuel representative of operation using natural gas. The European Community support of fuel cell and hydrogen R&D programs

approached approximately €250 million in 2005. SOFC and PEFC systems are
being developed for transportation applications planned by the EU. Figure 5
presents a schematic of future SOFC hybrids and coal use.

Future SOFC hybrids and coal use

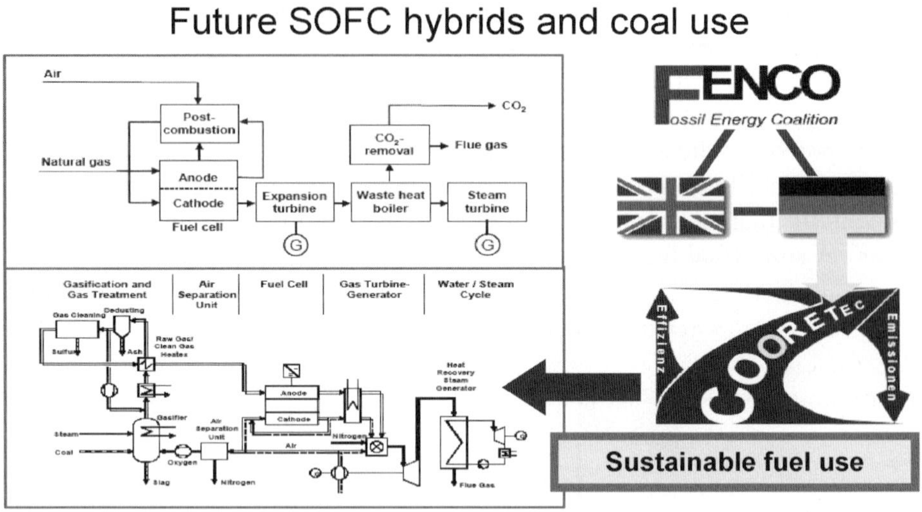

Figure 5. Future SOFC Hybrids and Coal Programs at EU

Fuel cells are being used for distributed generation (DG) while reducing
greenhouse gas emissions. Currently, these are small-in-size for the commercial
and small-scale industrial sectors. A comparison of fuel cells with other power
sources in DG mode is presented in Table 2. Table 3[8] contains a comparison of
SO_2, NO_x and CO_2 emissions from various power generation systems.

TABLE 2. Performance comparisons of selected distributed generation systems

Type	Size	Efficiency (%)
SOFC	5 KW – 3 MW	45–65
Proton exchange membrane	<1 KW–1 MW	34–36
Reciprocating engines	50 KW–6 MW	33–37
Micro turbines	10 KW– 00 KW	20–30
PAFC	50 KW–1 MW	40
Photovoltaics	1 KW–1 MW	NA
Wind turbines	150 KW–500 KW	NA

TABLE 3. Comparison of fuel cell hybrids emissions with other systems

Type	NO_2 (LBS/MWH)	SO_2 (LBS/MWH)	CO_2 (LBS/MWH)
Fuel cell hybrids	Negligible	Negligible	<750
Fuel cells	Negligible	Negligible	<1250
Combined cycle gt	<1	<1	<1000
Microturbine	1	1	<2000
US fossil fuel plant (2000)	>4	>8	>2000

Acknowledgment

The author wishes to acknowledge the US Department of Energy for the published information used in the generation of this document. The author is also grateful to Dr. Wolfgang Winkler of Germany for providing pertinent information on the EU fuel cell projects.

References

1. US Department of Energy. Fuel Cell Handbook Seventh Edition. November 2004.
2. A. Liang, Emerging Fuel Cell Developments at NASA for Aircraft Applications. 4th SECA Conference. 2003.
3. US Department of Energy. SECA 6th Annual Workshop. April 2005.
4. US DOE. Clean Coal Power Conference. November 2005
5. G. Agnew and Sy A. Ali, The Future, Right Here ... Right Now. The Development Programme for Pressurized Hybrid Fuel Cell Systems. Grove Fuel Cell Symposium, London. September 2001.
6. A. Layne and Sy Ali, Distributed Generation Market Potential for Fuel Cell Hybrids. PowerGen. Europe. July 2001.
7. W. Winkler, International Programs Europe. 6th ICEPAG Colloquium on Hybrid Fuel Cell Technologies. September 2005.
8. M. Williams. DOE. 6th ICEPAG Colloquium on Hybrid Fuel Cell Technologies. September 2005.

ACTIVE ELECTRIC CONTROL OF EMISSIONS FROM SWIRLING COMBUSTION

I. BARMINA, A. DESNICKIS, A. MEIJERE, M. ZAKE[*]
Institute of Physics, University of Latvia, Salaspils-1, Miera 32, LV-2169, Latvia

Abstract. The DC electric field effect on the nonpremixed swirling flame dynamics, flame temperature and composition profiles is studied experimentally with the aim of establishing the feasibility of electric control of swirling combustion and formation of polluting emissions. The results testify that the electric field-enhanced mass transfer inside the internal recirculation zone disturbs the balance between the axial fuel flows and the reverse axial motion of the combustion products. The DC field effect on the fuel/air mixing rate and the residence time of reactions is responsible for soot growth and carbon burnout, thereby providing electric control of swirling combustion and formation of polluting emissions. The influence of applied voltage and polarity of the axially inserted electrode on the swirling combustion and composition of polluting emissions is shown and analyzed.

Keywords: swirling combustion, electric control, soot formation, carbon capture

1. Introduction

The emission of greenhouse gases results in global warming, which leads to an increase in the average global surface temperature, a decrease in snow cover and a rise in sea levels. .During the 20th century, human activities have significantly increased the concentration of greenhouse gases, mainly, CO_2

---

[*] To whom correspondence should be addressed. Maija Zake, Institute of Physics, University of Latvia, Miera Street 32, Salaspils-1, LV-2169, Latvia; e-mail: mzfi@sal.lv

N. Syred and A. Khalatov (eds.), Advanced Combustion and Aerothermal Technologies, 405–412.

released into the atmosphere through the combustion of fossil fuels (oil, natural gas and coal). There are various methods of controlling carbon emissions during the combustion of fossil fuels[1,2]. One of them, the modification of primary fuel, resulting in fuel decarbonization, is a very promising technique[2]. Both elemental carbon and hydrogen, produced by thermal decomposition of hydrocarbons, have various technical applications[3] (e.g., the production of paints, fuel cells, etc.). Therefore, electric control of the nonpremixed swirling flame flow has intriguing consequences for industry[4]. The electric force can be used to control the swirl-induced recirculation of the products, providing DC field control of the mixing rate of the flame compounds and determining the combustion dynamics, fuel burnout, the residence time of soot formation, carbon capture and sequestration and composition of the products. Previous investigations have shown that reducing the rate of fuel/air mixing downstream of the nonpremixed strongly swirled flame flow can create favorable conditions for soot nucleation and production of nanoparticles of elemental carbon, whereas the electric field control enables the separation of carbon particles from the fuel-rich flame core[4]. The lower the amount of carbon burnt, the lower the greenhouse carbon emissions.

The observed field-induced variations of nonpremixed swirling combustion and the composition of polluting emissions require minimal electric power input (<1 W). However, they are very sensitive to variations in the applied voltage, field configuration and polarity of the axially inserted electrode. Therefore, detailed research into the primary mechanisms involved is necessary to develop towards the stable and repeatable control of nonpremixed swirling combustion, flame carbon and polluting NO_x emissions.

2. Experimental Apparatus

In the present study we use a swirl burner[4] with an axial fuel supply of $q_f \approx 0.5$–0.7 l/min and constant tangential air supply of $q_{air} \approx 17$ l/min, providing fuel-lean and stoichiometric combustion conditions at the burner outlet. For the given burner geometry and stoichiometric propane/air supply into the burner, the axial velocity of the swirling airflow (u) at the annular nozzle outlet does not exceed $u \approx 1$ m/s, while the azimuthal velocity component approaches $w \approx 5.9$ m/s. Hence, the swirling airflow has a relatively high swirl number ($S \approx 5$), determining the formation of low-temperature staged fuel combustion[5] with a reduced propane-air mixing rate downstream the swirling flame flow.

Downstream of the swirling flame flow, a steel electrode of 4 mm diameter 250 mm length is axially inserted. A high DC voltage source provides the required voltage between the axially inserted electrode and the burner surface. The bias voltage of the electrode can be varied in the range −3 to +3 kV. To restrict the effects of Joule dissipation and corona discharge on the combustion dynamics, the ion current in the flame is limited to 0.5 mA in this study. The field effect on the combustion dynamic is studied downstream the flame channel flow, when a swirling flame flow at the burner outlet is injected into a channel of diameter $D = 40$ mm and length $L = 335$ mm, composed of five sections. The investigation of the electric field effect on the combustion dynamics includes local measurements of the flame velocity, temperature and composition, using a Pitot tube and Pt/Pt-Rh thermocouples, and flame sampling, coupled with spectral analysis of the gaseous samples. The resulting electric field effect on the formation of polluting emissions is estimated from the measurements of the composition of the products. These measurements are carried out using a gas analyzer Testo 350XL. The electric field effect on the soot formation and carbon sequestration is valued from the gravimetric measurements of the sequestered soot mass.

3. Results and Discussion

Previous investigations confirm[4] that for the given swirl burner geometry, the nonpremixed swirling combustion develops at a reduced propane-air mixing rate downstream of the flame channel flow. Under such conditions a typical feature of the flame flow is the formation of low-temperature staged fuel combustion, developing at a relatively large distance from the burner outlet; up to $L/D \approx 7$. Local measurements of the flame temperature and composition have shown[4] that the swirling flame flow has a relatively cold ($T \approx 1,000$ K) fuel-rich ($\alpha < 1$) flame core ($r/R < 0.5$) surrounded by the annular reaction zone, where the peak value of the flame temperature approaches $T \approx 1,500$–$1,750$ K. Downstream of the fuel-rich flame core ($r/R < 0.5$) the formation of hydrocarbon species is detected, such as C_2H_2, CH_4, C_2H_4, etc., which play an important role in the primary reactions as ingredients in the formation of aromatics (PAH), soot nucleation and particle growth (Figure 1). The formation of the main products (CO, CO_2, and NO_2) is revealed downstream of the annular reaction zone and peaks at $r/R \approx 0.5$–0.6.

The formation of nonpremixed swirling flame velocity profiles is accompanied by recirculation of the products that counterbalances and gradually slows down the axial fuel flow from the burner outlet. Under undisturbed swirling flame conditions ($U = 0$) the balance between the axial fuel flow and the reverse axial flow of the products is found at a distance of $L \approx 40$–50 mm from the burner outlet, where the axial fuel flow from the burner outlet actually stops (stagnates). Since the residence time of soot growth is highly influenced by the axial fuel flow rate ($t_{res} \approx u$–1), the residence time of soot growth rapidly increases, when the recirculation of the products slows down the axial fuel flow from the burner outlet. Hence, at this distance ($L \approx 40$–50 mm) from the burner outlet, an intensive nucleation of the soot precursors is observed through the increase of the luminous flame emission in the visible spectrum range (550–600 nm), resulting in the formation and growth of primary soot nanoparticles of diameter $d \approx 20$–30 nm. Thermophoretic effects control the transport of primary soot nanoparticles to the surface of the central electrode, resulting in a mass growth of sequestered soot in the undisturbed ($U = 0$) swirling flame conditions.

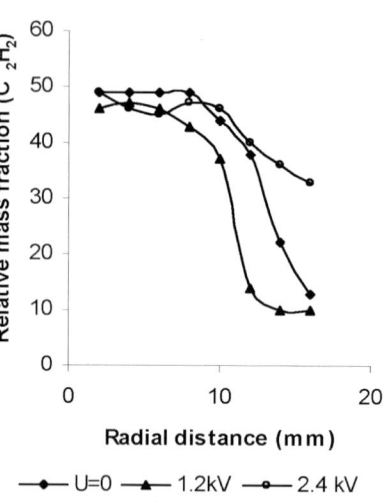

Figure 1. Typical shape of the relative mass fraction profiles of the formation of hydrocarbon fragments (C_2H_2) under the conditions of nonpremixed swirling combustion ($L = 30$ mm).

When the electric body force F is applied to the swirling flame flow, the field-enhanced heat and mass transfer processes vary the swirling flame flow field by enhancing or counterbalancing the axial fuel flow motion from the burner outlet and so varying the shape of the flame velocity profiles with transformation between spiral and bubble type vortex breakdown (Figure 2).

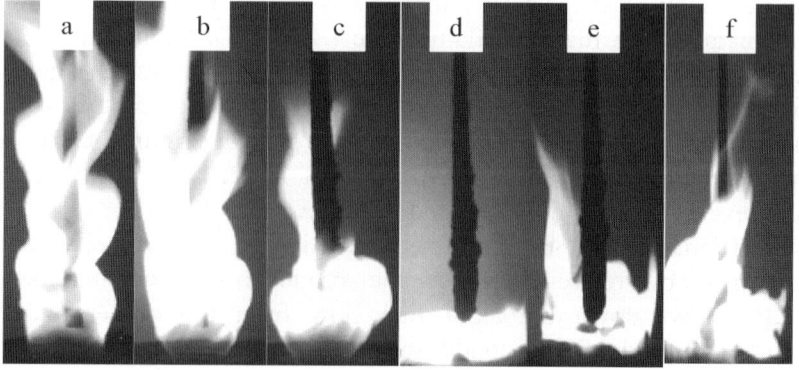

Figure 2. The electric field effect on the flame shape and length: (a) $U = 0$; (b) $U = +1.2$ kV; (c) $U = +1.5$ kV; (d) $U = +2.1$ kV; (e) $U = +3$ kV; (f) $U = -0.9$ kV.

In fact, by increasing the positive bias voltage of the central electrode, the field strengthens the reverse axial motion of the gaseous species and gradually slows down the axial fuel flow motion from the burner outlet (Figure 3), so increasing the residence time of soot formation and enhancing the growth and coagulation of the primary soot nanoparticles. The reverse field effect on the swirling flame flow is revealed for the negative bias voltage of the central electrode, when the field intensifies the axial fuel flow motion, decreasing the residence time of soot growth (Figure 3).

As soon as a soot particle is in contact with the weakly ionized flame flow ($n_e \approx n_i \approx 10^8-10^{10}$ cm^{-3}), is acquires a charge determined by the flux of either positively or negatively charged species (mostly electrons and positive ions) to the surface

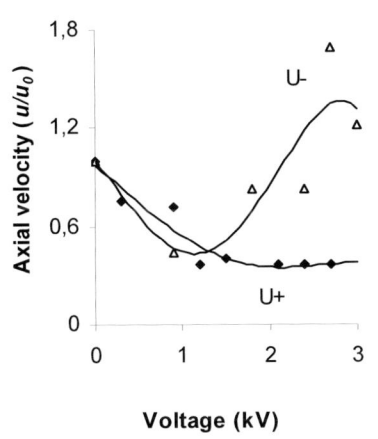

Figure 3. The electric field-enhanced relative variations of the axial velocity at the burner outlet (L = 30 mm) at the varying polarity and bias voltage of the central electrode.

of the particle. Once charged, the electric field promotes the electrophoretic transport of the particle to the surface of the central electrode and the surface growth of sequestered carbon particles proceeds (Figure 4a). Note that a positive bias voltage of the central electrode produces a stronger field effect on the carbon capture and sequestration than a negative bias voltage, when the field-enhanced growth and the electrophoretic transport of soot nanoparticles to the surface of the central electrode are the major mechanisms that determine the mass growth of sequestered soot. A correlating decrease in mass fraction of the main hydrocarbon species (C_2H_2, C_2H_4, CH_4) inside the fuel-rich flame core in a range of $U < 1.2$ kV shows that these species can be recognized as the main soot precursors (Figure 1). The peak value of the sequestered soot mass is registered at $U \approx +0.9-1.2$ kV – under conditions, when the field-enhanced reverse axial motion of the products completely locks up the axial fuel flow from the burner outlet.

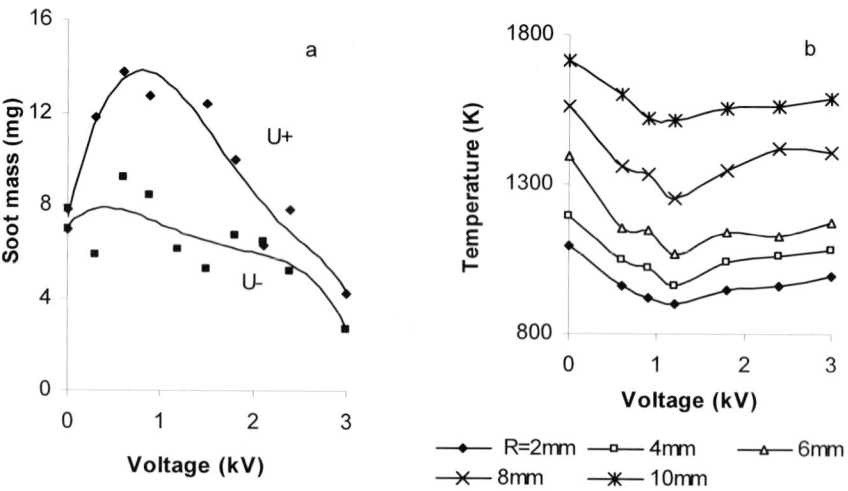

Figure 4. The effect of applied voltage and polarity of the central electrode on the mass growth of sequestered soot (a) and local variations of the flame temperature ($L = 30$ mm) (b)

Since the flame temperature is very sensitive to the amount of carbon burned, the field-enhanced mass growth of sequestered carbon results in a correlating decrease in the temperature inside both the fuel-rich flame core and the flame reaction zone (Figure 4b). As one can see, the minimum value of the temperature corresponds to the peak value of the sequestered soot mass (Figure 4a and b) and starts to increase in strong fields ($U > 1.2$ kV), where the field-enhanced recirculation actually discontinues the axial fuel flow from the burner outlet and advances the radial mass transfer of the hydrocarbon species by increasing the mass fraction of these species along the outside part of the flame reaction zone ($R > 10$–12 mm) (Figure 1). The field-enhanced radial mass transfer of the fuel compounds results in a more intensive fuel/air mixing, completing the fuel burnout, thereby increasing the flame temperature and efficiency of the fuel burnout. Conversely, it decreases the mass fraction of the soot precursors inside the flame core by restricting soot formation with a correlating decrease of the sequestered soot mass and visible flame length.

Measurements of the composition of the products, released during the swirling propane/air combustion have shown that the field-enhanced carbon capture and sequestration provide correlating variations in mass fraction of the polluting emission. In fact, by increasing the mass of sequestered carbon, the mass fraction of the greenhouse carbon emissions (CO_2) in the products can be decreased by 6–11% (Figure 5a), while the mass fraction of the free oxygen in the products increases by 14% (Figure 5a), and a field-enhanced 2.5% decrease in the fuel burnout efficiency is registered. This correlation testifies that the field-enhanced carbon capture and sequestration can be used as a tool to control the fuel burnout and the formation of greenhouse carbon emissions. Moreover, the correlating field-enhanced decrease in temperature downstream of the flame core and inside the flame reaction zone provides a corresponding decrease of the NO production (Figure 5b) that finally brings about a 16–20% decrease in the mass fraction of NO_x emissions. In fact, by increasing the positive bias voltage of the central electrode to $U \approx 0.9–1.2$ kV, the mass fraction of NO_x in the products can be lowered below 40 ppm at 7% CO_2 in the products (Figure 5a). Hence, cleaner fuel combustion can be achieved.

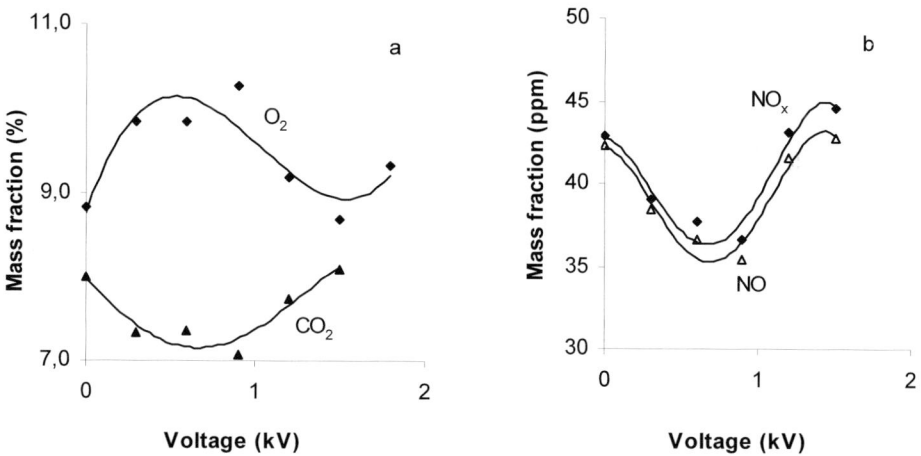

Figure 5. The electric field effect on the mass fraction of free oxygen and carbon emissions in the products (a) and the formation of polluting NO and NO_x emissions (b).

References

1. H.J. Herzog, E.E. Adams, M. Akai, G. Alendal, L. Golmen, et al., Update on the international experiment on CO_2 ocean sequestration, *Proc. the 5th International Conference on Greenhouse Gas Control*, 399–405 (2000).
2. R. Sokolov, Fuels decarbonization and carbon sequestration: Report of a Workshop. PU/CEES Report N302, Center for Energy and Environmental Studies, School of Engineering and Applied Science, Princeton University, Princeton, NJ 08544, 1–47 (1997).

3. M. Steinberg, J.F. Cooper, N. Cherepy, High efficiency carbon and hydrogen fuel cells for CO_2 mitigating power, *Report UCRL-JC-146774, Lawrence Livermore National Laboratory,* Livermore, CA, 1–4 (2002).
4. M. Zake, I. Barmina, D. Turlajs, M. Lubāne, A. Krūmiņa , Swirling Flame, Part 2, Electric field effect on the soot formation and greenhouse emissions, *Magnetohydrodynamics*, **40**(2), 183–202 (2004).

INFLUENCE OF GAS DISCHARGE PLASMA ON COMBUSTION OF A HIGH-SPEED HYDROCARBON FLOW

V. M. SHIBKOV[*], V. G. GROMOV, R. S. KONSTANTINOVSKIJ
*Physical Faculty of M.V.Lomonosov Moscow State University,
119992 Moscow, Russia*

Abstract. A direct current discharge, a transversal in relation to the gas flow pulse-periodic electrode discharge, a freely localized microwave discharge and a surface microwave discharge under condition of a high-speed flow are considered. It is shown, that all types of discharge result in a reliable ignition of hydrocarbon fuel. In order to determine the influence of different channels of energy transfer on the ignition of combustible mixtures in a high speed flow, the kinetic model of ignition of hydrocarbon–air mixtures was developed, taking into account the influence of the electric field on molecule dissociation and the creation of active radicals i.e., excited and charged particles (electrons, positive, and negative ions), under conditions of nonequilibrium plasma in the gas discharge. Mathematical modeling has revealed the strong influence of the reduced electric field on the induction period.

Keywords: ignition, combustion stabilization, high-speed hydrocarbon fuel flow, gas discharge, low-temperature plasma, numerical modeling, diagnostics

1. Introduction

Advances in aviation technology call for research and development aimed at creating new efficient means of reducing the ignition time and controlling the combustion of a high speed fuel flow. A new solution to these problems is the application of different gas discharges. This has given rise to a new field of research in plasma physics and supersonic plasma aerodynamics, which is currently progressing rapidly[1,2]. Specifically, it is conjectured that the use of a

---
[*]To whom correspondence should be addressed. Valery M. Shibkov, Moscow State University, Physical Faculty, 119992 Moscow, Russia. e-mail: shibkov@phys.msu.ru

N. Syred and A. Khalatov (eds.), Advanced Combustion and Aerothermal Technologies, 413–423.

nonequilibrium gas-discharge plasma will reduce the fuel ignition time in a high speed stream.

The mechanism of gas-phase oxidation of various combustible gases, including hydrocarbons and hydrogen, has been thoroughly investigated[3-10,] with the emphasis on their ignition mechanism. The great majority of publications in this field have dealt with factors determining the induction period preceding the ignition event. In recent decades, there have been several publications discussing the possibility of effectively controlling combustion processes by various physical means. For example, Semenov[11] studied the broadening of the inflammability range of the hydrogen–oxygen mixture under the action of short-wave radiation or oxygen atoms. Furthermore, it has been experimentally demonstrated that under the action of ultraviolet radiation (λ = 175 nm), the inflammability "peninsula" broadens and shifts to lower temperatures[12]. A number of studies, have suggested initiating ion-molecule and ion-atom reactions using low-temperature gas-discharge plasma[1,2]. In the same studies, plasma jets and laser radiation are discussed as possible ignitors for supersonic hydrocarbon streams. Numerical analysis and experimental studies of the ignition of H_2–O_2, H_2–air, and CH_4–O_2 mixtures with a nanosecond high-voltage discharge have revealed marked distinctions between the equilibrium and nonequilibrium excitations of the mixtures[13]. The effects of the initial concentration of free radicals (H and O atoms) and of the radiolysis rates of dihydrogen and dioxygen on the ignition limits of the stoichiometric hydrogen–oxygen mixture have been studied by numerical simulation[14]. Numerical simulation predicts that singlet oxygen $O_2(a^1\Delta_g)$ will cause an increase in the hydrogen–oxygen flame velocity[15]. The ignition of various hydrocarbon-containing combustible mixtures by laser heating or laser-induced breakdown has been studied as a function of gas pressure and laser wavelength[16-18]. The ignition of a high-speed propane-air flow by microwave discharges has been studied as a function of gas pressure, pulse duration, microwave power, equivalent ratio, etc.[19-26].

A cursory survey of the literature has demonstrated that there are numerous ways of intensifying the chain combustion of hydrocarbons. However, the ignition kinetics are not completely understood, even for the rather simple model system of hydrogen–oxygen under low-temperature gas-discharge plasma conditions, which are established at large values of the reduced electric field. The study of the ignition and combustion of hydrogen-containing mixtures under low-temperature plasma conditions is of importance from various standpoints: it is necessary to carry out both fundamental research in the mechanism and kinetics of atom-molecule reactions in a strong electric field and an analysis of a variety of applied problems, including the optimization of plasma chemical processes. One practical challenge is the development of the physical principles of combustion of high speed hydrocarbon–air flows. In this case, it is necessary to ensure a rapid space ignition of the high-velocity fuel flow. To do this, it is necessary to minimize the induction period. It is known that ignition of combustible gaseous mixtures can occur either through heating of the gas to high temperature (thermal autoignition), or as a result of radicals and active particles. Determining the mechanisms responsible for the ignition of nonequilibrium low-temperature plasma of the gas discharge, at high values of the reduced electric field, is one of the principal goals of the investigations.

The results of research into low-temperature nonequilibrium plasma in high-speed streams of a hydrocarbon–air fuel, which have been conducted at the Physical Faculty of the Moscow State University[1,2,19–31] during the last several years, are submitted in this paper.

2. Experimental Results

The experimental installation consisted of a vacuum chamber, a receiver of a high pressure of air, a receiver of a high pressure of propane, a system for mixing propane with air, a system for producing a high speed gas flow, two magnetron generators, two systems for delivering microwave power to the chamber, cylindrical and rectangular aerodynamic channels, two sources of high-voltage pulses, a synchronization unit, and a diagnostic system. The basic component of the experimental setup was an evacuated metal cylindrical chamber, which enabled high speed flow creation, and contained the combustion products. The inner diameter of the vacuum chamber was 1 m, and its length was 3 m. A high speed flow was produced by filling the vacuum chamber with air through a specially profiled Laval nozzle mounted on the outlet tube of an electromechanical valve. In our experiments, we used cylindrical and rectangular nozzles designed for a Mach number of $M = 2$.

The microwave source was a pulsed magnetron generator operating in the centimeter wavelength range. The parameters of the magnetron generator were as follows: the wavelength was $\lambda = 2.4$ cm, the pulsed microwave power was $W_p < 200$ kW, the pulse duration was $\tau = 1$–100 μs, and the period-to-pulse duration ratio was $Q = 1,000$. The magnetron was powered from a pulsed modulator with a partial discharge of the capacitive storage. Microwave power was delivered to the discharge chamber through a 9.5×19 mm^2 rectangular waveguide. The input microwave power was measured with the help of a directional coupler installed in the waveguide so that a fraction of microwave power was directed to the measuring arm containing an attenuator and a section with a crystal detector. All the components of the microwave transmission line were sealed. To avoid electric breakdowns inside the waveguide, it was filled with an insulating gas (SF_6) at a pressure of 4 atm. The vacuum system of the chamber enabled pressure variationover a wide range from 10^{-3} to 10^3 torr.

The ignition of a high speed stream was detected as a glow in the aerodynamic channel downstream of the discharge section. No glow was observed either when a gas discharge was generated in an air flow, a high-speed propane-air flow, although its properties (pulse duration, discharge current, electric field strength in the plasma, and the electric power deposited into discharge) were inappropriate for ignition, or when the mixture was far from stoichiometric. Induction time was simultaneously derived from different measurements: (1) the minimum microwave pulse duration, resulting in a glowing flame in the aerodynamic channel downstream of the discharge section; (2) the time taken by the intensity of the molecular band of the excited CH^* radical (the (0;0) band of the $A^2\Delta \rightarrow X^2\pi$ transition) with a wavelength of $\lambda = 431.5$ nm, to achieve the maximum growth rate; (3) the time taken by the signal from the double probe to achieve the maximum growth rate; and (4) the time taken by the current through the plane capacitor at the outlet of the aerodynamic channel to achieve the maximum growth rate. The ignition of a high speed flow was also detected as an increasing output signal from an acoustic noise meter, as a sharp change of general view of plasma radiation spectrum and as a sharp increase of gas temperature.

In the overwhelming majority of experimental and theoretical research devoted to the application of various types of gas discharges for ignition of fuel mixtures, the gas discharge was considered only as a source of thermal energy entered into system. But various gas ionization degrees are achieved for different types of gas discharge at the same applied specific power. Hence, the electric energy input is differently distributed according to the internal degrees of freedom of molecular gas. This distribution is strongly dependent on the reduced electric field which, in turn, is determined by the electrodynamics of the discharge. A different result can be received at the same power of the energy source. For example, for ignition of a high-speed propane-air flow with Mach number $M = 2$ a DC electrode discharge pulse-periodic transversal electrode discharge and freely localized and surface microwave discharges were used. Figure 1 presents the induction period as a function of the reduced electric field under these conditions. For example, the ignition of a propane-air mixture with the help of the surface microwave discharge was took place after a pulse duration of just $\tau = 5$–20 μs, whereas at the same conditions the pulse-periodic transversal electrode discharge only resulted in ignition for $\tau \geq 150$–200 μs. This indicates that there are more charged and active particles in a microwave discharge plasma than in a pulse-periodic electrode discharge plasma. Therefore from our point of view it is necessary to investigate in detail the influence of the charged and active particles on reducing the ignition delay and increasing the completeness of propane–air mixture combustion.

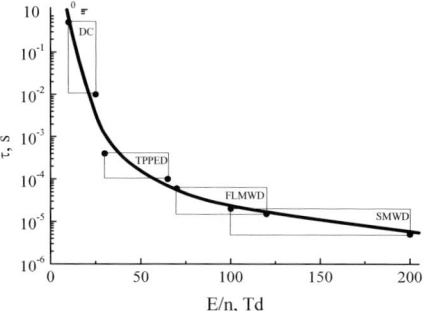

Figure 1. Induction period as a function of reduced electric field under conditions of a high-speed propane-airstream with Mach number $M = 2$. DC – direct current electrode discharge; TPPED – transversal pulse-periodic electrode discharge; FLMWD – freely localized microwave discharge; SMWD – surface microwave discharge

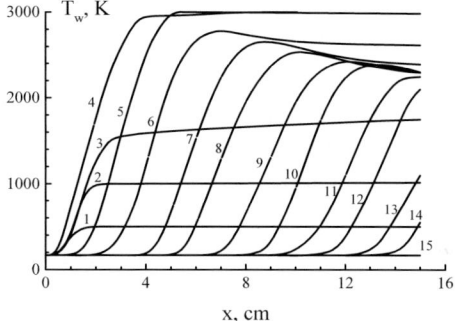

Figure 2. Time history of wall temperature distribution. t, μs: 1–25, 2–50, 3–75, 4–100, 5–150, 6–200, 7–250, 8–300, 9–350, 10–400, 11–450, 12–500

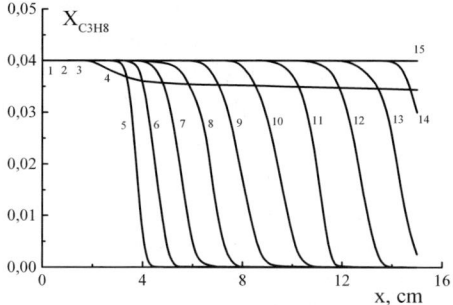

Figure 3. Time history of wall C_3H_8 mole fraction distribution. t, μs: 1–25, 2–50, 3–75, 4–100, 5–150, 6–200, 7–250, 8–300, 9–350, 10–400, 11–450, 12–500

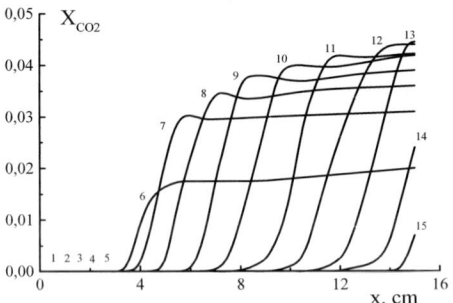

Figure 4. Time history of wall CO$_2$ mole fraction distribution. *t*, μs: 1–25, 2–50, 3–75, 4–100, 5–150, 6–200, 7–250, 8–300, 9–350, 10–400, 11–450, 12–500

3. Mathematical Modeling

A numerical model has been developed to study C$_3$H$_8$–air thermal autoignition by near wall heat deposition. Computations have been performed for supersonic flow of stoichiometric propane–air mixtures over a blunted flat plate 15 cm long and 0.02 cm thick for a free stream velocity of 519 m/s, pressure of 98 torr, and temperature of 167 K. The gas phase model, capable of dealing with 30 species (C$_3$H$_8$, O, H, O$_2$, N$_2$, H$_2$, CO, OH, H$_2$O, HO$_2$, H$_2$O$_2$, HCO, CO$_2$, CH, CH$_2$, CH$_3$, CH$_4$, C$_2$H, C$_2$H$_2$, C$_2$H$_3$, C$_2$H$_4$, C$_2$H$_5$, C$_2$H$_6$, C$_3$H$_5$, C$_3$H$_6$, i-C$_3$H$_7$, n-C$_3$H$_7$, CH$_2$O, CH$_2$OH, CH$_3$OH) and 70 chemical reactions, was employed to describe the ignition process. Adiabatic, full noncatalytic modeling of gas-wall thermal and chemical interactions was used. Calculations were carried out for conditions similar to experimental conditions created by a surface microwave discharge. It was assumed that heat is transferred uniformly during a time period of 100 μs with constant total power $P_d = 2.5$ kW/cm into a near wall region of dimensions 0.5 cm $\leq x \leq$ 15 cm, 0 cm $\leq y \leq$ 0.1 cm. The heat supply is inserted into an initially steady-state turbulent stream. Time history of predicted gas temperature and wall C$_3$H$_8$ and CO$_2$ mole fraction distributions along the plate surface are shown in Figure 2–4. The data presented indicate that for the considered conditions, ignition of the fuel mixture occurs post heat supply pulse at a distance of about 4 cm from the plate heading. Up until this time, the gas temperature near the wall reaches ~3,000 K. Subsequently, ignition front burning moves downstream at a rate of about 300 m/s. The temperature of the gas decreases and completeness of the combustion increases. In the gas mixture behind the front, the fraction of final combustion products H$_2$O and CO$_2$ rise, whereas the fraction of oxygen and the intermediate products of combustion drop. Numerical analysis has also shown that for a shorter heating pulse or a lower heating power, when the temperature does not reach ~3,000 K ignition does not occur. These results contradict the experimental data (executed at the same initial conditions) which show (see Figure 1), that ignition of a supersonic propane-airstream occurs much faster ($t = 5$–25 μs) and at a lower temperature ($T_g \sim 1{,}000$ K). This is connected to the fact that in calculations the surface microwave discharge was considered as the source of thermal energy inserted into the boundary layer, and plasma effects were not examined.

The kinetic model was developed to determine the influence of different channels on the ignition of combustible mixtures. As an example, mathematical modeling was performed for a motionless hydrogen–oxygen mixture. The model has been developed to define the influence of various channels on the ignition of a gas mixture, including 29 components and 241 direct and reverse reactions[24]. Such components as stable particles H_2, O_2, H, O, OH, HO_2, H_2O_2, H_2O, O_3, the excited molecules of oxygen $O_2(a)$, $O_2(b)$, positive ions O^+, O_2^+, O_4^+, H^+, H_2^+, H_3^+, H_5^+, OH^+, H_2O^+, H_3O^+, $O_3H_2^+$, negative ions O^-, O_2^-, O_3^-, O_4^-, H^-, OH^- and electrons (e) were considered during modeling. The system of equations describing the process of oxidation in such mixtures, including the equation for energy, the equation for concentration of particles and the equation of state have been combined to form the basis of the model, Eq. (1) and (2)

$$\frac{dH}{dt} = 0, \quad H = N \cdot \sum_{i=1}^{M_1} \gamma_i \left(h_{0i}[T_0] + \int_{T_0}^{T} C_{Pi} dT \right), \quad \frac{d\gamma_i}{dt} = G_i - \gamma_i \sum_{k=1}^{M_1} G_k, \qquad (1)$$

$$G_i = \sum_{q=1}^{M_2} \frac{\alpha_{iq}^- - \alpha_{iq}^+}{N} \left[R_q^+ - R_q^- \right], \quad R_q^{\pm} = k_q^{\pm} \prod_{j=1}^{n_q^{\pm}} \left(N\gamma_{n_{jq}^{\pm}} \right)^{\alpha_{jq}^{\pm}}, \quad \frac{dN}{dt} = N \sum_{k=1}^{M_1} G_k, \qquad (2)$$

where γ_i – mole fraction of corresponding component i; N [mol/cm^{-3}] – total concentration, c_{pi} – molar heat capacity of species i at $p = $ const, $h_{0i}[T_0]$ – molar enthalpy of formation of species i at $T_0 = 298$ K, M_1 – full number of components, $k_q^{\pm}$ – rate constant with participation of species i, M_2 – full number of such reactions, α_{iq}^+ and α_{iq}^- – stoichiometric coefficients. Constants for the rate of reactions are presented by Konstantinovskii et al.[24] and in the references quoted within.

Ignition of a hydrogen–oxygen mixture under conditions of nonequilibrium low-temperature plasma was modeled, taking into account dissociation of molecules and creation of active radicals and charged particles. Since the gas temperature increases at a rate of about 10^2 K/μs under discharge conditions, it was considered that at power-up the gas temperature instantly increases up to some initial value T_0. Calculations were carried out without considering losses of energy and diffusion of active particles in surrounding space. Calculations have been performed at an initial pressure $p = 0.1$ MPa for H_2–O_2 mixture for a range of initial gas ($T_0 = 800$–$1{,}200$ K) and electron ($T_e = 0.1$–1.6 eV) temperatures and stoichiometric fractions ϕ. The stoichiometric fraction ϕ is the ratio of consumption of hydrogen in mixture to its share in stoichiometric mixture. Modeling of the kinetics of autoignition of a hydrogen–oxygen mixture demonstrated that the ignition occurs at time ~1 ms at $T_0 = 900$ K At the initial stage ($t = 0$–1 ms), the gas temperature rises very slowly and then during the moment of ignition suddenly increases to its final value ~3,000 K.

The dependence of the ignition time of a hydrogen–oxygen mixture on hydrogen consumption was evaluated. It was shown, that the induction time depends on both the mixture ratio, and the initial gas temperature. At that, the ignition time decreases with increasing temperature for any ratio of components in a mixture. At the same time, the ignition time of a mixture grows at any temperature for both an increase and a reduction in the fraction of hydrogen in a mixture compared to $\phi = 1$.

The time dependence of the mole fraction at ignition of a stochiometric H_2–$O_2 = 2$–1 mixture under conditions of nonequilibrium gas discharge plasma is represented in Figure 5. The mechanism of ignition represented in Figure 5 is distinct from the mechanism of autoignition of the hydrogen–oxygen mixture. So, for example, the creation of radicals and active particles is essentially accelerated in the presence of the discharge due to quantitative and qualitative changes from the simple formation of active radicals as a result of the interaction of stable particles, to the creation of active radicals as a result of interaction with charged particles. As a consequence of this fact, the mechanism of ignition varies also. In this case the creation of active components occurs generally due to the discharge.

The rate of creation of radicals and charged particles increases with the growth of the reduced electric field E/n at electron temperature T_e. This essentially results in the reduction of the induction time. It was shown too, that with increase in electronic temperature, i.e., the reduced electric field, the ionization frequency sharply grows. At $T_0 = 900$ K the delay time of avalanche development also decreases from ~200 μs at $T_e = 1.35$ eV to ~0.5 μs at $T_e = 1.6$ eV. At $T_e = 1.35$ eV the concentration of metastable molecules $O_2(a^1\Delta_g)$, $O_2(b^1\Sigma_g^+)$, atoms H and O, and also radicals at stages of avalanche development increase by one order of magnitude, whereas at $T_e = 1.6$ eV the increase in the concentration of these particles reaches 5–6 orders of magnitude. Naturally, this results in the significant reduction of the induction period. Recombination and recharged collisions of ions, processes of attachment and detachment result in the establishment of a balance of concentration of the charged particles. Thus at a stage prior to ignition of a hydrogen–oxygen mixture, the H_5^+ and O_2^+ ions are the basic positive ions in the discharge, and the ion O_4^- is the basic negative ion.

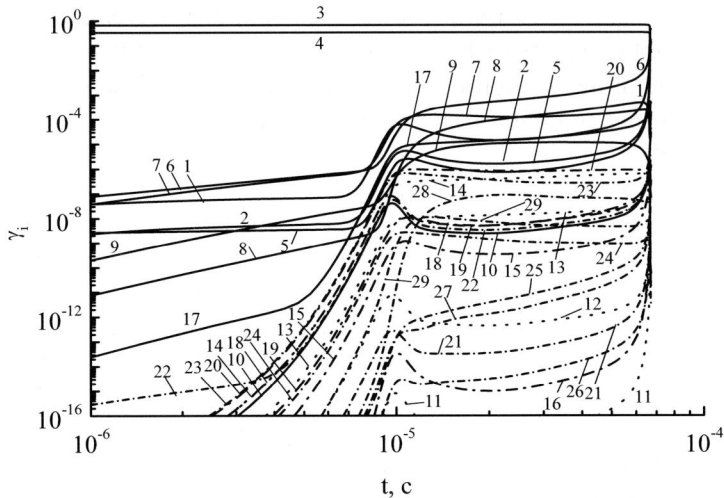

Figure 5. Time dependence of mole fractions of ignition of stochiometric H_2–O_2=2–1 mixture under condition of nonequilibrium plasma of gas discharge at electron temperature T_e = 1.4 eV, T_0 = 900 K, N_0 = 4.4·10^{-5} mole/cm^{-3}. 1 – H, 2 – O, 3 – H_2, 4 – O_2, 5 – OH, 6 – H_2O, 7 – HO_2, 8 – H_2O_2, 9 – O_3, 10 – e, 11 – H^+, 12 – H_2^+, 13 – H_3^+, 14 – H_5^+, 15 – O^-, 16 – O^+, 17 – $O_2(a^1\Delta_g)$, 18 – O_2^-, 19 – O_3^-, 20 – O_4^-, 21 – H^-, 22 – $O_2(b^1\Sigma_g^+)$, 23 – O_2^+, 24 – O_4^+, 25 – OH^+, 26 – H_2O^+, 27 – OH^-, 28 – H_3O^+, 29 – $O_3H_2^+$.

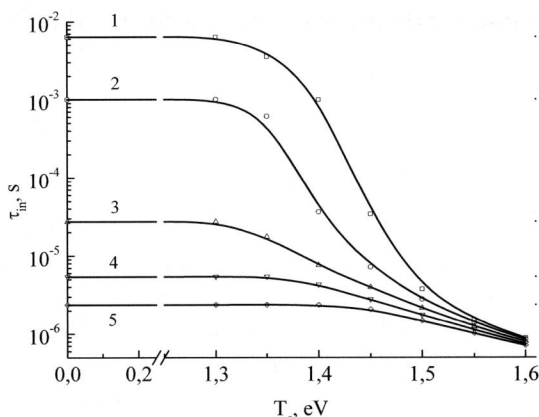

Figure 6. Ignition time of H_2–O_2 stoichiometric (67–33%) mixture vs electron temperature at p = 1 atm and at instant gas heating to various initial gas temperatures T_0, K: 1 – 800; 2 – 900; 3 – 1000; 4 – 1100; 5 – 1200.

The influence of gas discharge on the induction period of a stoichiometric (67–33%) mixture of H_2–O_2 is shown in Figure 6. The duration of the ignition delay, under conditions of the gas discharge, decreased to of the order of value associated with low gas temperature. Calculations have also shown that increasing the percentage of hydrogen in a H_2–O_2 mixture, causes a greater decrease in the ignition delay time compared to a stoichiometric mixture, whereas the influence of the gas discharge on the ignition of poor mixtures is less significant.

Acknowledgment

The work was supported by the Russian Foundation of Basic Research (grant #05-02-16532).

References

1. *International Space Planes and Hypersonic System and Technologies Conference. Workshop on Weakly Ionized Gases.* Proceedings AIAA, USA, Colorado-1997; Norfolk-1998, 1999; Anaheim-2001; Reno-2002, 2003, 2004, 2005, 2006.
2. *The International Workshops on Magneto- and Plasma Aerodynamics for Aerospace Applications.* Proceedings IVTAN, Russia, Moscow, 1999, 2000, 2001, 2002, 2003, 2005.
3. N.N. Semenov, *Tsepnye reaktsii* (Chain Reactions), Moscow: Nauka, 1986.
4. B. Lewis and G. Von Elbe, *Combustion, Flames and Explosions of Gases*, New York: Academic, 1961.
5. T.R. Coffee, *Combustion and Flame*, 1984, **55**(2), 161.
6. M. Frenklach and D.E. Bornside, *Combustion and Flame*, 1984, **56**(1), 1.
7. D.J. Seery and C.T. Bowman, *Combustion and Flame*, 1970, **14**, 37.
8. V.V. Azatyan, *Doctoral (Chem.) Dissertation*, Moscow: Inst. of Chemical Physics, 1978.
9. N.G. Dautov and A.M. Starik, *Kinetics and Catalysis*, 1997, **38**(2), 207.
10. A.M. Starik, N.S. Titova, and L.S. Yanovskiy, *Kinetics and Catalysis*, 1999, **40**(1), 11.
11. N.N. Semenov, *O nekotorykh problemakh khimicheskoi kinetiki i reaktsionnoi sposobnosti* (Some Problems of Chemical Kinetics and Reactivity), Moscow: Akad. Nauk SSSR, 1958.
12. A.B. Nalbandyan, *Zh. Fiz. Khim.*, 1946, **20**, 1259.
13. S.A. Bozhenkov, S.M. Starikovskaya, and A.Y. Starikovskii, *Combustion Flame*, 2003, **133**, 133.
14. A.A. Seleznev, A.Yu. Aleinikov, and V.V. Yaroshenko, *Khim. Fiz.*, 1999, **18**(5), 65.
15. V.Ya. Basevich and A.A. Belyaev, *Khim. Fiz.*, 1989, **8**(8), 1124.
16. M. A. Tanoff, M.D. Smooke, K.E. Teets, and J.A. Sell, *Combustion Flame*, 1995, **103**(4), p.253.
17. J. X. Ma, D.R. Alexander, and D.E. Poulain, *Combustion Flame*, 1998, **112**(4), 492.
18. M.H. Morsy, Y.S. Ko, and S.H. Chung, *Combustion Flame*, 1999, **119**(4), 492.
19. V.M. Shibkov, A.F. Alexandrov, A.V. Chernikov, et al. *AIAA Paper* No. 2001-2946.
20. I.I. Esakov, L.P. Grachev, and K.V. Khodataev. *AIAA Papers* No. 2001-2939 and No. 2006-1212.
21. V.M. Shibkov, V.A. Chernikov, A.P. Ershov, et al. *AIAA Papers* No. 2004-0513 and 0838.
22. V.M. Shibkov and R.S. Konstantinovskii, *AIAA Papers* No. 2005–0779 and 0987.
23. V.M. Shibkov, A.F. Aleksandrov, V.A. Chernikov, et al. *AIAA Paper* No. 2006–1216.

24. R.S. Konstantinovskii, V.M. Shibkov, and L.V. Shibkova, *Kinetics and Catalysis*, 2005, **46**(6), 775.
25. V.M. Shibkov, A.F. Aleksandrov, A.P. Ershov, et al. *Plasma Physics Reports*, 2005, **31**(9), 795.
26. V.M. Shibkov, A.F. Aleksandrov, A.P. Ershov, et al. *Moscow University Physics Bulletin*, 2004, **59**(5), 64.
27. A.S. Zarin, A.A. Kuzovnikov, and V.M. Shibkov, *Svobodno lokalizovannyi SVCh razryad v vozdukhe* (Freely localized microwave discharge in air). Moscow: Neft i Gaz, 1996.
28. V.M. Shibkov, D.A. Vinogradov, A.V. Voskanyan, et.al. *Moscow University Physics Bulletin*, 2000, **55**(6), 80.
29. V.M. Shibkov, A.P. Ershov, V.A. Chernikov, and L.V. Shibkova, *Technical Physics*, 2005, **75**(4), 455.
30. V.M. Shibkov, S.A. Dvinin, A.P. Ershov, and L.V. Shibkova, *Technical Physics*, 2005, **75**(4), 462.
31. S.A. Dvinin, V.M. Shibkov, and V.V. Mickeev, *Plasma Physics Reports*, 2006, **32**(6).

VARIABLE FUEL PLACEMENT INJECTOR DEVELOPMENT

K. D. BRUNDISH*, M. N. MILLER
QinetiQ, Cody Technology Park, Farnborough, Hants, GU14 OLX, UK

L. C. MORGAN, A. J. WHEATLEY
Smiths Aerospace Components Burnley Ltd, Burnley, 1 Bentley Wood Way, Network 65 Business Park, Burnley, BB11 5TG, UK

Abstract. The operational uncertainty of lean premixed prevaporized technology coupled with the weight and complexity penalties associated with double annular aerocombustors has led to the emergence of the single annular, lean-burn variable fuel placement fuel injector. QinetiQ, in partnership with Smiths Aerospace, are involved in the development of an in-burner-staged airblast fuel injector, which produces two autonomous recirculation zones allowing for low NO_x control over a wide turndown range. Various prototype fuel injectors have been tested in an angled effusion cooled combustor liner at up to 13 bar and 900 K. An idle power efficiency of over 99.5% was achieved with less than 10 $EINO_x$ at scaled climb-out power. In addition, smoke numbers were consistently below 5 SAE, although further optimization for idle blow-out stability and traverse quality is required.

Keywords: variable fuel placement, low NO_x, lean burn, single annular, in-burner staging, airblast, fuel injector, turndown, aero, combustor, stability, efficiency, fuel–air ratio

1. Introduction

Emission legislation coupled with the need for increased thrust has forced gas turbine manufactures to move from traditional designs of combustion systems. Single annular systems are still desired to save weight, however the fuel

---
* To whom correspondence should be addressed. Kevin Brundish, Energy & Materials Centre, QinetiQ, Cody Technology Park, Ively Road, Farnborough, Hants, GU14 OLX, United Kingdom. e-mail: kdbrundish@QinetiQ.com.

N. Syred and A. Khalatov (eds.), Advanced Combustion and Aerothermal Technologies, 425–443.

injector must accept as much as 70% of the combustor air to ensure adequate high power emissions. New technology is required to ensure the operating envelope is preserved with current safety margins whilst meeting these emissions targets. One such technology is variable fuel placement injectors.

The variable fuel placement injector described here features four concentric coswirling air swirlers and two concentric coswirling "airblast" fuel filmers (pilot and main). Atomizer aerodynamics have been developed that produce two different airflow recirculating regions within the combustor (Figure 1). A concentric fuel filmer feeds each of these. By staging the fuel into each flame recirculation zone the variation of local "fuel-air ratio" can be more accurately controlled. In a similar manner to a double annular system one zone can be configured to meet low power stability, whilst the other is optimized for high power emissions performance.

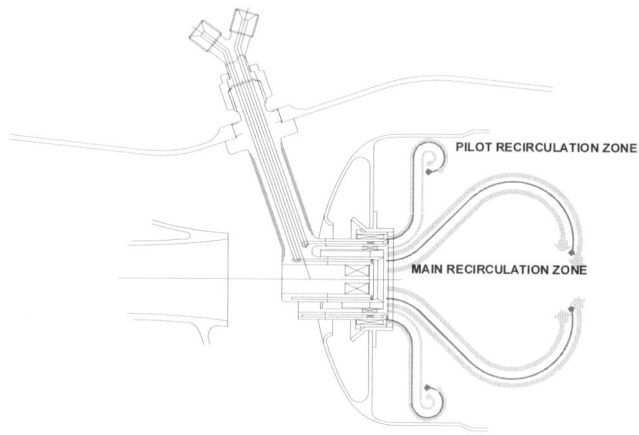

Figure 1. Desired flowfield

Target performance data for the design has been derived from anticipated future engine cycles, and translates into:

- Acceptable fuel injector stability limit (250:1 AFR at flight idle)
- Over 99.5% idle efficiency
- $EINO_x$ values of less than 10 at full power
- Smoke levels less than 10 at full power

This paper describes continued development of the variable fuel placement injector. A combination of experimental and CFD methods has been used to study two potential designs.

2. Nomenclature

AFR   Air/fuel ratio

ACd   Effective area

CO   Carbon monoxide

CO_2 Carbon dioxide

$EINO_x$ NO_x emissions index

FAR   Fuel/air ratio

HC   Unburnt hydrocarbons

LTO   Landing and take-off

NO   Nitrogen monoxide

NO_x Oxides of nitrogen

SFC   Specific fuel consumption

SN   SAE smoke number density

θ Swirler blade angle

3. Fuel Injector Configurations

The fuel injectors have been designed to accept 70% of the total combustion air based on a 32:1 pressure ratio engine cycle. The cycle is representative of a future medium thrust engine of approximately 29,000 lbs thrust. Each fuel injector is designed to have an overall effective area (ACd) of 700 mm^2.

Figure 2 shows the two fuel injector configurations examined. These devices are a development of an injector configuration (shown in Figure 1) detailed in a previous paper (Morgan et al., 2001).

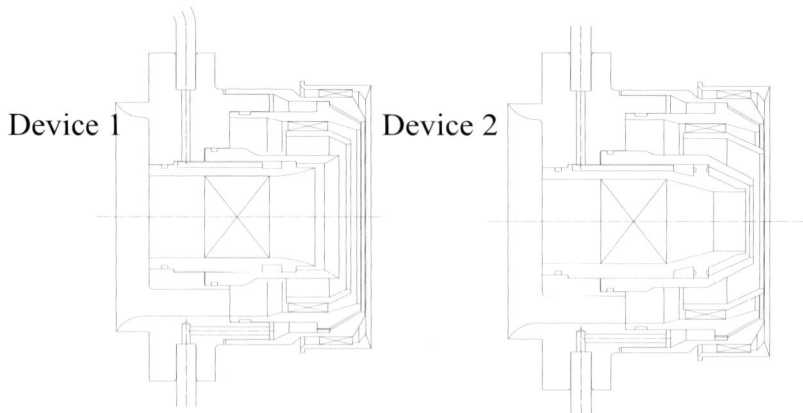

Figure 2. Devices 1 and 2 configuration

Four axial air passages are configured radially, namely the inner, axial curtain, outer and the dome. The inner swirler provides the core flow for the main zone. Limited in swirl, and hence recirculation, it is surrounded by the air curtain which contains no swirl but provides a barrier between the pilot and main zones. The highly aerated low swirl main provides a lean, low residence time zone aimed at low emissions at high power.

The outer and dome swirlers make up the pilot zone. Highly swirled, they provide an outer recirculation zone aimed at low power stability. The configurations were developed from a combination of bench testing and CFD modeling (Morgan et al., 2001). In all cases, the swirlers are corotating, although counter rotating swirlers have been examined.

The difference between the two devices is that the main zone geometry of device 2 is nozzled whereas the main zone geometry of device 1 is open and divergent. Table 1 below shows the measured airflow distribution between the four air passages. Swirl number is included and was calculated in accordance with the Beer and Chigier method (1972).

TABLE 1. Measured fuel injector airflow distribution

Fuel nozzle type	Inner swirler		Axial air		Middle swirler		Outer swirler		Combined
	AC_D (MM^2)	S_N	AC_D (MM^2)	S_N	AC_D (MM^2)	S_N	AC_D (MM^2)	S_N	AC_D
Device 1	217	0.7	288	–	86	2.5	91	2.5	723
Device 2	134	0.7	323	–	91	2.5	100	2.5	644

4. CFD Evaluation

Insight to airflow visualization was provided by constructing a three-dimensional (3D)-axisymmetric model of each of the fuel nozzles configured to a simplified cylindrical combustor primary zone. In order to model the near fuel injector region and accurately predict separation of the two flow fields, a sector of the fuel injector was taken which includes the air swirler blades.

The CFD code used in this study was CFX 5.5, a commercial code supplied and developed by AEA Harwell. Boundary conditions were set by calculating the ACd of each air swirler passageway and then setting flow velocities upstream to give a peak blade passage exit velocity in accordance with the combustor pressure drop at atmospheric conditions. It is intended in future to use pressure boundary conditions. CFD predictions were based on a 3D, cylindrical coordinate system and the $k-\varepsilon$ RNG turbulence model set with incompressible flow and no heat transfer. Neither reacting flow nor liquid fuel interactions were considered in these models, the analysis consisted solely of airflow pattern within a simple cylindrical combustor. The grid system used was an unstructured tetrahedral/prismatic volume mesh and comprised approximately 1,000,000 computational cells for each model.

The two concepts were examined isothermally to determine the aero-dynamic distribution. Figures 3 and 4 show the results for devices 1 and 2, respectively.

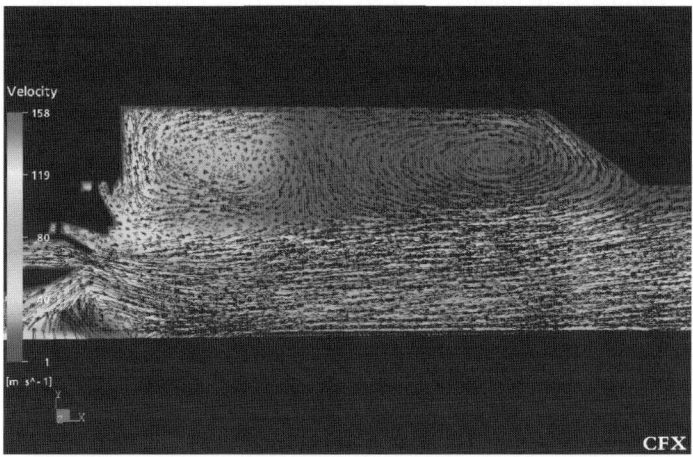

Figure 3. Device 1 CFD flowfield prediction

It is clear from Figure 3 that device 1 forms two separate zones in the upstream section of the combustor, although there is some mixing between zones at downstream locations. A good recirculation zone exists in the pilot whilst a small recirculation is present in the main. The presence of the two separate zones was further proven in bench testing (Morgan et al., 2001). Device 1 was configured to give approximately 26% of the fuel injector airflow into the pilot zone.

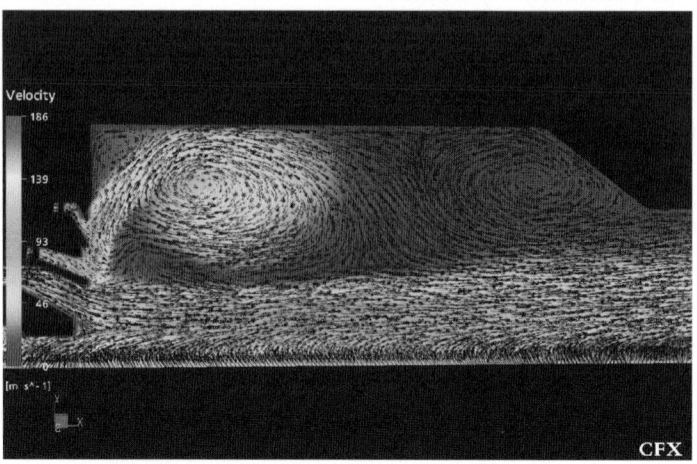

Figure 4. Device 2 CFD flowfield prediction

Figure 4 shows that device 2 also forms two separate zones. The highly recirculating pilot zone is completely separated from the main zone compared to device 1. No recirculation is present in the main zone for device 2 and there is little evidence of downstream mixing. Device 2 was configured to give approximately 29% of the fuel injector airflow into the pilot zone.

From the initial development work, device 1 was thought to be more promising as it offered better mixing and some recirculation for the main zone. The two devices were evaluated experimentally for low power stability/ efficiency and high power emissions performance.

5. Experimental Evaluation

5.1 ATMOSPHERIC PRESSURE TEST RIG HARDWARE CONFIGURATION

Combustion stability was measured in an atmospheric pressure test facility at AIT. The hardware configuration is shown in Figure 5 below, and comprises a simple airbox arrangement with quartz tube to confine the flame and simulate an injector/combustor system.

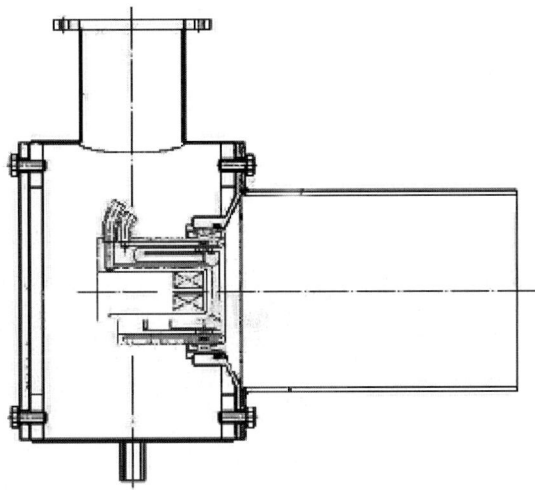

Figure 5. Fuel injector stability test rig

The stability test rig was fed from a compressed air supply. Kerosene fuel was fed to the injector and metered using calibrated rotameters. Airbox pressure differential was measured with a calibrated digital manometer.

5.2 HIGH-PRESSURE TEST RIG HARDWARE CONFIGURATION

The combustion facility at QinetiQ was used for the test program. Separate control over temperature (up to 900 K, nonvitiated), mass flow (up to 5 kg/s) and pressure (up to 16 bar) is available, allowing a wide variation of combustor inlet conditions to be achieved. An outline of the experimental rig, designed and manufacture at AIT is given in Figure 6.

The test combustor was cylindrical and designed with 70% air through the fuel injector and 30% through an angled effusion cooled combustor liner. Calibrated instrumentation was utilized to measure inlet air and fuel mass flows, inlet air temperature and inlet air pressure. The associated error with mass flow measurement is $\pm5\%$, whilst the error with pressure and temperature measurements is $\pm2\%$. The combustion system was instrumented with surface thermocouples to determine heat loading to the combustor wall during operation. Additionally, dynamic pressure measurements were made to determine if any onset of combustion driven oscillations occurred.

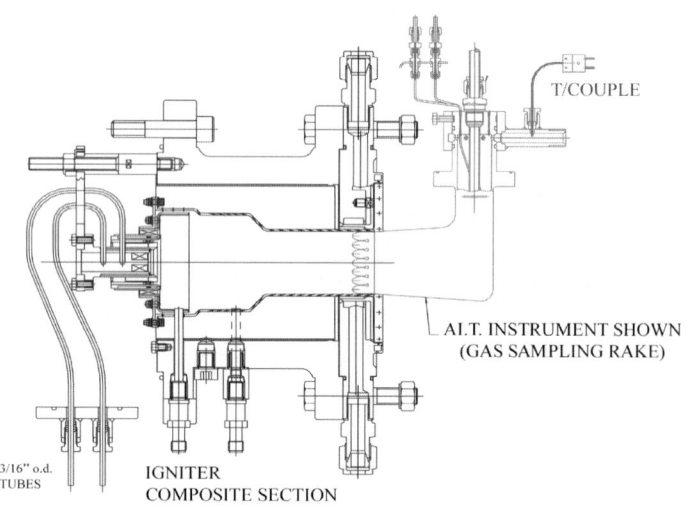

Figure 6. Experimental setup

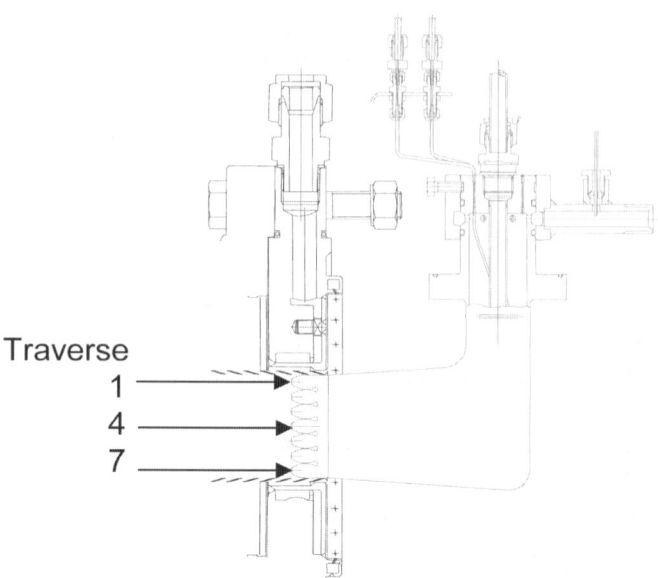

Figure 7. Gas analysis probe

An intrusive emissions measurement probe is included as part of the configuration (Figure 7). The probe has inlet ports aligned on radii of equal area to give a single diametrical traverse. Measurements reported were made for each individual inlet port of the probe to determine traverse data (with traverse location identified at 1–7, Figure 7), and with the individual ports gauged to give averaged data. The sample probe orifice is very small and designed so that the sample enters at a high velocity and is quenched as it enters the relatively cool environment. It is important that condensation does not take place at any point in the system as species of interest may be condensed out and lost on the walls of the sample line. Conditioning water not only provides cooling to enable survival in the harsh environment, but also ensures that the sample does not condense whilst within the probe.

Additional instrumentation was present during the testing, such as surface thermocouples and acoustic pressure measurements. No significant acoustic pressures were detected. Wall temperature limitations occurred when pilot fuel flows were high, and limited the split between pilot and main fuel.

5.3 GAS ANALYSIS CONFIGURATION AND FUNCTIONALITY

A heated stainless steel sample line maintained at a temperature of 423 ± 15 K was connected to the sample probe which was used to convey the sample from the combustor to the gas analysis system. This sample acquisition system incorporates high-pressure back purge air, which can be maintained at a pressure of 1,500 Pa. This air is used to stop fuel flowing down the sample line at light-up.

At a short distance from the probe, the sample line is split into two separate channels. One of the sample lines leads to the smoke analysis equipment, which consists of both a filter stain, and an optical measuring technique. The other sample line conveys the sample to a filter oven to remove any particulate matter and then to the gas analysis suite.

Gas analysis instrumentation contained within the facility was used to determine the concentration of CO, CO_2, NO, NO_2, total hydrocarbons, and O_2. CO_2 and CO was measured using heated nondispersive infrared (NDIR) analyzers and O_2 by a heated paramagnetic detector. NO and NO_x were measured using the chemiluminescent technique, which relies on the reaction between nitric oxide (NO) and ozone (O_3). NO_x is measured after converting NO_2 to NO in a converter built into the analyzer. Thus two measurements are made, NO and NO_x. The NO/NO_x analyzer is a dual channel system that allows both species to be measured simultaneously. As is typical in such analyzers, the converter not only converts NO_2, but also any other higher oxides of nitrogen (e.g., N_2O_5) or acids (e.g., HNO_2, HNO_3) and therefore actually makes a measurement of NO_y. Total unburned hydrocarbons were measured by a heated flame ionization detector (FID). This measurement is reported as a carbon equivalent volume fraction i.e., ppmC (methane equivalent). The total error associated with the gas analysis measurements is about $\pm 2\%$ of reading with the exception of smoke, which is ± 3 SAE.

The NO/NO_x analyzer, and the total hydrocarbon analyzer measure the species concentration using a wet sample, whilst for the rest of the analyzers the sample is dried. The dew point of the inlet air is measured so as to determine the concentration of ambient water, and the dried sample dew point is also measured so that volume corrections may be applied.

The measurements performed conformed to ARP 1256 for gaseous species and ARP 1179 for smoke. The raw data was post-processed to ARP 1533. This includes corrections for inlet air humidity, cross interference effects and dried sample corrections. The post processing also includes calculations of AFR, combustion efficiency and gas temperature.

5.4   TEST CONDITIONS

The atmospheric testing was performed over a range of combustor pressure drops, vented to atmosphere. The airflow was at ambient conditions with an inlet temperature of approximately 293 K.

The test conditions for high power were scaled by $M\sqrt{T}/P$ from a future medium thrust engine cycle (29,000 lbs). Two conditions representing idle and climbout were tested. The test conditions are shown in Table 2.

TABLE 2. Test conditions

	Idle (7%)	Climb-out (85%)
Temperature (K)	526	793
Pressure (bar)	5.67	13.7
$\Delta P/P$ (%)	3	3
AFR	52	28

6. Experimental Results and Discussion

6.1   COMBUSTION STABILITY PERFORMANCE

The main aim of the atmospheric testing tests was to examine weak stability characteristics. Optimization of ignition characteristics has not been performed, although the data is reported for completeness. Further work would be required in order to improve ignition characteristics, such as optimization of igniter location.

Figure 8 shows the stability results for device 1 plotted as extinction FAR against fuel injector pressure drop. Figure 8 shows that the injector has the potential to achieve 0.01 Fuel injector FAR, or 100:1 AFR, prior to extinction. The area within the loop shows where successful ignition was achieved. Ignition performance is clearly poor as the system has a very narrow operating envelope.

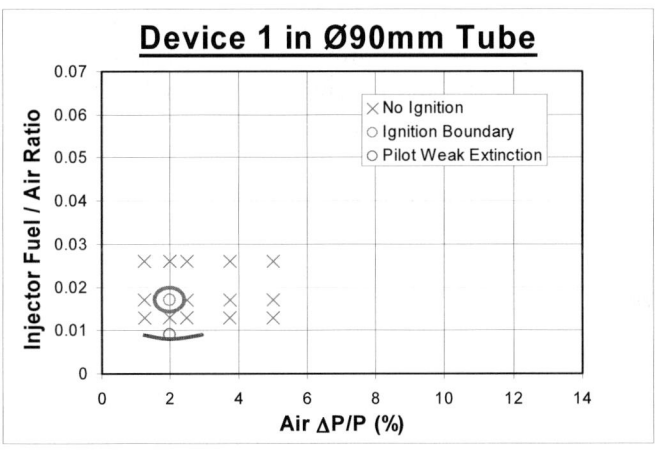

Figure 8. Fuel injector stability performance, device 1

The results for device 2 can be seen in Figure 9. Device 2 has significantly improved operability with stability results better that 0.01 fuel injector FAR (100:1 AFR) and blowout limit better that 12% $\Delta P/P$. Ignition has also improved, with a larger loop showing increased regions where ignition could be achieved.

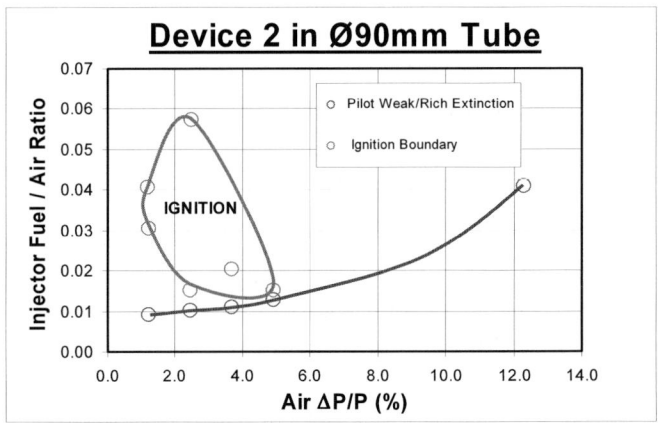

Figure 9. Fuel injector stability performance, device 2

With a fuel injector AFR of 100:1, at 70% of the combustion air, this translates to a combustor AFR of approximately 143:1. With cold airflow and atmospheric pressure this result was considered adequate.

6.2 HIGH-PRESSURE RESULTS DEVICE 1

Results for the idle condition can be seen in Table 3 for 100% pilot fuel flow. It is clear that the efficiency was unacceptable. Lean stability limits were also poor at idle with extinction at approximately 80:1 combustor AFR at idle pressures and temperatures. This was not apparent in the atmospheric testing, and there was some concern that the flowfield had altered with the elevated pressure.

TABLE 3. Idle performance data

CO (ppm)	1,328
HC (ppm)	1,270
Efficiency (%)	93
NO_x (ppm)	32

Figure 10 shows $EINO_x$ values against main fuel flow split for the scaled climb-out power condition. It was anticipated that minimum NO_x values would occur at a similar fuel split to the airflow split, that being 26% pilot and 74% main. However, it is clear that the minimum NO_x occurs with a main fuel of around 30%. With further fuel added to the main zone the NOx emissions increased. It was clearly evident that the design airflow splits had not been achieved at elevated pressure. It was probable that some of the main air had attached itself to the pilot zone airflow thus explaining the poor idle efficiency/stability and the unexpected NOx emissions results. Lower main zone fuel splits could not be achieved due to temperature limitation when pilot fuel was increased.

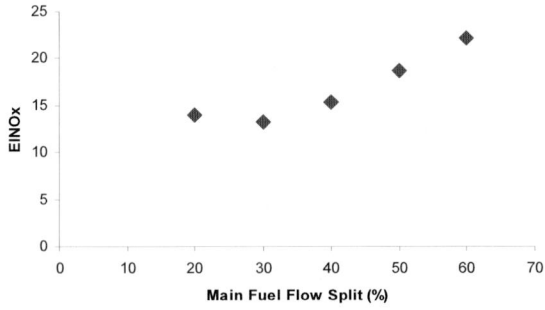

Figure 10. $EINO_x$ vs Main Zone fuel split at scaled climb-out conditions, device 1

To examine this possibility further, individual traverse points were taken. This allowed the NO_x emissions to be measured at different radial positions, and so identify where the NO_x production zones were located. Figure 11 shows $EINO_x$ values against traverse location for a range of main zone fuel splits.

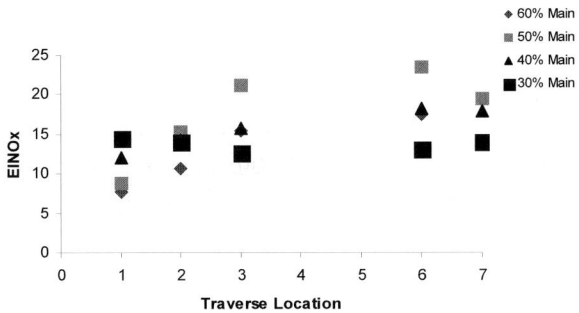

Figure 11. $EINO_x$ vs traverse location at scaled climb-out conditions, device 1

When considering the 30% main fuel split result, it can be seen that the profile is fairly flat indicating an equal contribution from both zones. With respect to the 40% main, a degree of asymmetry is present which could indicate a poor distribution of fuel on the prefilmer at this condition but may also be as a result of the aerodynamic flowfield. With 50% main zone fuel flow, NO_x emissions are shown to be high at traverse location 3 and 6 suggesting an equal contribution from both zones. These two locations also show the greatest NO_x concentrations. At 60% main zone fuel the NO_x emissions are seen to reduce at these two locations. Overall the maximum amounts of NO_x are found in the center rather than at the outer extremities, suggesting the main zone is the dominant NO_x production zone. This supports the suggestion that the airflow splits have not remained as designed.

The optimum NO_x result was 15 $EINO_x$ at scaled climb-out power conditions. Figure 12 shows that the NO_x emissions were achieved with efficiencies over 99.9%. Although it did not meet the target $EINO_x$ value of 10, it was a promising result at such an early stage in the development program. Target smoke levels were achieved with less that 10 SAE at the same main zone fuel split as the minimum NO_x result (Figure 13).

It was clear from the results that the design airflow splits had not been achieved at high pressure. However, CFD prediction for device 2 had shown a clearer separation of the two zones. Device 2 was therefore also tested at idle and scaled climb-out power conditions.

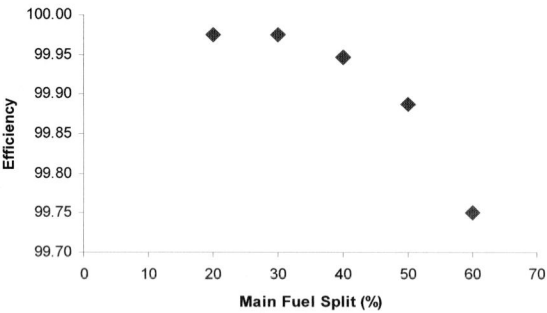

Figure 12. Efficiency vs main zone fuel split at scaled climb-out conditions, device 1

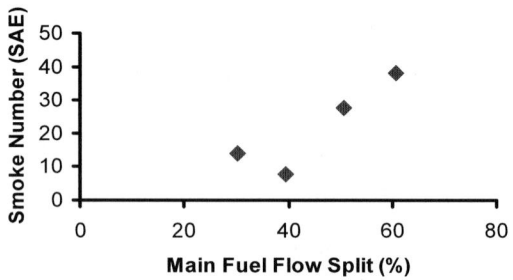

Figure 13. Smoke number vs main zone fuel split at scaled climb-out conditions, device 1

6.3   HIGH PRESSURE RESULTS DEVICE 2

The idle results for device 2 are shown in Table 4 for 100% pilot fuel flow split. The target of over 99.5% efficiency was achieved at the idle condition.

TABLE 4. Idle performance data

CO (ppm)	53
HC (ppm)	5
Efficiency (%)	99.87
NO_x (ppm)	54

EINO$_x$ emissions against main zone fuel flow split for scaled climb-out power conditions are shown in Figure 14. NO_x emissions were minimum at 80% main zone fuel for device 2. Higher main zone fuel splits were not achieved as the flame extinguished probably due to the lean pilot zone. However, EINO$_x$ results were close to the target values although it is recognized that these will increase at take-off power conditions due to the increased inlet

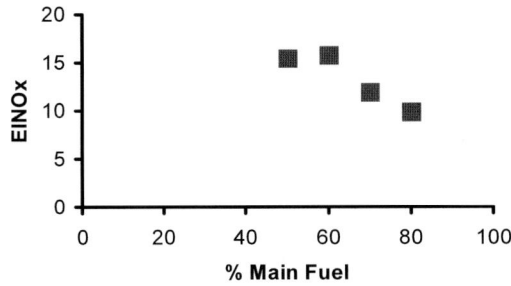

Figure 14. EINO$_x$ vs main zone fuel split at scaled climb-out conditions, device 2

temperature and pressure. Main zone fuel splits lower than 50% could not be achieved due to temperature limitation when pilot fuel was increased.

Figure 15 shows the EINO$_x$ values at each individual traverse location, for a range of main zone fuel splits. With 50% and 60% main zone fuel, the pilot zone can be seen as the dominant contribution to the NO$_x$ emissions. EINO$_x$ values are relatively even for both zones at 70% main zone fuel, but the main zone is clearly dominant at 80% main zone fuel split.

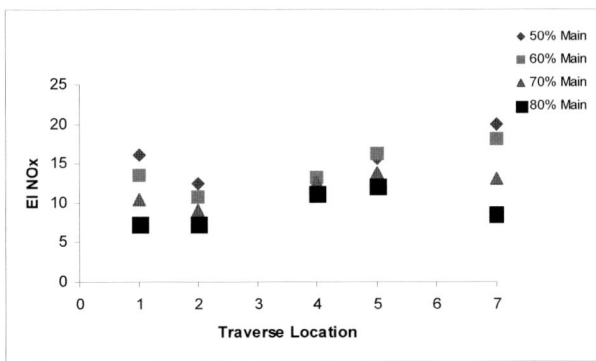

Figure 15. EINO$_x$ vs traverse location at scaled climb-out conditions, device 2

In all cases, the efficiency was above 99% as shown in Figure 16. Although not ideal, as 99.99% would be desirable at such high operating powers, the results were promising at such an early stage in the development program.

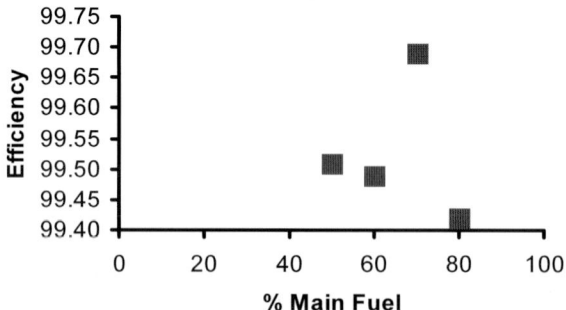

Figure 16. Efficiency vs main zone fuel split at scaled climb-out conditions, device 2

Smoke levels were below the target value for all fuel flow splits, as shown in Figure 17. The measured levels are within the error for smoke measurements (±3 SAE) and hence no further interpretation of the data is reported.

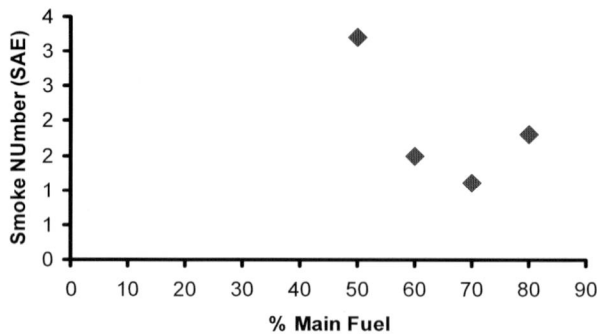

Figure 17. Smoke number vs main zone fuel split at scaled climb-out conditions, device 2

6.4 COMPARISON OF TEMPERATURE TRAVERSE

Although emissions performance has been assessed for each design, another important feature is combustor exit traverse. With the presence of two separate zones there is a risk of temperature traverse distortion. Figure 18 shows the temperature traverse where the temperature at each location has been nondimensionalized to the mean temperature. The data shown is for the point at which minimum NO_x emissions were measured, 40% main zone fuel for device 1 and 80% main zone fuel flow for device 2.

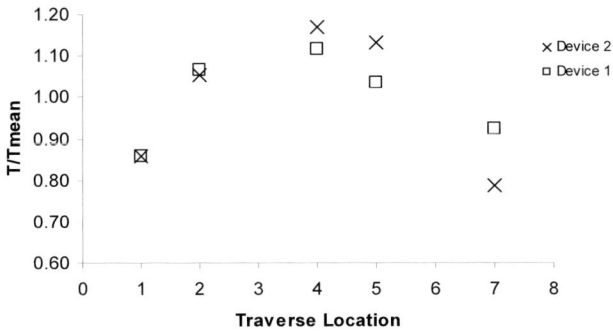

Figure 18. Temperature exit traverse, scaled climb-out conditions

When considering device 1, the temperature increased from the outboard pilot to the main zone. Examination of the temperature traverse was performed using an RTDF type factor defined by Eq. (1). The RTDF for device 1 was 20%, which although high, was acceptable.

$$RTDF = \frac{(T\max - T\mathrm{mean})}{(T\mathrm{mean} - T\mathrm{inlet})} \tag{1}$$

When examining the traverse for device 2, again peaks in the temperature occurred in the main zone. The RTDF for device 2 is 30%, somewhat higher than acceptable. The mixing process between the main zone and the pilot zone is insufficient at this stage. Figure 4 suggested that the temperature traverse may be problematic, as the two zones are distinctly separate and little expansion of the main zone occurs. In order to improve traverse, a higher degree of swirl is required for the main zone. This will expand the main zone and promote inter zone mixing as well as improve the mixing within the main zone mixture.

7. Summary

In summary, device 1 did not meet the target values for low power efficiency or high power emissions. The flowfield structure altered at high pressure, leading to a merging of the air curtain with the pilot zone.

Device 2 met target efficiency performance at low power, although emissions were slightly higher than target values at high power. Exit traverse was also unacceptable at high power. Although the design has shown promising performance at such an early stage in the development, some design improvement is required. The percentage of air through the main zone should be increased to further reduce emissions. Additionally increasing the swirl in the main zone is required to promote better mixing and reduce exit temperature traverse.

8. Conclusions and Recommendations

Two fuel injector configurations have been identified which create a pilot zone optimized for stability, and a main zone optimized for high-power emissions. In addition to atmospheric pressure evaluations the two fuel injector configurations were examined at two conditions: idle and climbout (scaled).

Although device 1 showed good stability at atmospheric pressure, its performance at elevated pressures was poor. Further evaluation suggested that the airflow split between the pilot and main zones changed at elevated pressures and performance was affected.

Device 2 achieved 10 $EINO_x$ at scaled climb-out power and over 99.5% efficiency at idle power.

Although the target values were not met, device 2 performed better than device 1 and the results were encouraging at such an early stage in the development program.

The potential of the concept as part of a low emissions combustion system for future cycles has been proven.

The airflow splits between the pilot and main for device 2 need to be further optimized in order to achieve target emissions values. Additionally the main zone swirl requires further optimization to improve combustor exit traverse.

Further CFD will be performed, at more representative conditions.

Development testing at simulated take-off power conditions is required, followed by validation tests at 32 bar to fully prove the fuel injector performance at take-off power.

Acknowledgments

This work was carried out in a collaboration between AIT and QinetiQ with funding from AIT and the UK DTI

The authors would like to acknowledge Rolls-Royce Deutschland Ltd & Co KG, Rolls-Royce plc. and UMIST for their support in this work.

References

Beer J.M., Chigier N.A., 1972, *Combustion Aerodynamics*, Applied Science, London.

Morgan L.C., Brundish, K.D., Wheatley A.J., 2001, *Development of a Variable Fuel Placement Airblast Atomiser*, ASME-2001-GT-0377.

SAE Aerospace Recommended Practice 1256, *Procedure for the Analysis and Evaluation of Gaseous Emissions from Aircraft Engines*, issued March 2001.

SAE Aerospace Recommended Practice 1179, *Aircraft Gas Turbine Engine Exhaust Smoke Measurement*, Revised October 1997.

SAE Aerospace Recommended Practice 1533, *Procedure for the Analysis and Evaluation of Gaseous Emissions from Aircraft Engines*, issued March 2001.

COMPARATIVE STUDY OF EQUILIBRIUM
AND NONEQUILIBRIUM EVAPORATION MODELS
FOR VAPORIZING DROPLET ARRAYS AT HIGH-PRESSURE

G. LAMANNA*, H. SUN, M. SCHÜLER, B. WEIGAND
*Institute of Aerospace Thermodynamics, University of Stuttgart,
70569 Stuttgart, Germany*

D. MAGATTI, F. FERRI
*Department of Physics and Mathematics, University of Insubria
and INFM, 22100 Como, Italy*

Abstract. Combustion performances and emissions are mainly influenced by atomization, evaporation of fuel droplets and mixing of fuel and air. In particular, vaporization is often the rate-controlling process and is directly related to the specific fuel consumption of the engine and to pollutant formation. The present work focuses on two competing factors, which strongly influence the vaporization process, specifically: drop–drop interactions and high-pressure effects. Numerical predictions from equilibrium and nonequilibrium vaporization models are compared with the experimentally determined evaporation rate of droplet arrays at high-pressure conditions. The objective is to improve the predictive capabilities of spray models by introducing drop interaction effects in the modeling and by verifying which formulation of heat and mass transfer is able to model adequately their mutual interdependence at high-pressure.

Keywords: vaporization at high-pressure, drop–drop interaction, nonequilibrium evaporation models, light scattering

---

* To whom correspondence should be addressed. Grazia Lamanna, Institute of Aerospace Thermodynamics, University of Stuttgart, Pfaffenwaldring 31, D-70569 Stuttgart, Germany. e-mail: grazia.lamanna@itlr.uni-stuttgart.de

N. Syred and A. Khalatov (eds.), Advanced Combustion and Aerothermal Technologies, 445–455.

1. Introduction

Liquid droplet vaporization in a high-pressure/temperature environment is of relevance to combustion science and technology, due to the need for developing high-pressure combustion devices such as liquid-propellant rockets, gas turbines and diesel engines. In order to calculate the evaporation rate of such droplets, equations are required to describe the mass, momentum and energy transfer between the droplet and its surroundings. Theoretical analysis is difficult because the equations must take into account drop–drop interaction and nonequilibrium effects. The latter have become increasingly important in modern propulsion systems due to the fact that vaporization occurs under conditions of large departure from equilibrium, measured in terms of temperature and/or pressure difference between the droplet and the carrier gas. The literature in this area is now extensive, addressing these problems at various levels of complexity (Imaoka and Sirignano, 2005; Leiroz and Rangel, 1995; Wong and Lin, 1992; Sangiovanni and Labowsky, 1982). However, as yet no conclusive evidence has been achieved due to a lack of reliable experimental data. This paucity of experimental data is attributable to the difficulties of performing controlled experiments on free droplets of micrometer size in a high pressure/high-temperature environment. As a first step in the solution of this challenging problem, the vaporization rate of droplet arrays of micrometer size under high-pressure conditions has been derived experimentally (Sun, 2006). From these measurements, an empirical correlation for the drag coefficient of interacting droplets was obtained. The latter is of relevance for the development of semiempirical models, aimed at providing an accurate and comprehensive prediction of the individual droplet drag, Nusselt and Sherwood numbers for the case of interacting vaporizing droplets. These global parameters together with the instantaneous droplet locations, size, and speed are important for less computer-intensive spray simulations.

Due to the complexity of spray calculations, the traditional modeling approach consists of specifying the governing equations for a single, isolated droplet including drag, convective heat, and mass transfer. The derived equations, corrected for drop–drop interaction effects, are then used for a subset of statistically representative droplets in various forms of computational spray models (Crowe et al., 1996). Concerning the vaporization of an isolated droplet, the classical approach uses an equilibrium evaporation model, which leads to the so-called d^2-law. Over the years, significant advances have been made to the classical method to include the effects of variable thermophysical properties (Yuen and Chen, 1976), transient liquid heating (Law and Sirignano, 1977) and gas-phase convection (Ranz and Marshall, 1952). Despite the noteworthy progress, the above-mentioned models cannot overcome the limitations inherent in the approach based on equilibrium thermodynamics. Alternatively, more advanced evaporation models have been developed based on the Langmuir–Knudsen approach, where nonequilibrium effects can be relatively easily incorporated by specifying an appropriate molecular velocity distribution function within the Knudsen layer. Different formulations of nonequilibrium evaporation models have been proposed in literature (Bellan and Summerfield, 1978; Bellan and Hardstad, 1987; Young, 1993), characterized by a different degree of complexity in the choice of the molecular velocity distribution function.

This paper presents the preliminary results of an ongoing systematic evaluation of evaporation models capable of describing the vaporization of small hydrocarbon droplets in a high-temperature/pressure environment, as found in modern propulsion systems. The models considered here include three versions of the heat-mass transfer analogy models and two nonequilibrium models based on the Langmuir–Knudsen approach. The models are evaluated through comparison with experiments. First, isolated hydrocarbon droplets vaporizing in low and high-temperature environments are considered. Subsequently, the comparison is extended to droplet arrays vaporizing in high-pressure convective surroundings.

2. Model Description

The set of governing equations describing the temporal evolution of a single droplet embedded in a carrier gas is reported below. In order to facilitate model comparison, the system of equations is cast in a very general form and the model dependency is incorporated into the specific expressions employed for the momentum F, mass $\dot{M}$ and thermal energy $\dot{Q}$ transfers between the droplet and the carrier gas:

$$
\begin{aligned}
\frac{dx}{dt} &= \Delta v_d \\
-\frac{d\Delta v_d}{dt} &= \frac{F}{m_d} - g \\
4\pi r_d^2 \rho_d \frac{dr_d}{dt} &= -\dot{M} \\
m_d c_{vd} \frac{dT_d}{dt} &= \dot{Q} + \dot{M} L - Q_{\Delta T}
\end{aligned}
\tag{1}
$$

where m_d is the droplet mass, c_{vd} the liquid heat capacity, ρ_d the liquid density, g the gravitational acceleration, L the latent heat of evaporation, and $Q_{\Delta T}$ represents any additional term used to incorporate nonuniform internal temperature effects (e.g., finite liquid thermal conductivity). These equations

represent a system of first-order ordinary differential equations for the droplet
position x, relative velocity Δv_d, droplet radius r_d, and temperature T_d. As
starting conditions for the time integration, the initial droplet radius, position,
temperature, and relative velocity are assigned. A real gas equation of state and
generalized multiparameter estimation methods are employed for the determination of the thermophysical properties of the liquid droplet, vapor phase,
and ambient gas (Sun, 2006).

2.1 MODELING OF TRANSFER RATES

The equations describing the quasisteady mass, momentum, and energy transfer
between a single droplet and a carrier gas can be written as follows:

$$\begin{aligned}
\dot{M} &= 2\pi r_d D_m \rho_m Sh f_M \\
\dot{Q} &= 2\pi r_d k_m (T_\infty - T_{dr}) Nu f_Q \\
F &= \pi r_d^2 \rho_m \Delta v_d^2 c_D
\end{aligned} \qquad (2)$$

where D is the diffusion coefficient, k the thermal conductivity, c_D is the drag
coefficient, Nu and Sh, represent the Nusselt and Sherwood numbers,
respectively and are empirically modified to account for convective effects. The
subscript m refers to the reference state, which uses free stream conditions for
the gas density and the so-called "1/3 rule" for all other properties. The factor f_M
is directly related to the driving potential for mass transfer; f_Q represents, the
correction to heat transfer due to the complementary role between heat and
mass transfer. Their specific expressions can be found in Table 1, which
identifies each evaporation model.

   Five different models are selected for comparison. Among the three
equilibrium formulations (termed MA1, MA2, and MA3, respectively), MA3
incorporates a correction for mass transfer accounting for higher departures
from equilibrium. When the temperature difference between the drop and its
surroundings is considerable, the rate of vaporization is a logarithmic function
of the difference in vapor pressure at the drop surface and at infinite distance
(Godsave, 1953). Additionally, models MA3 and MA1 also contain a correction
for heat transfer (i.e., $f_Q \neq 1$) derived on the basis of film theory (Spalding,
1953), which accounts for the cross-influence of mass transfer on heat transfer.
As nonequilibrium formulations, two versions of Bellan's Langmuir–Knudsen
model are considered: LK1 simulates heat transfer inside the droplet using the
infinite liquid conductivity hypothesis; LK2 includes finite liquid conductivity
effects. Detailed descriptions of the models can be found in Bellan and Harstad
(1987) and Miller et al. (1998) and hence, they are not repeated here.

TABLE 1. Expressions for the evaporation corrections f_Q, internal temperature gradient $Q_{\Delta T}$ and mass transfer potential f_M from various models

Model	Name	f_M	f_Q	$Q_{\Delta T}$
MA1	Mass analogy I	B_M	$\ln(1+B_M)/B_M$	0
MA2	Mass analogy II	B_M	1	0
MA3	Mass analogy III	$\ln(1+B_M)$	$\ln(1+B_M)/B_M$	0
LK1	Langmuir–Knudsen	$\ln(1+B_{M,neq})$	G	0
LK2	Langmuir–Knudsen	$\ln(1+B_{M,neq})$	G	$(2\beta\theta_1/3\,\tau_d Pr)\,(T_{dr}-T_{di})$

Nomenclature is as follows: Y_{vr} is the vapor mass fraction at the droplet surface (the subscript ∞ refers to free steam conditions), θ_1 is the ratio of gas heat capacity to that of the liquid phase $\theta_1 = c_{p,g}/c_{vd}$, Pr is the gas phase Prandtl number, τ_d is defined as $\tau_d = \rho_d D^2/18\mu_g$, the subscripts dr and di refer to conditions at the droplet surface and inside, respectively. The nonequilibrium Spalding transfer numbers for mass (B_M) and energy (B_T) are defined as

$$B_{M,(n)eq} = \frac{Y_{vr,(n)eq} - Y_{v\infty}}{1 - Y_{vr,(n)eq}}, \qquad B_T = \frac{c_{pm}(T_\infty - T_{dr})}{L} \qquad (3)$$

where L is the latent heat of vaporization. For the calculation of the parameters ($Y_{vr,eq}$, $Y_{vr,neq}$, $B_{M,eq}$, $B_{M,neq}$) and for the definition of the function G, the reader is referred to the work of Miller et al. (1998). Here, we simply recall that the function G provides an analytical expression for heat transfer derived from the quasisteady solution of the gas field and contains explicitly the dependency on mass transfer. Essentially, this implies that for models LK1 and LK2 it is not necessary to employ empirical correlations to simulate the cross-influence between heat and mass transfer. However, the same consideration does not hold for the mass analogy models, where empirical correlations are required for the Nusselt and Sherwood numbers to reintroduce the mutual interdependency between heat and mass transfer into the modeling. For the present work, the Renksizbulut and Yuen (1983) correlations have been used

$$Nu = (2 + 0.57 Re_m^{1/2} Pr_f^{1/3})(1 + B_T)^{-0.7}$$

$$Sh = (2 + 0.87 Re_m^{1/2} Sc_f^{1/3})(1 + B_M)^{-0.7} \qquad (4)$$

$$c_D = c_D^{iso}(Re_m)(1 + B_T)^{-0.2}$$

where Re is the Reynolds number ($Re_m = 2\,\rho_g\,r_d v_d/\mu_m$), Sc the Schmidt number ($Sc_f = \mu_m/\rho_m D_m$) and Pr the Prandtl ($Pr_f = c_{pm}\mu_m/k_m$). The drag coefficient of an isolated droplet is estimated using the correlation proposed by Virepinte et al. (1999). Note that, for models LK1 and LK2, the factor $(1+B)^\alpha$ in Eq. (4) is retained only in the correlation for the drag coefficient, as the two models were initially developed for droplets evaporating in a stagnant environment and thus do not take into account the influence of heat transfer on momentum transfer.

Interaction effects are modeled by means of empirical correlations, expressing the reduction of mass transfer and drag coefficient in terms of the nondimensional interdroplet distance ($\Delta s/D_0$). For the mass transfer, the correlation proposed by Atthasit et al. (2005) is employed:

$$Sh^{prx} = Sh\left[1 - 0.57\left(\frac{1 - e^{-0.13(\Delta s/D_0 - 6)}}{1 + e^{-0.13(\Delta s/D_0 - 6)}}\right)\right] \tag{5}$$

The latter is valid for $2 < \Delta s/D_0 < 16$ and $12 < Re < 25$. For the drag coefficient, instead, the correlation proposed by these authors is used. For a discussion on its derivation and accuracy, the reader is referred to Sun (2006)

$$c_D^{prx} = c_D^{ev}\left[1 - a\exp(-b\frac{\Delta s}{D})\right] \tag{6}$$

where

$$a = -0.9139 + 0.2955\exp(-0.0244Re)$$
$$b = -0.0302 - 0.0634\exp(-0.0132Re) \tag{7}$$

3. Results and Discussion

The predictive capabilities of the different models are best evaluated through comparison with droplet vaporization experiments occurring in high-temperature/pressure convective environments, as those are more representative of the conditions normally encountered in modern propulsion systems. In an attempt to evaluate separately the accuracy of the models in capturing correctly both nonequilibrium and interaction effects, isolated droplet experiments are considered first followed by monodisperse droplet streams. Although the extent of experimental data is too limited to draw definite conclusions on model quality, it was possible to identify some specific trends.

3.1 ISOLATED DROPLET EXPERIMENTS

Figure 1a shows the temporal evolution of the relative surface area for a single water droplet ($D_0 = 1.1$ mm and $T_{d,0} = 282$ K) evaporating in stagnant air at $T_\infty = 298$ K. As can be immediately seen, all models predict identical evaporation histories and agree with the experiments. The temperature evolutions are also basically identical for all models and approximately equal to the wet-bulb temperature, predicted by Miller's formula (1998). As expected, for a small departure from equilibrium, all mass analogy models perform rather well. This result contradicts those of Miller et al. (1998), who still observed some slight differences in the temperature evolutions when employing the Ranz and Marshall (1952) correlation. In the present work, however, Renksizbulut and Yuen's (1983) correlation is used, where the *Nu* and *Sh* numbers are nonlinear functions of the respective transfer numbers. This indicates that the latter is more capable of capturing the mutual interdependency between heat and mass transfer for low evaporation rates.

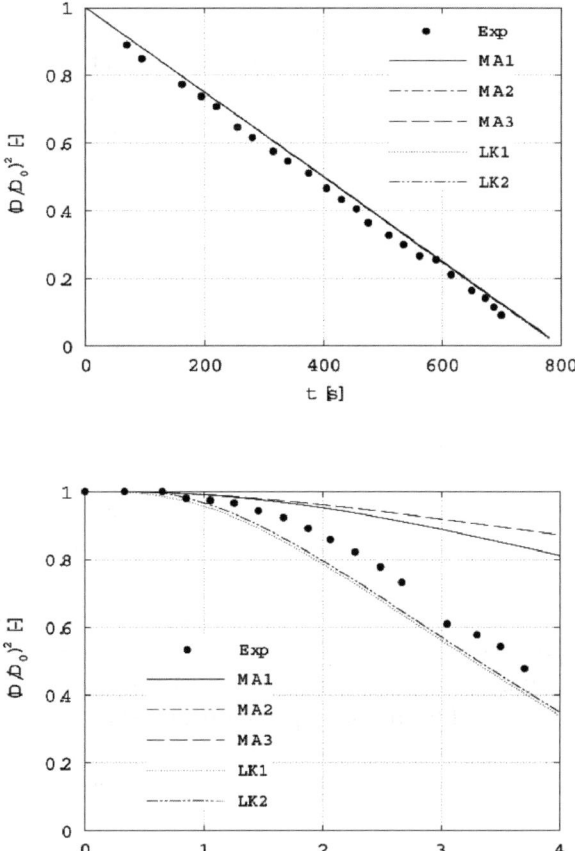

Figure 1. Numerical predictions of the nondimensional droplet surface area compared to the experimental results from (a) Ranz and Marshall (1952). Initial conditions: $T_\infty = 298$ K, $T_{d,0} = 282$ K, $D_0 = 1.1$ mm and $Re_d = 0$; (b) Wong and Lin (1992). Initial conditions: $T_\infty = 1000$ K, $T_{d,0} = 315$ K, $D_0 = 2.0$ mm and $Re_d = 17$

Variations among the model predictions start to appear as soon as the evaporation process occurs under conditions of high departure from equilibrium as in the experiments by Wong and Lin (1992). Their experiments consisted of a decane droplet with initial size $D_0 = 2.0$ mm and temperature $T_{d,0} = 315$ K evaporating in a convective airstream at $T_\infty = 1,000$ K. Figure 1b clearly reveals that in the initial phase of the vaporization process, corresponding to the heat up transient stage, the d^2-law is invalid. The results shown in Figure 1b suggest that the droplet size evolution is best modeled using either of the two nonequilibrium models. However, the mass analogy models clearly overestimate the droplet size. This is due to the fact that the droplet temperature is largely underestimated. The heat transfer reduction through the factor $(1 + B_T)^{-0.7}$ in Eq. (4) provides a relatively strong contribution, thus resulting in an underestimation of the droplet temperature. In model MA1 and MA2, the size prediction is somewhat improved by the stronger mass potential B_M (when compared to ln (1 + B_M) of model MA3), which compensates somewhat for the reduced heat transfer, thus resulting in a faster evaporation rate.

3.2 INTERACTING DROPLET EXPERIMENTS

As a final verification, the numerical predictions from the various models are compared with our own experimental results (Sun, 2006). The objective is twofold. First, we intend to determine the accuracy and range of validity of each model with respect to high pressure effects. Second, we want to verify whether the proposed empirical correlations for Sh^{prx} and c_D^{prx} are indeed capable of simulating proximity effects among the droplets correctly. The results can be summarized as follows. Except for model MA3, the numerical results agree fairly well with the experiments as the chamber pressure is raised from 5 to 40 bar. For example, Figure 2 shows the temporal evolution of the nondimensional droplet surface area for two different chamber pressures. At low pressure, the larger deviations observed for model MA3 are due to the concomitant effect of the factors $f_M = \ln (1 + B_M)$ and $f_Q = \ln (1 + B_M)/B_M$, which results in a higher quenching of the mass transfer. For models MA2 and MA3, this effect is counterbalanced by the stronger mass potential $f_M = B_M$. At higher pressures (i.e., higher droplet Reynolds number), convective heat transfer becomes the dominant effect in controlling the rate of vaporization, thus overshadowing completely the role of the correction factor f_Q. The macroscopic result is no difference in the model predictions, as shown in Figure 2b. Both LK1 and LK2 perform very well at all chamber pressures without the need to employ any correction factors for the heat and mass transfer.

Figure 3 depicts the nondimensional droplet speed for the same experimental conditions shown in Figure 2. The results clearly show that its temporal evolution is well captured by all models at all pressure levels. This indicates that the proposed correlation for the drag coefficient (Eq. (6)) together with the correction for the vaporizing case (Eq. (4)) is indeed able to model correctly the momentum transfer process in vaporizing arrays. The same conclusion holds for the proximity correction, proposed by Atthasit et al. (2005), for mass transfer, since it is demonstrated that the good performance of the correlation is independent from the specific vaporization model chosen.

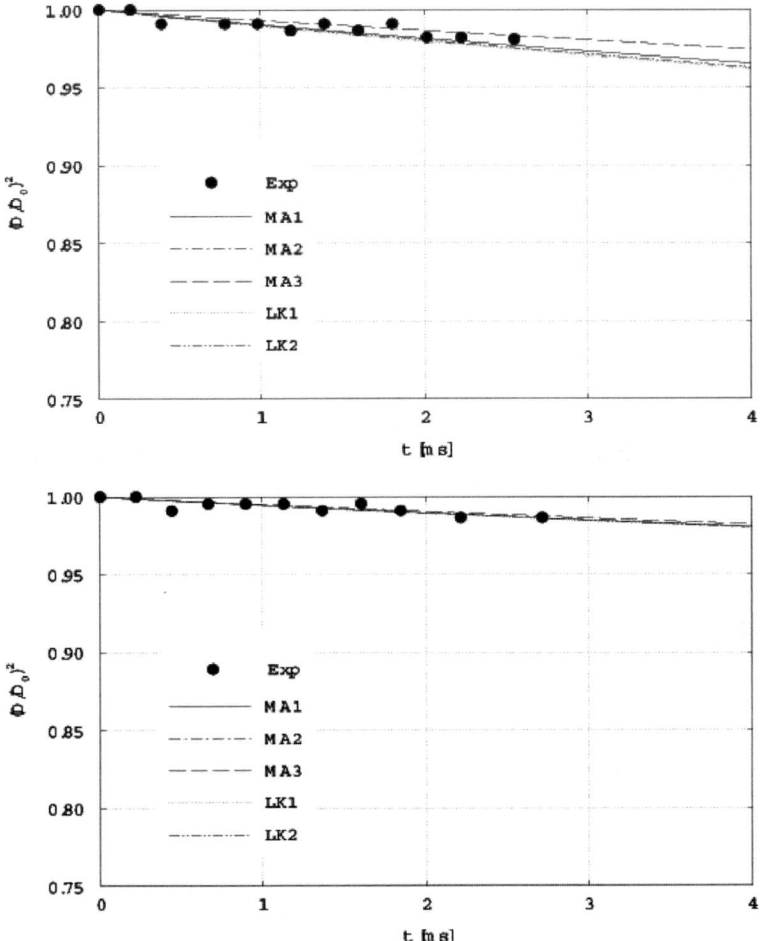

Figure 2. Temporal evolution of the nondimensional droplet diameter squared: comparison between experimental results and numerical predictions for different chamber pressure (a) $P_\infty =$ 10 bar and (b) $P_\infty = 40$ bar. Experimental conditions: $D_0 = 88$ μm $T_\infty = T_{d,0} = 315$ K, $\Delta s/D_0 = 3.2$, and $36 < Re_{d,0} < 236$

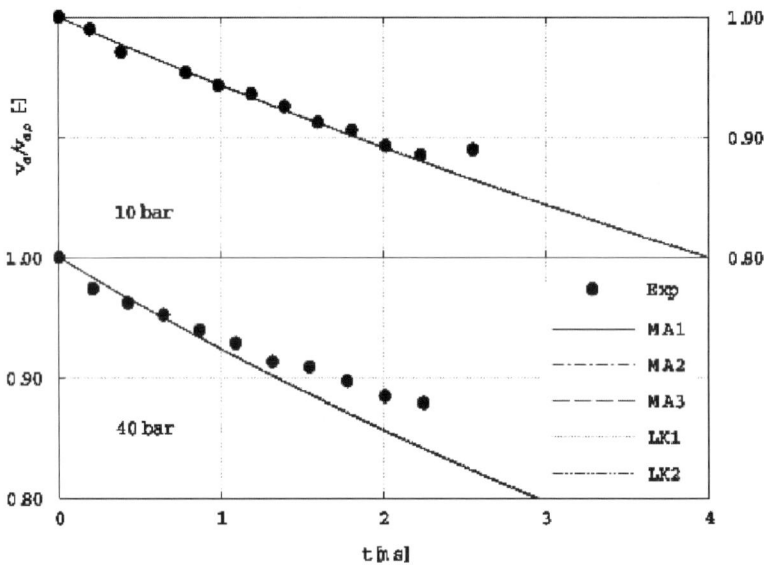

Figure 3. Temporal evolution of the nondimensional droplet speed: comparison between experimental results and numerical predictions for different chamber pressure (a) $P_\infty = 10$ bar and (b) $P_\infty = 40$ bar. Experimental conditions: $D_0 = 88$ μm $T_\infty = T_{d,0} = 315$ K, $\Delta s/D_0 = 3.2$, and $36 < Re_{d,0} < 236$

References

Atthasit, A., Doué, N., Rosa, N. G., Biscos, Y., Lavergne, G., 2005, Influence of Droplet Concentration on the Dynamics and Evaporation of a Monodisperse Stream of Droplets, *Proc. Int. Symp. on Heat and Mass Transfer in Spray Systems*, Antalya, Turkey, June 5–10.

Bellan, J. and Summerfield, M., 1978, Theoretical Examination of Assumptions Commonly Used for the Gas Phase Surrounding a Burning Droplet, *Comb. Flame*, **33**:107–122.

Bellan, J. and Hardstad, K., 1987, The Details of the Convective Evaporation of Dense and Dilute Clusters of Drops, *Int. J. Heat Mass Transfer*, **30**(6):1083–1093.

Crowe, C.T., Troutt, T.R., Chung, J.N., 1996, Numerical Models of two-phase Turbulent Flows, *Ann. Rev. Fluid Mech.*, **28**:11–43.

Godsave, G.A.E., 1953, Studies on the Combustion of Drops in a Fuel Spray: The Burning of Drops of Fuel, *Proc. 4th Int. Symp. on Combustion*, Combustion Institute, Baltimore, MD, pp. 818–830.

Imaoka, R.T. and Sirignano, W.A., 2005, A generalized Analysis for Liquid-Fuel Vaporization and Burning, *Int. J. Heat Mass Transfer*, **48**(21–22):4342–4353.

Law, C. K. and Sirignano, W.A., 1977, Unsteady Droplet Combustion with Droplet-Heating-II: Conduction Limit, *Combust. Flame*, **28**:175–186.

Lamanna, G., Sun, H., Weigand, B., Magatti, D., and Ferri, F., 2005, Measurements of Droplet Vaporization by means of Light Scattering, *Colloids and Sufaces A: Physicochem. Eng. Aspects*, **261**:153–161.

Leiroz, A.J.K. and Rangel, R.H., 1995, Interaction Effects during Droplet-Stream Combustion, *Proc. 8th Int. Symp. on Transfer Phenomena in Combustion*, San Francisco, USA, pp. 1–12.

Miller, R.S., Hardstad, K., and Bellan, J., 1998, Evaluation of Equilibrium and Non-Equilibrium Evaporation Models for Many-Droplet Gas-Liquid Flow Simulations, *Int. J. Multiphase Flow*, **24**:1025–1055.

Ranz, W.E., and Marshall, W.R., 1952, Evaporation from Drops, PART I, *Chem. Eng. Prog.*, **48**(3):143–146.

Renksizbulut, M. and Yuen, M.C., 1983, Experimental Study of Droplet Evaporation in a High-Temperature Air Stream, *Int. Heat Transfer*, **105**:384–388.

Sangiovanni, J.J., and Labowsky, M., 1982, Burning Times of Linear Droplet Arrays: A Comparison of Experiment and Theory, *Combust. Flame*, **47**:15–30.

Spalding, D.B., 1953, The Combustion of Liquid Fuels, *Proc.4th Int. Symp. on Combustion*, Combustion Institute, Baltimore, MD, pp. 847–864.

Sun, H., 2006, *On the Vaporization of Droplets at Elevated Pressure*, Dr. Hut Verlag, Muenchen. ISBN 3-89963-300-8.

Virepinte, J.F., Adam, O., Lavergne, G., and Biscos, Y., 1999, Droplet Spacing on Drag Measurement and Burning Rate for Isothermal and Reacting Conditions, *J. Propul. Power*, **15**(1):97–102.

Wong, S.C. and Lin, A. R., 1992, Internal Temperature Distributions of Droplets Vaporizing in High-Temperature Convective Flows, *J. Fluid Mech.*, **237**:671–687.

Young, J.B., 1993, The Condensation and Evaporation of Liquid Droplets at Arbitrary Knudsen Number in the Presence of an Inert Gas, *Int. J. Heat Mass Transfer*, **36**:2941–2956.

Yuen, M. C. and Chen L.W., 1976, On Drag of Evaporating Droplets, *Comb. Sci. Tech.*, **14:** 147–153.

RESULTS OF ENVIRONMENTAL RETROFIT FOR GAS PIPELINE GAS-PUMPING UNITS

A. SOUDAREV*
Research-Engineering "Ceramic Heat Engines" Center named after A.M. Boyko, Ltd, Polyustrovsly Av., 15, block 2, 195221, St. Petersburg, Russia

S. VESELY
EKOL, Ltd, Brno, Czech Republic

E. VINOGRADOV, Y. ZAKHAROV
NPP "EST", Ltd, St. Petersburg, Russia

Abstract. Provision of an environmental friendliness is an important global challenge. This is why the improvement of the environmental characteristics have involved the development of the gas turbine engines (GTE) applied to the gas pipelines as a natural gas blower drive. Stationary GTEs have been run on gas pipelines for some decades. During this time, the environmental norms have become much more demanding. In connection with this, an environmental update of the operated gas-pumping units becomes necessary. Tasks of the environmental update as applied to the existing gas-pumping units are, in principle, different from those confronting the engineers involved in development of improved machines. The paper presents some engineering approaches developed at the NPP "EST" (Russia) and put into effect as a result of joint efforts undertaken by the "EKOL" (Czech Republic), including solutions of the research-engineering concepts, combustor designs and test findings for the updated gas-pumping units of the following types:: GTK-10 («Nevsky Works», 10 MW), GT-750-6 («Nevsky Works», 6 MW), MS-3002 (General Electric, 10 MW), KWU VR-438 (Siemens, 10 MW). The experience of the environmental retrofit for over 500 GPUs on the compressor stations of Russia, Ukraine, Kazakhstan, Czech Republic, Slovak Republic, Germany, and Hungary is described.

Keywords: emissions, environment, combustor, burner, GTU, GPU

---
*All correspondence should be addressed to Soudarev Anatoly, "NIZ KTD", Ltd, Polyustrovsky Av.15, block 2, 195221, St.Petersburg, Russia. e-mail: soudarev@boykocenter.spb.ru

N. Syred and A. Khalatov (eds.), Advanced Combustion and Aerothermal Technologies, 457–471.
© 2007 *Springer.*

1. Introduction

A main environmental challenge encountered with the operation of the gas turbine plants is the atmospheric pollution by the toxic hydrocarbon fuel combustion products. First and foremost, it concerns the nitrogen oxides as having the highest toxicity and the greatest volume of exhaust. Next are the carbon oxides and heavy hydrocarbons. The sustained and ISO supported trend towards making the environmental legislation tougher in time puts limits to the admissible level of toxic emissions into atmosphere. Improvement of the environmental values has assumed lately a status of a main trend in developments and update of the gas turbine engines, especially those applied as drives for the natural gas blowers on the gas pipelines.

On the basis of the analysis carried out in the reference[1], it emerges that only in Russia there are over 3,000 stationary gas turbines, the bulk of which have been operated for more than 15 years. As it is shown in this reference, their environmental characteristics are notably below the current environmental norms. Despite a sustained activity in terms of update of the gas turbine stock, many units will operate dozens of years longer. Therefore, along with developments of new generation low-toxic GTUs it is necessary as well to carry out a regular environmental update of the existing stock. The task of environmental update of turbine combustion chambers for the operated gas turbines is principally different from that of development of novel low-toxic machines. Considering that the life of these units is more or less depleted, expenses on the update of the older GTUs must be minimum; otherwise, the update will be economically unjustified.

First, to make the environmental update economically viable , the expenses on the update must be as low as possible. Therefore, there must be no changes introduced into the shroud and the combustor's main systems design:

- Solid shroud
- Fuel supply–discharge systems, automatic control systems
- Governing and control system

Furthermore, the operation properties of units, i.e., the main characteristics of the updated combustor (combustion completeness, hydraulic resistance, temperature pattern nonuniformity for gases downstream of the combustor, the maximum temperature of the hot elements metal, reliability of ignition of the fuel–air mixture at startups, limits for the "lean" and "rich" blowouts) must be close to the characteristics of a conventional design combustor.

Then, the unit outage at retrofit must be minimum. To ensure this, simplicity of the updated design must be available as well as technological capabilities to carry out the retrofit on site.

The authors accumulated a vast expertise in the solution of these tasks. The present paper gives an insight into some results of activities dedicated to the environmental retrofit of the stationary gas turbine combustion chambers on the basis of application of developed and commercialized low-toxic hydrocarbon fuel combustion techniques.

2. The Air Blow-in Local Dosing Technique

2.1 THE TECHNIQUE'S RATIONALE

The combustors of the gas turbine engines (GTE) operated for over 15–20 years have been developed on the basis of design concepts used in the 1970s when there had been no norms that would regulate the toxic components concentrations in the exhaust gases. The main task of a combustor design was to ensure the maximum possible combustion completeness, the latter achieved at the expense of reducing the excess-air coefficient inside the primary combustor up to 1.2–1.3 at the rated conditions of the GTE operation with the relevant mean-mass temperature of the flame being 1,700–1,800°C. Therefore, the NO_x emissions amounted to 600–1,000 mg/Nm^3 and higher.

It is natural that the highest NO_x generation occurred within high-temperature zones of the flame whose location was dependent on the hydrodynamic flow patterns in the combustor.

The core of the technique of the local dosed blow-in, as described in detail in the patent, is to lower the flame temperature within the highest temperature zones of the flame through a jet supply of the cold air into them. It means a reduction of the dwell time for the burning particulates at very high temperatures with the subsequent suppression of the CO emissions at the minimum operation duties of a GTE.

The examples of a concrete engineering implementation of the local dosed blow-in technique as applied to different combustor designs are provided below.

3. Retrofit of GT-750-6 Combustion Chamber

The GT-750-6 combustor (Figure 1) is the typical large size-type combustor with a separate cylindrical shroud connected at the bottom to the turbine casing. The combustor dome contains seven register burners, six thereof constitute a circular row, while one burner is the pilot one placed along the chamber axis.

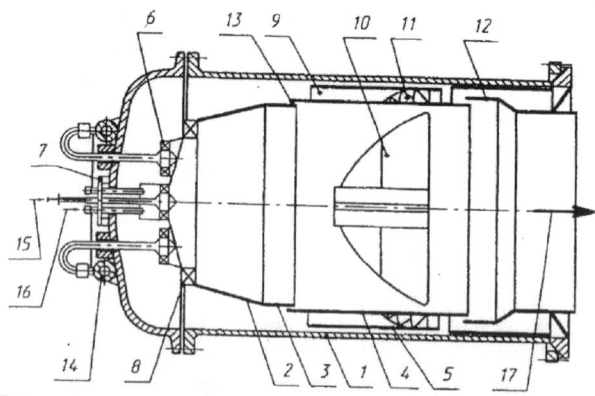

Figure 1. Design scheme of seven-burner combustor. Designations: 1 – shroud, 2 – dome, 3 – liner, 4 – mixer, 5 – screen, 6 – main burner, 7 – pilot burner, 8 – swirler, 9 – annular mixer passage, 10 – mixer opening, 11 – mixer guide vanes, 12 – output section shell, 13 – circular slot, 14 – main fuel manifold, 15 – fuel supply to pilot burner, 16 – fuel supply to igniter, 17 – combustion products discharge to turbine.

Outside, the circular row of burners is embraced by the vane swirler of a large diameter intended to cool the liner. The mixer is an annular passage formed by the liner and the screen closed on one end and communicated with the combustion space by two mixer openings located one against another in the liner.

The combustion zone structure (configuration and sizes, fuel burnout behavior and the temperature pattern, sizes, and form of the back streams) are determined, primarily, by the interaction of the burner flames one with another and, also, by the interaction of the burger flames with the annular airstream of the opposite twist that flows from the large size swirler.

On the basis of the gas temperature patterns within the combustor fire space (Figure 2) and the accurate three-dimensional localization of the high-tempe-rature zones, the jet directions for the additional air to reduce the NO_x emissions were identified.

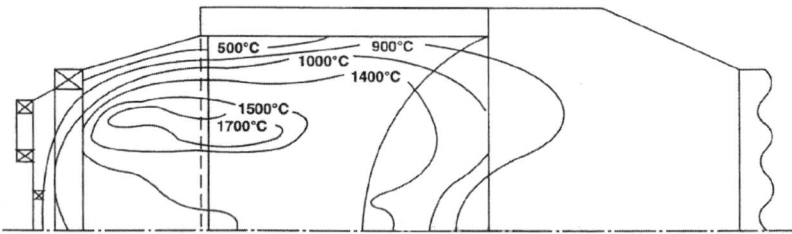

Figure 2. Gas temperature pattern within the fire space of seven-burner combustion chamber

The combustor retrofit included the mounting of two rows of air-guide pipe branches of various diameter on the dome; their aim was to supply the air jets into the maximum temperature zones and, also, to reduce the flow area of the annular passage in the mixer. The scheme and the general view of the retrofitted dome is shown in Figure 3.

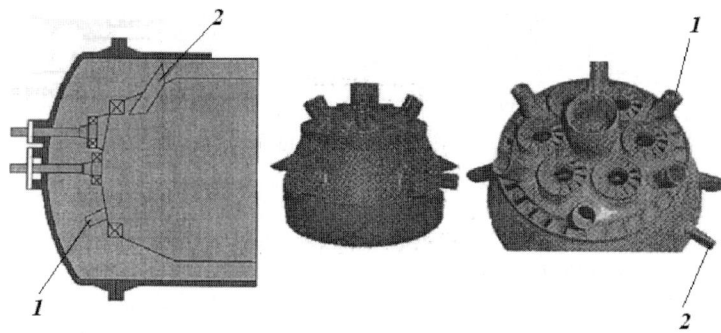

Figure 3. Scheme and general view of retrofitted dome for GT-750-6 combustor. Designations: 1, 2 – first and second row nozzles.

TABLE 1. The main characteristics of the GT-750-6 combustor

	Characteristics	Numerical value			
1	GTE	GT-750-6	GTK-10	KWH VR-438	MS 3002
2	Diameter of liner (mm)	1,000	1,195		
3	Number of liners (pc)	1	1	1	6
4	Temperature of air at combustor inlet, K(°C)	665(392)	676(403)	541(268)	543 (270)
5	Combustion products temperature upstream of HPT, K(°C)	1,023(750)	1,053(780)	1,053 (780)	1,216 (943)
6	Air pressure at combustor inlet (MPa)	0.475	0.440	0.76	0.72
7	Air flow across combustor (kg/s)	56.75	83.33	33.84	51.00
8	Total excess-air coefficient	6.78	6.50	4.77	3.60
9	Reduced concentration of NO_x at 15% O_2 (mg/Nm^3)	800 135	580 15	340 140	232 150
10	Reduced CO concentration at 15% O_2 (mg/Nm^3)	0 35	0 60	218 27	58 58
11	Number of retrofitted units (Czech Republic, Slovak Republic)	150	353	1	6

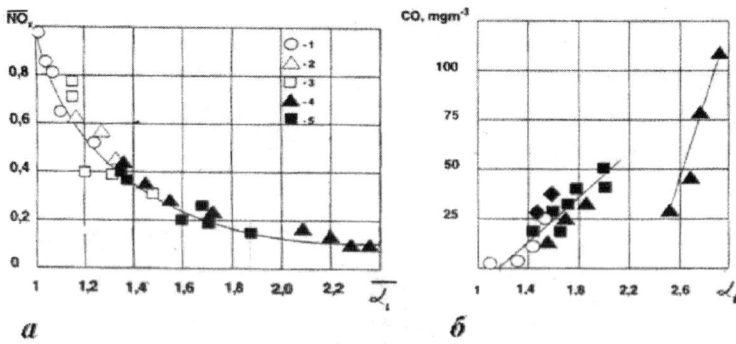

Figure 4. Test results for standard (1,2,3) and retrofitted (4,5) GT-750-6 combustor as a function of: a – relative emission $\overline{NO_x}$ on the relative primary air coefficient $\overline{\alpha}_1$; b – CO emission (mg/m³) on the excess-air coefficient α_1

   In the process of the experimental development of the retrofitted combustor, the diameters and the number of the air-guide pipe branches protruding into the fire space of the pipe branch areas, their cooling systems and other design details were derived.

   As follows from references[3,4], the test results for the retrofitted GT-750-6 combustor showed that the nitrogen oxide emissions decreased by as much as 6–8 times (Figure 4a), from 620–1,050 mg/Nm³ to 105–30 mg/Nm³ with no substantial carbon oxide emissions increase (Figure 4b). The graph below shows the average NO_x and CO emissions values for all updated GT-750-6 units. Other operation characteristics of the updated combustors, not related with the exhaust toxicity, are not inferior to those of the conventional design combustors. The main characteristics of the GT-750-6 combustor are summed up in the Table 1.

   This retrofit was relatively simple, inexpensive and easily realized at the compressor station conditions. The CO content in the combustion products did not exceed standard levels valid in Russia (300 mg/Nm³ at 15% O_2).

   At $\alpha_1 > 2.6$, an intensive CO emissions increase takes place with the excess-air coefficient rise. This is due to differences in the actual state of separate elements and devices of the GTU, whose guaranteed life is, mainly, depleted and, also, because of a difference in the individual flow characteristics for the combustor paths as well as for the compressor air heaters, etc.

   The environmental update of 150 GPUs was mainly carried out on the gas pipelines "Tranzit" (Czech Republic) and "Slavtransgaz" (Slovak Republic).

4. Retrofitting of GTK-10 Combustor

Over 3700 GPUs are in service on the territory of the former USSR (C.I.S. countries) with the total power nearing 40 million KW, almost 25% of these are the "Nevsky Works" production GTK-10 plants of the 10 MW power range with the NO_x concentration at 15% O_2 in the exhaust gases up to 670 mg/Nm^3. As it is stressed in the reference[4], this is thrice exceeding the specified level from the GTK-10 unit and is virtually identical to that of the GT-750-6 unit combustor (Figure 1). Its main parameters are presented in Table 1.

The results presented in this paper were obtained on the basis of the analysis relative to velocity, temperature and concentration patterns during the tests of the full-scale combustors. The maximum temperature areas (Figure 2) are placed opposite to the axis of the main burners at the distance 1–3 times as much as the swirler diameter. The effect of the air supplied through the peripheral swirler is expressed as a deviation from the high-temperature end areas in the direction of twist of the swirler.

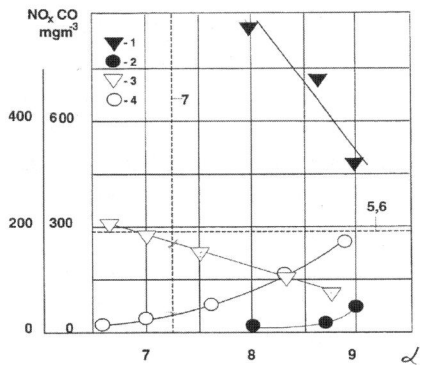

Figure 5. Comparison of NO_x and CO emissions at 15% O_2 for conventional and retrofitted GTK-10 combustors Designations: 1, 2 – NO_x and CO emissions for conventional combustor; 3, 4 – the same for retrofitted combustor; 5, 6 – the same in compliance with norms valid at present in Russia; 7 – design duties; α – total excess-air coefficient

The NO_x content in the combustion products (Figure 5) compared with a conventional combustor decreased by nearly five times; the CO emissions, though increased somewhat, remains still at a relatively low level, namely close to 60 mg/ Nm^3 which gives the combustion completeness of 99.9%. Comparison of the temperature patterns for both the conventional and retrofitted combustors indicates that the maximum peripheral nonuniformity decreased from 116°C to 85°C. Due to this, the power and efficiency of GTU tends to increase since the mean-mass temperature of the working media at the high pressure turbine inlet is almost 15°C higher with the reliability maintained.

The NO_x concentration scatter in the combustion products of the first series units with conventional combustors under design conditions and at 15% O_2 was 635–1,230 mg/ Nm^3. So high a spectrum of the NO_x concentration scatter shows a need for an individual approach at the update of each combustor. Thanks to such updates, the NO_x, concentration decreased by 4–5 times amounting to (160 ± 29) mg/Nm3. The CO concentration in all combustors used was low and did not exceed 60 mg/Nm3 at 15% O_2.

For the gas-pumping compressor stations of the "Tyumentransgaz" enterprise, over 350 GPUs were thus retrofitted.

5. Retrofitting of KWU VR-438 Combustor

The KWU VR-438 combustor installed at the compressor station «Emsburen» (Ruhrgas) includes two sections on the either side of the turbo unit. The general view of the KWU VR-438 combustor is shown in Figure 6.

The liner and burner (1) are mounted inside the solid shroud.

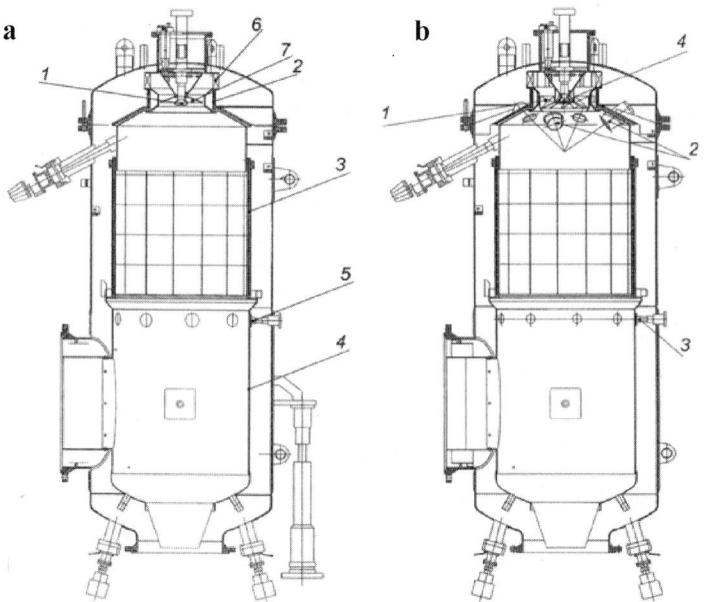

Figure 6. General view of the KWU VR-438 combustor installed at the compressor station "Emsburen" (Ruhrgas): a – conventional; b – retrofitted

The liner includes three sections: upper section – dome (2), middle high-temperature section (3) and bottom section – mixer (4). The inner surface of the middle high-temperature section of the liner is lined with ceramic plates.

The mixer section has nine holes to supply air to the mixing zone. Three of these are equipped with special valves (5) that allow variation of the mixer flow area without dismantling the combustor. The valve transfer is carried out outside using the thread spindle. The air valves enable the adjustment of the mixer direction aimed to provide the maximum possible uniform temperature pattern upstream of the turbine. Such adjustment is typically carried during the scheduled overhauls.

The burner has rather an original design. Here, to stabilize the flame, the airstream twist is applied using the radial blade swirler (6) as well as the conical perforated flame holder of a large diameter (7).

The following features of the operation process inside the VR-438 combustor should be singled out:

- A complex air supply scheme in the burner which hinders an accurate identification of the air flow to the burner by a computational technique

- A relatively high diameter of the conic flame holder in combination with a weakly twisted jet that flows around it generating a lengthy and large volume counterflow zone

- Compared with the diameter of the conic flame holder, width of the annular twisted jet cannot sustain the turbulent exchange conditions within the counterflow zone at a sufficiently high intensity level

- A consequence of the above is a lengthy and "feeble" flame, a long dwell of the combustion products within the high-temperature zone, increased nitrogen oxide emissions

- An insufficient intensity of the turbulent mixing processes within the combustion zone is likely to be a cause of a relatively high carbon monoxide emission: due to a longer process of the fuel mixing with air, the combustion reactions fail to be completely accomplished in the high-temperature liner section and the combustion products are "frozen" by the cold air jets in the mixer.

On the basis of the operation analysis as applied to the conventional VR-438 combustor, an environmental update technique was worked out including as follows (Figure 6b):

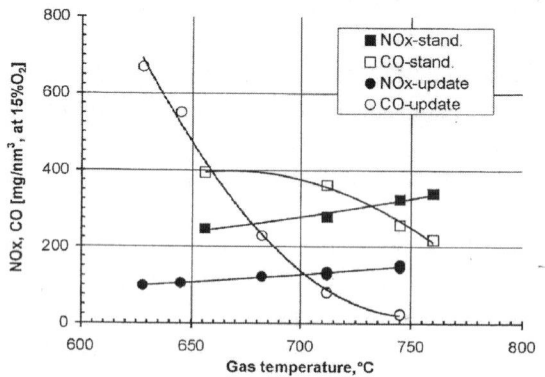

Figure 7. Test results for conventional and retrofitted KC VR-438

- Replacement of the conic perforated flame holder with the axial vane air swirler (1)
- Mounting of four air-guide pipe branches (2) on the dome cone
- Diminishing of the flow area for the mixer (3)

The results presented in references[4,5] give rise to the following conclusions.

Through minimum variations of the conventional design combustor with use of the local dosing air blow-in technique, the toxic emissions at the rating could be decreased by as much as:

- 2.5 times for nitrogen oxides
- 8.1 times for carbon oxide

Thus, the GT-750-6 combustor, the essential reduction in the emissions of both NO_x and CO was, thus, achieved. This evidently paradoxical result was obtained due to intensification of the fuel/air mixing process. The intensification was achieved by replacement of the conical flame holder with a register burner which gave notably changes to the temperature within the combustion zone and the aerodynamic flow pattern; at the same time, with the update of the GT-750-6 combustor, the flow hydrodynamics changes were insignificant though the temperature within the combustor notably decreased.

The other operation characteristics of the retrofitted combustor are not inferior to those featured by a conventional design combustor.

6. Retrofitting of General Electric MS3002 Combustor

The General Electric MS3002 gas turbine, known also as Frame-3, is distinguished by its excellent reliability and good operation qualities. No wonder that it is widely applied all over the world. There are over 1,000 machines of this series that are being operated, mostly as a part of the electricity generation stations to drive the electric generators.

At the same time, the environmental records of these machines especially the nitrogen oxide emissions, do not meet the current norms. Bearing in mind that the Frame-3 life is essentially over 100,000 h, a need arises to reduce the NO_x emissions for the machines with a sufficiently high life to the levels admissible for the currently valid environmental norms.

The low-toxic fuel combustion techniques, e.g., DLN, proposed by General Electric, based mainly on burning of the premix lean fuel/air mixture, is too complicated for its control and too costly (about $1 million per one unit). For the units, which have been in service for a long time and for which life is partially depleted, the application of such costly technique does not seem to be economically justified. So, there must be cheaper and simpler engineering approaches available.

The development of techniques included the following stages:

- Analysis of design and operation process for a conventional combustor
- Design and manufacture of versions of full-scale prototypes for low-toxic combustors
- Bed tests of prototypes and options for an optimum version
- Design and manufacture of a full-scale combustor
- Tests of updated combustor as a part of a unit at the compressor station aiming to identify its main characteristics
- Trial-industrial operation of the updated combustor aiming to verify its reliability

The MS3002 combustor includes tubular sections placed in separate parallel shrouds, three on each side of the machine normal to its axis.

The air for combustion is supplied via the annular channel formed by the liner and shroud, by the counterflow scheme. The liner design is typical for the combustors with staged air supply into the combustion zone (ref. Figure 8). The liner sleeve has five rows of holes (two rows for primary air supply, two rows of afterburn holes and one row of the mixer holes) placed at different distances from the dome. The dome is a cone with a developed perforation as scale-shaped stamp-outs – louvers for cooling. A similar cooling system was applied on the liner sleeve at the area from the dome to the mixer holes.

The gas burner is secured on the shroud cover and has gas distribution jet-type head pieces located along the dome axis. The operation process in the combustors with the staged air supply is well studied and it has some special features. The combustion process intensity and the fuel burn-out length in the combustors of this type are mainly determined by the diameter and location of the primary air supply holes and the afterburn holes in the liner.

As for the combustor under examination, it features low stream velocity and small, close to the unit, excess-air coefficients in the head area and a larger counterflow zone volume; all these result in a longer time of dwell for the combustion products within the high-temperature zone and cause higher nitrogen oxide emissions.

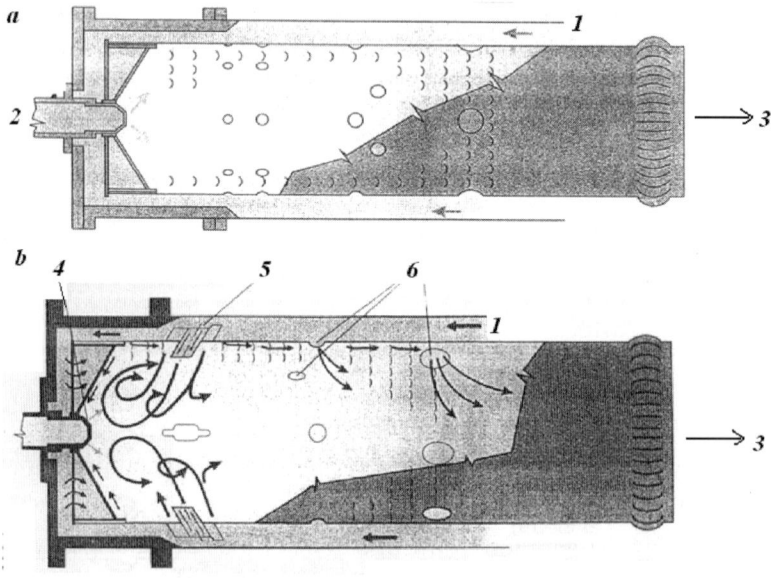

Figure 8. The MS-3002 combustor: a – conventional; b - updated. Designations: 1 – air, 2 – fuel gas, 3 – combustion products, 4 – burner with varied geometry of the burner head piece, 5 – air blow-in nozzles, 6 – holes with reduced flow area

As was stressed earlier, the main factors that determine the formation of the nitrogen oxides are the combustion zone temperature and the dwell time of the combustion products within the high-temperature zone. It is also known that these factors affect the CO emissions though with an opposite effect. So, a straightforward approach that would pursue a simple increase of the excess-air coefficient inside the primary zone at the expense of variation of the number and diameter of the holes within the liner looks to be leading to nowhere since it causes the CO emissions increase and, also, can aggravate the combustion stability, causes pressure fluctuations, reduction in the ignition reliability and other disturbances. The air blow-in local dosing technique was used as the basis for designing experimental versions of the MS-3002 low-toxic combustor. An optimal version, based on the experimental results, is shown in Figure 8b.

The distinguishing features of the newly developed combustor (Figure 8b) are an increased flow for the air supplied to the primary zone, its accurate dosing and a strict orientation of the air jets to supply it to the local high-temperature zones. This was achieved using a special system that includes the air-guide pipe branches welded to the liner (Figure 8c). Along with this, some changes were introduced into the air distribution system and the angle of the gas outflow from the burner was also changed.

The above changes introduced into the liner design result in the formation of the highly turbulent toroidal vortex within the primary zone which promotes as follows:

- A faster fuel/air mixing
- A reduction in the average temperature of the flame
- An increase in the stream velocity within the recirculation zone and, therefore, a decrease in the time of dwell for the combustion products within the high-temperature zones
- An intensification of the fuel combustion process

Totally, eight different models of low-toxic combustors were designed, manufactured and tested on the test bed. They differed one from another by their mounting seat, direction, number and configuration of the air-guide pipe branches, the angle of the fuel jet outflow from the burner holes, diameters of the afterburn and mixing holes.

On the basis of the test results of the combustor models, an optimum low-toxic combustor design was selected (Figure 8b) that allows reduction in the NO_x emissions by as much as 1.8 time compared with a conventional design combustor. The CO emissions at the rated load remained close to 0 (Figure 9).

The complex tests of the combustor as a part of the unit confirmed the conclusions made on the basis of results of the test bed refinement of the combustor, in particular:

- 100% ignition reliability and functioning of the cross-fire tubes at startups
- High stability of the flame and a reliable functioning of the flame detectors over the entire operating condition range
- High-fuel combustion completeness which as in the base-line combustor exceeded 99.9%
- Absence of the working media pressure fluctuations
- A considerable decrease of the pressure losses within the combustor which leads to a reduction in the fuel consumption

Comparison of emissions from a conventional design combustor with that from a retrofitted low-toxic combustor is shown in Figure 9.

The tests performed demonstrated that the nitrogen oxides from the updated combustor is 1.6 times lower that from the conventional design combustor and does not exceed 140 mg/Nm^3. These characteristics meet the requirements of the environmental norms of Germany TA Luft.

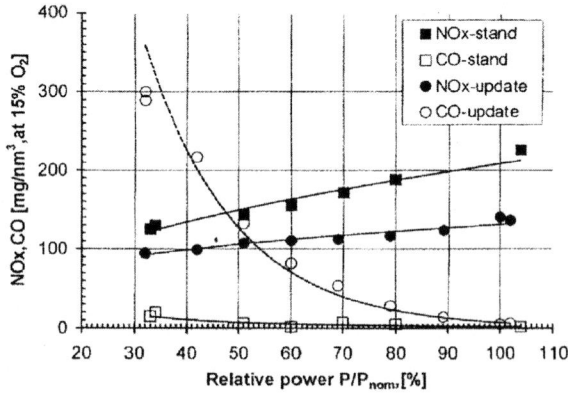

Figure 9. Cogeneration of NO$_x$ and CO in the combustion products vs relative GPU load

The complex test results analysis for the lox-toxic combustor revealed that its main operation characteristics are not inferior to those of the conventional design combustor or even superior to them.

After the short-term complex tests were over, the trial-industrial operation of the low-toxic combustor was launched, its operation time is now over 5 years of successful operation. To this date, the environmental update was conducted on more than four «Frame3» units operated in Germany and Hungary.

7. Summary

On the basis of the available environmental update experience, the following conclusions could be made:

- The local dosing air blow-in technique is an effective means of toxic emissions reduction that allows the lowering of the emissions to the level that would be admissible for the current environmental norms and regulations valid in the industrially developed countries.

- The local air blow-in technique is sufficiently universal and applicable to the combustors of essentially different designs.

- Use of this technique at the environmental update of combustors does not need the introduction of any changes into the solid shroud, fuel system design and the automatic control system.

- Thanks to the inherent simplicity and adaptability to manufacture for the examined updated designs, they differ advantageously from the engineering approached used by the original manufacturers in terms of their low cost, high reliability and a short outage at retrofit.

References

1. V. Tschurovsky. State of production and experience of maturing of new generation GTUs for compressor stations application, *Paper presented at the gas turbine commission of the Russian Academy of Sciences*, 15.02.2006, Moscow.
2. E. Vinogradov, Y. Zakharov, A. Soudarev, A. Bystrov. Combustion Chambers, Patent of Russia, No. 2027114 of 20.01.95, Priority claimed 13.01.92.
3. S. Vesely, S. Parizek. Experience of environmental update of gas-pumping units, ASME Turbo Expo 2000. May 8–11, 2000, Munich, Germany, *Paper 2000-GT-89*, 9 p.
4. A. Soudarev, et al. Gas Turbine units for pipeline compressor stations: environmental update problems, *ASME – Paper 94-GT-497*, 5 p.
5. A. Soudarev, et al., Experimental refinement of technologies for environmental update of gas turbine units applied to electro-generator driving, *ASME – Paper 96-TA-45*.

APPLICATION OF ADVANCED TECHNOLOGIES IN PRODUCTION OF LOW-EMISSION COMBUSTORS AND TURBINE COMPONENTS

N. TUROFF*

President of Advanced EDM Automation, Inc., San Diego, USA

Abstract. This chapter focuses on the advanced manufacturing methods used in the fabrication of gas turbine fuel injectors, combustor liners, and turbine nozzles. Uniform distribution and proper mixture of the air and fuel is essential for effective combustion process. High-precision sizing of the multiple cooling holes in combustor liners and turbine nozzles is critical for engine emission performance and durability of these components. These requirements led to increased industry standards for significantly tighter tolerances. This chapter describes the newest machinery and manufacturing processes developed for fabrication of hot section components, reviews actual examples of such components and describes some concepts for advanced technological fixtures.

Keywords: advanced fabrication methods, fixtures, combustor liners, injectors, turbine nozzles

1. Discussion

Most advanced low-emission combustors rely heavily on advancements in manufacturing processes with particular emphasis on tight manufacturing tolerances. This is true for both fuel injectors and combustor liners. The precise ratio of the fuel-to-air mixture supplied by the multiple fuel injectors within a single combustor determines the flame temperature variation and corresponding combustion emission levels. For this reason, high-precision fuel and air supply orifices are required with consistent dimensions throughout all of the injectors. The same requirement extends to the cooling holes in the combustor liners since

---

*To whom correspondence should be addressed. Norm Turoff, Advanced EDM Automation, Poway, CA, USA

N. Syred and A. Khalatov (eds.), Advanced Combustion and Aerothermal Technologies, 473–480,
© 2007 *Springer.*

their durability depends on uniform distribution and precision of the dimensions. The following sections focus on the advanced manufacturing techniques
and processes that help in meeting these requirements. The chapter is structured
around examples of the actual combustor hardware for advanced industrial gas
turbines.

Many of the fuel and air passages in turbine components are relatively deep
and have small diameters, often at an acute angle to the surface where they
begin (see examples in Figures 1 and 2). Conventional drilling of such holes is
often impossible without spot facing. Moreover, the required accuracy usually
precludes conventional drilling altogether, gun drilling being the only accurate
enough conventional method.

Figure 1. Typical examples of combustor fuel injectors

Precision is only one of the manufacturing challenges presented by
advanced turbine components. Accessibility of features for conventional
machining is another. Advanced design requirements often necessitate features
that are simply not accessible by conventional machine tools. Cleanliness is yet
another one. Air and fuel passages in the turbine components often take the

shape of an intricate labyrinth consisting of both cast and machined sections. During conventional machining, chips inevitably accumulate in this labyrinth and are virtually impossible to completely remove. Just one such chip, carried by the air or fuel flow of the fully assembled operational turbine and lodged in one of the precision orifices downstream can mean an extensive overhaul and associated downtime.

Figure 2. Steep-angled holes in fuel injector and turbine nozzle

These challenges are made even tougher by the materials typically used in the combustor and injector components. Machinability of stainless steels and high-temperature alloys by conventional methods leaves much to be desired.

All these issues are being addressed by electrical discharge machining (EDM). The only fundamental limitation of the EDM process is the requirement that the material machined be electrically conductive. Regular steels, aluminum, stainless steels, and high-temperature alloys are machined at comparable speeds. Only parts manufactured by powder metallurgy require significantly slower speeds due to variations in the material conductivity inherent in the porous powder metal.

Since no cutting forces (or high temperatures in the case of laser machining) are applied to the machined components, the stresses introduced to them by EDM process are minimal. So is the warping after machining, provided that the part did not have residual stresses from prior machining operations.

Since no force is applied to the tool (electrode) in the process of machining, the creation of holes at acute angles to starting surface is simplified and their accuracy is tremendously improved.

For the same reason, part holding fixtures for EDM machining can be significantly less expensive than their conventional counterparts: they need to be just as accurate, but only sturdy enough to hold the part in place. They do not need to accommodate the cutting forces of a conventional machine tool.

EDM machines do not leave any chips inside the part: the by-products of the electrical discharge erosion are microscopic and easily washed away by water or oil circulating through the machined area. Thus, the cleanliness is always assured.

EDM machines fall into one of the three general types.

Wire EDM machines allow feature accuracy of ±0.0001″ (±2.5 μm). They are, however, mostly limited to two- or three-axis machining.

Sinker EDM machines achieve similar accuracy and are capable of three or more axis machining. They also allow access to hard-to-reach features. In any conventional drilling or milling machine, the cutting tool itself is a direct extension of the machine's spindle. This severely limits the ability of such machine tools to "reach around the corner." A sinker EDM machine, on the other hand, is able to reach around such corners and machine features at an arbitrary angle to any of the machine's axes of motion (Figure 3). Naturally, this requires custom designed electrodes and electrode holders. In the case of several parallel holes, a sinker EDM machine is able to make all these holes in one setup with a group of electrodes held in a custom holder. Figure 3 provides an example of a part with multiple small-diameter, high length-to-diameter ratio holes, which are inaccessible by a conventional machine tool. A sinker EDM machine creates all five holes in one "5-up" operation using simple cylindrical electrodes in a custom holder.

Figure 3. Application of a sinker EDM machine for a multihole machining

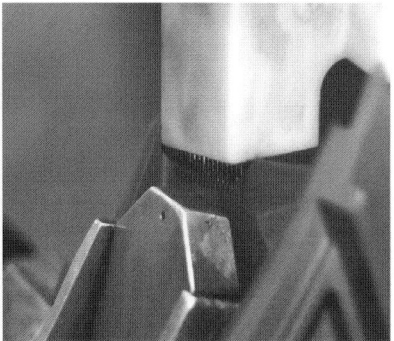

Figure 4. Fabrication of nozzle cooling holes

Figure 4 illustrates a nozzle setup in a sinker EDM machine. A pattern of small-diameter holes entering the nozzle's vane at an acute angle is being made in one setup.

EDM drills, also known in the industry as hole pop machines, provide an alternative to conventional and laser drilling. Depending on the type of material and depth of the hole, they can achieve accuracy from ±0.002″ (±50 μm) to ±0.0005″ (±12.7 μm). While not as accurate as wire or sinker EDM machines, EDM drills are much faster. Unfortunately, they do not allow for simultaneous drilling of several holes, which limits their speed advantage to "one-up" drilling operations. On the other hand, many groups of holes in combustor and injector components have their axes arranged in some sort of a circular, rather than parallel, pattern. This makes "multiple-up" machining impossible anyway. In such situations EDM drill equipped with an indexing table is clearly the best choice (Figure 5).

A few examples of practical application of this equipment are presented in Figures 6 and 7.

Figure 5. One of the EDM drilling machines installed at advanced EDM

Figure 6. A Typical EDM drilling setup for a low-emission fuel injector

Figure 7. A Combustor liner being drilled on a customized EDM drilling machine

EDM drilling also has several advantages over the method of laser drilling. One major drawback of laser drilling is the fact that it is limited to through holes. Even a hole that is technically through but with some feature of the part behind (like a through hole in one wall of a tube) is impossible for the laser drill. An EDM drill, on the other hand, has no problems with blind holes. Holes produced by EDM drills typically have much better repeatability than holes produced by laser drilling. Also, for holes of any significant depth, the cylindrical shape of an EDM drill-produced hole is much more consistent than a laser drill-produced hole.

These features make EDM process uniquely suited to the needs of designers and manufacturers of the highly sophisticated modern day turbines.

References

1. B. N. Raghunandan, R. K. Sullerey, Charlie Oommen, Indian Institute of Technology (Kanpur, India). Air Breathing Engines and Aerospace Propulsion: Proceedings of NCABE 2004, 5–7 November 2004 – pp. 381.
2. Institute of Metals, American Society for Metals, Metals Society. Metals Abstracts. 1996.
3. Aviation Week and Space Technology. 1916.
4. Inc Cambridge Scientific Abstracts, Materials Information (Information service), Internet Databases Service. Engineered Materials Abstracts. 1995 – pp. 6.
5. K. U. Kainer. Magnesium Alloys and Technology. Wiley-VCH, Weinheim, Germany. 2003– pp. 380.